Igapó (Black-water flooded forests) of the Amazon Basin

Randall W. Myster
Editor

Igapó (Black-water flooded forests) of the Amazon Basin

Springer

Editor
Randall W. Myster
Biology Department
Oklahoma State University
Oklahoma City, OK, USA

ISBN 978-3-319-90121-3 ISBN 978-3-319-90122-0 (eBook)
https://doi.org/10.1007/978-3-319-90122-0

Library of Congress Control Number: 2018951548

This Springer imprint is published by the registered company Springer Nature Switzerland AG
The registered company address is: Gewerbestrasse 11, 6330 Cham, Switzerland

Dedicated to the memory of Al Gentry.

"Every man dies, but not every man truly lives." – William Wallace

Foreword

The western Amazon forest is truly the world's most precious treasure of biodiversity. I have been fortunate to explore this region for some 37 years as a field biologist, an ecotourism owner, and creator/manager of the Area de Conservacion Regional Comunal de Tamshiyacu-Tahuayo, Loreto Region, Peru. This has included a 20+ year collaboration with Dr. Myster. This book is the fruit of those efforts and includes chapters from the premier authors doing current research in igapó forest, most of them residing in South American countries. Some of these authors are of limited means, and so this is an important vehicle for the publication of their work. It is the first book to focus on igapó forest exclusively and is a comprehensive compilation. It includes all aspects of igapó research: spatial and temporal scales, light, water, soil and the carbon cycle, litter, fungi, plants, both invertebrate and vertebrate animals, and management. In addition to editing, Dr. Myster provides both an excellent introduction, which summarizes what we know about igapó forests in the Amazon and motivates the chapters that follow, and a concluding chapter that points out key results, synthesizes, and suggests future research avenues.

I believe it is critical that such work continues in the future, and our study sites, along with my efforts as scientific coordinator, are ready to help facilitate that research. I hope this book can serve as a significant part of those studies, both as a reference and as an organizational tool, as we learn more about the critically important igapó forest.

Miami, Florida, USA Paul Beaver

Preface

After rereading the prologue to my first book on the Amazon, I felt that I should – in this second book – be more forthcoming about some of the challenges I have faced doing research there, especially deep in the Amazon under primitive conditions, so that those who wish to follow may be more prepared, in so far as that is possible. You may have to – as I did – live on an uncovered wooden platform, sleeping on one end and cooking on the other. You may be asked to – as I was – live in a canvas tent so hot, even at night, that I actually entertained the idea of offering myself up to the clouds of insects outside. Whatever your accommodations, the forest will be your toilet where you should learn to finish quickly before the ants can find you. I would also suggest that you learn how fast to walk through that same forest, faster than the mosquitos can fly but slow enough to allow the snakes time to get out of the way. A bite from one of the many poisonous snakes could ruin your whole day (life). With time, your body may come to resemble mine where the mites, ticks, gnats, wasps, biting flies, leeches, etc. have all left their marks. Then there are the diseases, some of which you can prepare for with vaccinations and pills (e.g., yellow fever, malaria), but other bacterial, viral (welcome, Zika!), and "God knows what" infections you cannot and so must just lay back and enjoy them, playing the perverse game of describing their symptoms as you come down with them. You may also try to temporarily boost your immune system with gamma globulin injections (It doesn't work. You get sick anyway).

If nothing else, the Amazon is a place where the usual calm and confidence afforded to us by Western medicine may seem to break down, leaving you with the feeling that you are at the mercy of forces you cannot understand or control. Indeed, this sheer "wildness" of the Amazon can be both exhilarating and frightening. You just never know what may come walking out of the rainforest at you. I have had primal, intense experiences there, ranging from wonder all the way to horror. You may also.

And so, after 20+ years of working in the Amazon, I believe I can say that there are at least two character traits you will need if you want to do research there: (1) you must love science and the beauty of nature and (2) you must be willing to pay for it with your body.

R. W. M.

Contents

Contributors

Arcilan Assireu Instituto de Recursos Naturais, Universidade Federal de Itajubá, Itajubá, Minas Gerais, Brazil

Adrian A. Barnett Department of Zoology, Universidade Federal do Amazonas, Manaus, AM, Brazil

Amazon Mammal Research Group, Biodiversity Studies, Instituto Nacional de Pesquisas da Amazonia, Manaus, AM, Brazil

Centre for Evolutionary Anthropology, University of Roehampton, London, England

Wanderley Rodrigues Bastos Universidade Federal do Rondonia, Porto Velho, RO, Brazil

Reinaldo Luiz Bozelli Universidade Federal do Rio de Janeiro, Rio de Janeiro, RJ, Brazil

Ángela Cano Departamento de Ciencias Biológicas, Laboratorio de Ecología de Bosques Tropicales y Primatología, Universidad de los Andes, Bogota, Colombia

Danielle Justino Capossoli Deloitte, Rio de Janeiro, RJ, Brazil

Sasha Cárdenas Departamento de Ciencias Biológicas, Laboratorio de Ecología de Bosques Tropicales y Primatología, Universidad de los Andes, Bogota, Colombia

Luisa F. Casas Departamento de Ciencias Biológicas, Laboratorio de Ecología de Bosques Tropicales y Primatología, Universidad de los Andes, Bogota, Colombia

Janice Chism Department of Biology, Winthrop University, Rock Hill, SC, USA

Diego F. Correa School of Agriculture and Food Sciences, The University of Queensland, Brisbane, Queensland, Australia

Rosecelia Moreira da Silva Castro Museu Paraense Emilio Goeldi, Belém, PA, Brazil

Denise de Andrade Cunha Instituto Federal do Pará, Castanhal, PA, Brazil

Francisco de Assis Esteves Núcleo de Pesquisas Ecológicas de Macaé, Universidade Federal do Rio de Janeiro, Rio de Janeiro, RJ, Brazil

Maria de Lourdes Pinheiro Ruivo Museu Paraense Emilio Goeldi, Belém, PA, Brazil

Rita Denize de Oliveira Universidade Federal do Pará, Belém, PA, Brazil

André Tavares Corrêa Dias Departamento de Ecologia, Universidade Federal do Rio de Janeiro, Rio de Janeiro, RJ, Brazil

João Henrique Fernandes Amaral Instituto Nacional de Pesquisas de Amazonia, Manaus, AM, Brazil

Helena do Amaral Kehrig Universidade Estadual do Norte Fluminense, Campos dos Goytacazes, RJ, Brazil

Rodil Tello Espinosa Biology Department, Facultad de Ingenieria Forstal, Universidad Nacional de la Amazonia Peruana (UANP), Iquitos, Peru

G. Wilson Fernandes Ecologia Evolutiva e Biodiversidade/DBG, ICB/ Universidade Federal de Minas Gerais (UFMG), Belo Horizonte, Minas Gerais, Brazil

Marcos Paulo Figueiredo-Barros Núcleo de Pesquisas Ecológicas de Macaé, Universidade Federal do Rio de Janeiro, Rio de Janeiro, RJ, Brazil

Bruce Rider Forsberg Instituto Nacional de Pesquisas de Amazonia, Manaus, AM, Brazil

Richard L. Jackson Jr Department of Biology, Winthrop University, Rock Hill, SC, USA

Thays Jucá Amazon Mammal Research Group, Biodiversity Studies, Instituto Nacional de Pesquisas da Amazonia, Manaus, AM, Brazil

Genimar R. Julião Coordenação de Ecologia, Instituto Nacional de Pesquisa da Amazonia (INPA), Manaus, Amazonas, Brazil

Fiocruz Rondônia, Laboratório de Entomologia, Porto Velho, Rondônia, Brazil

Daniele Kasper Universidade Federal do Rio de Janeiro, Rio de Janeiro, RJ, Brazil

John L. Koprowski The University of Arizona, School of Natural Resources and the Environment, Tucson, AZ, USA

Darley C. Leal Matos Museu Paraense Emilio Goeldi, Belém, PA, Brazil

Elessandra Laura Nogueira Lopes Universidade Federal do Pará, Belém, PA, Brazil

Juan Celidonio Ruiz Macedo Herbarium Amazonense (AMAZ), Centro de Investigacion de Recursos Naturales (CIRNA) de la Universidad Nacional de la Amazonia Peruana (UNAP), Iquitos, Peru

Olaf Malm Universidade Federal do Rio de Janeiro, Rio de Janeiro, RJ, Brazil

Randall W. Myster Oklahoma State University, Oklahoma City, OK, USA

Maria Fernanda Quintela Souza Nunes Departamento de Ecologia, Universidade Federal do Rio de Janeiro, Rio de Janeiro, RJ, Brazil

Walter Palacios Universidad Tecnica del Norte, Ibarra, Ecuador

Rosa R. Palmer The University of Arizona, School of Natural Resources and the Environment, Tucson, AZ, USA

Dario Dávila Paredes Facultad de Ingenievia Forestal, Universidad Nacional de la Amazonia Peruana (UANP), Iquitos, Peru

Pedro Henrique Medeiros Rajão Departamento de Ecologia e Evolução, Universidade Estadual do Rio de Janeiro, Rio de Janeiro, RJ, Brazil

André Reis Instituto de Recursos Naturais, Universidade Federal de Itajubá, Itajubá, Minas Gerais, Brazil

Fabio Roland Departamento de Biologia, Universidade Federal de Juiz de Fora, Juiz de Fora, MG, Brazil

Lidianne Salvatierra State University of Roraima, Postgraduate Program in Education, Boa Vista, Roraima, Brazil

Jerônimo Boelsums Barreto Sansevero Departamento de Ciências Ambientais, Universidade Federal Rural do Rio de Janeiro, Seropédica, RJ, Brazil

Fabio Rubio Scarano Fundação Brasileira para o Desenvolvimento Sustentável, Departamento de Ecologia, Universidade Federal do Rio de Janeiro, Rio de Janeiro, RJ, Brazil

Larissa Schneider Instituto Nacional de Pesquisas da Amazônia, Manaus, Brazil

Archaeology and Natural History, Australian National University, Canberra, ACT, Australia

Veridiana Vizoni Scudeller Departamento de Biologia – ICB, Universidade Federal do Amazonas – UFAM, Manaus, Brazil

Pablo Stevenson Departamento de Ciencias Biológicas, Laboratorio de Ecología de Bosques Tropicales y Primatología, Universidad de los Andes, Bogota, Colombia

María Natalia Umaña School of Forestry and Environmental Studies, Yale University, New Haven, CT, USA

Eduardo M. Venticinque Departamento de Ecologia, CB/Universidade Federal do Rio Grande do Norte, Campus Universitário, Lagoa Nova, Brazil

Boris Villanueva Ciencias Forestales – Grupo de Investigación en Biodiversidad y Dinámica de Ecosistemas Tropicales, Universidad del Tolima, Ibagué, Tolima, Colombia

Richard C. Vogt Instituto Nacional de Pesquisas da Amazônia, Manaus, Brazil

Luiz Roberto Zamith Departamento de Biologia Geral, Universidade Federal Fluminense, Niterói, RJ, Brazil

Fabrício Berton Zanchi Centro de Formação em Ciências Ambientais-CFCAm, Universidade Federal do Sul da Bahia-UFSB, Porto Sequro, BA, Brazil

Mauricio Camargo Zorro Instituto Federal de Educação, Ciência e Tecnologia da Paraíba, Belem, Brazil

Chapter 1
Introduction

Randall W. Myster

1.1 Rationale

Within many plant communities, the action of aqueous solutions such as rain, dew, mist, and fog easily leaches various compounds out of plant biomass and necromass (mainly fallen leaves; Tukey 1970) creating "black-water." Indeed, this leaching process is analogous to boiling leaves for black tea. The leached compounds are both organic and inorganic and include carbohydrates (most common), organic acids, pectic compounds, minerals, growth hormones, alkaloids, and phenolic compounds (Pallardy 2008). The leached "black-water" is so common that it can lead to specific plant species adaptations and plant species associations. In the United States alone, associations include the flatwoods of longleaf pine and turkey oak and the savannas of longleaf pine (both in Virginia), Blackwater Creek Nature Preserve and Blackwater River State Park (in Florida), and plant communities found in the states of Maine, New Hampshire, Florida, Alabama, Maryland, Missouri, and Massachusetts (Burke et al. 2003).

In this book I focus on black-water in the Amazon Basin – the watershed of the Amazon River – which predates the separation of South America from Africa some 110 million years ago (MYA; Junk et al. 2010) and is generally found below 200 m above sea level (a.s.l.). Up to 6 m of rain falls there every year, mainly during the rainy season (October to May) and beginning in the south. Of that rainfall, 25.6% returns to the atmosphere by evaporation, 45.5% is taken up by plants and transpired, and the rest can be absorbed into the soil or run off into rivers and streams (Salati 1985). Leaching is intense, mainly occurring during the rainy season and when human activity increases erosion.

The black-water in the Amazon Basin is transparent due to its low amount of dissolved and suspended matter (sediments), its low amount of solutes, and its low

R. W. Myster (✉)
Oklahoma State University, Oklahoma City, OK, USA
e-mail: myster@okstate.edu

R. W. Myster (ed.), *Igapó (Black-water flooded forests) of the Amazon Basin*,
https://doi.org/10.1007/978-3-319-90122-0_1

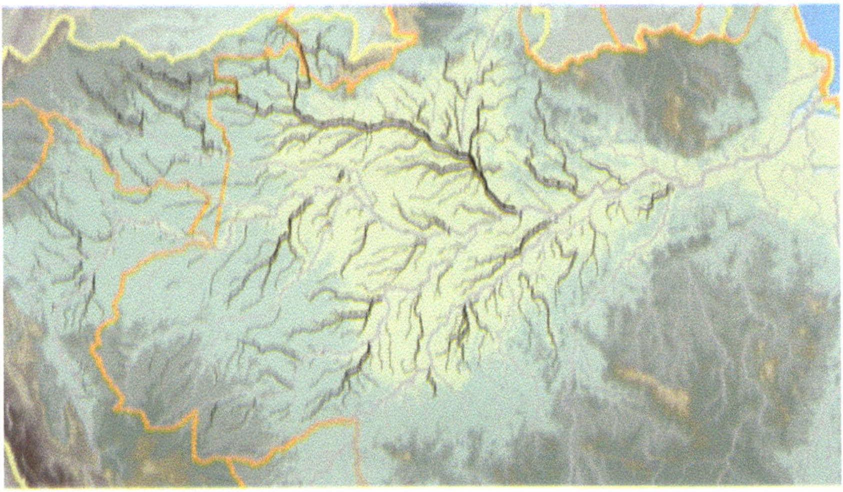

Fig. 1.1 Map of the black-water rivers in the Amazon Basin. The Rio Negro is the largest black-water river (middle-top), and its largest tributary is the Rio Branco. The boundaries of those South American countries that are part of the Amazon Basin are outlined in orange

amount of dissolved nutrients. It has a high amount of humic acid that results in a pH between 4 and 5 and an electrical conductivity <20 μS per cm. In addition the black-water can have 612 mg per kg of organic carbon, 5.1 g per kg of N, 0.21 g per kg of P, 0.10 Cmol per kg of Na, 0.27 Cmol per kg of K, 0.31 Cmol per kg of Mg, and 0.31 Cmol per kg of Ca, with dominant ions SO_4 and Cl (Wittmann et al. 2010b). Black-water and soils of igapó forests in Western Amazonia, however, can have twice or more electrolytes and nutrients compared to igapó forests in central and eastern Amazonia (F. Wittmann, "personal communication"). Similar to black-water, there is clear-water, which is transparent but greenish, with also a low amount of sediment and dissolved solids (Junk and Furch 1985). Clear-water pH varies between 5 and 6 in large rivers, and its electrical conductivity is in the range of 20–40 μS per cm but can decrease to 5 in low-order streams (Junk et al. 2011). Clear-water rivers are of intermediate fertility compared to the low fertility of black-water rivers (Junk et al. 2000). References to igapó in this book will refer to forests flooded with black-water, unless otherwise noted. Finally igapó in the Western Amazon, perhaps as a result of their geographic proximity to the Andes, may have more nutrient-rich soils compared to black-water forests farther east in Amazonia.

Both black-water and clear-water may stay locally on long-term flooded or waterlogged habitats, creating igapó swamp forest called chavascal (dominated by shrubs with a few trees; Ayres 1993). More likely, however, these waters flow into streams and rivers that drain the Archaic or Precambrian shields of Guyana and Central Brazil. Dominant black-water rivers include the Negro, Tefé, and Uatumã (Fig. 1.1) which have large areas of white sand. Both black-water and clear-water rivers transport mostly a sandy quartz bed load and a small fraction of low-fertility kaolinite (Junk et al. 2011). While this sandy soil is low in nutrients (e.g., phosphorus, potassium, calcium, and magnesium; Junk 1997), a high infection rate of AM fungi and a relatively high amount of fine roots in the upper layer of the soil help these forests get nutrients in the poor nutrient conditions where decomposition can take several years in the deep litter layer (Junk et al. 2011). The sandy soil is not just nutrient poor but does not hold water well and so often leads to drought after the flooding is over.

Regular, periodic, and seasonal flooding of black-water rivers creates a floodplain that covers an area of approximately 100,000 km^2. Forests within that floodplain are generally called igapó and have existed since the early Cretaceous (145–66 MYA; Junk et al. 2010). Black-water river floodplains also occur on alluvial paleo-várzea substrates, occurring along rivers such as Coari, Jutaí, and Tefé and along lakes in the central and western Amazon (Junk et al. 2011). This kind of flooding has been referred to as a "flood pulse" (Junk 1989) which has a regime of amplitude, duration, frequency (almost always once a year [monomodal]), timing, water quality, shape, and predictability (Myster 2009; Junk et al. 2000, 2010). The high point of these fluctuations, the amplitude, is greatest in the central part of the Amazon Basin (up to 10 m) and declines in the eastern and western parts of the basin (e.g., 8 m near Iquitos). The highest point for black-water is at the lower Negro River near its confluence with the Solimões River, leading to the greatest extent of igapó forest having a distinct flora with high levels of endemism. Flooding depth affects hydrostatic pressure, light intensity, and oxygen concentration in deeper water layers, and flooding duration is critical in determining much of the structure of igapó forests (Ferreira 1997).

Flooding creates a gradient of black-water starting at the river's edge and continuing into the forest (Junk et al. 2000; Myster 2001, 2007b, 2010, 2015b) where it interacts with the topographic gradient so that tree species grow in elevation zones, directly related to period and duration of inundation. These tree species zonational distributions (Wittmann et al. 2010a) may be largely determined by the submergence tolerance of their seedlings (Parolin et al. 2004) and how far sunlight filters down the water column (2.0–2.5 m; Junk et al. 2011). Variation within the flooding regime greatly affects the distribution and abundance of plant species (Junk 1989; Lamotte 1990; Ferreira and Stohlgren 1999). Indeed, flooded forest types, vegetation formations, and plant communities lie on a continuum defined by the duration of the aquatic and terrestrial phases of the annual cycles and the physical stability of the habitat influenced by sedimentation and erosion processes (Junk et al. 2010). The predictability of the flood pulse can lead to species adaptations, including flood-resistant plant ecotypes of upland species and the coevolution of tree seeds and the fish that eat and disperse them (at least 100 species of fish species eat fruit; Junk 1997). Adaptations can be to periodic drought, as well as to the flooding, and can include synchronization of plant reproduction cycles and phenology. Common physiological plant adaptations to flooding include internal aeration, anoxia tolerance, and ability to prevent or repair oxidative damage during reaeration. Adaptations may be remnants of preadaptations from the non-flooded *terra firme* species in which floodplain trees originated (Kubitzki 1989).

Particularly for trees growing in igapó forests, (1) after fruit maturation, which occurs at high water levels, seeds fall into the water and may float and/or be submerged for several weeks. That is, seed production coincides with the flood pulse, in order to facilitate dispersal by water flow and by fish. Indeed morphological adaptations that enhance floatation, such as spongy tissues or large, air-filled spaces, are often present; (2) seed losses to predators and pathogens are intense (my own experimental results are presented later); (3) seeds germinate, and seedlings establish after flood waters recede; and while flooding may return after a few months, mortality of waterlogged and submerged seedlings is generally low compared to tree seeds in *terra firme* forest, and overall height growth and new leaf production are not severely affected by waterlogging in most species. The seed mass for trees in igapó is higher for trees compared to other flooded forests, and seeds of large mass enable rapid height growth of seedlings, allowing escape from submergence. Fast germination and growth may be crucial for survival of tree species, especially those that do not tolerate submersion and must grow tall quickly in order to maintain some leaves above the water surface when the flood comes. (4) Seedlings may be completely submerged for several months, and many species tolerate several weeks of submergence in a state of rest. Soon after the water recedes, leaves resprout and seedlings have a high ability to compensate for the period of rest induced by submergence; vigorous resprouting after seedling damage through rotting or mechanical injury may be considered further adaptations to effective establishment, and (5) for mature trees, the flood pulse triggers vegetative phenology; for leaves, common leaf adaptations include large epidermal cells, thick outer epidermis walls, thick cuticle, compact spongy parenchyma with only few and small intercellular spaces, sunken stomata, and transcurrent vascular bundles with a strong sclerenchymatous bundle sheath; for stems, periodical growth reductions as a consequence of flooding are reflected by the formation of increment rings and by periodic shoot elongation; and for roots, common root adaptations include hypertrophy of lenticels, formation of adventitious roots, plank buttressing and stilt rooting, development of aerenchyma, and the deposition of cell wall biopolymers such as suberin and lignin in the root peripheral cell layers (Parolin et al. 2004). Most species grow during the flooded times of the year and flower/fruit when the waters start to subside (Junk 1997; Junk et al. 2010). Moreover, there is an absence of both fast-growing tree species such as *Cecropia* spp., *Pseudobombax munguba*, and *Salix numboldfiana* (Junk et al. 2000) and aquatic macrophytes (but slow-growing sedges may be present). These are fragile ecosystems, which need protection and good management of human impacts, such as fishing and tourism. Books such as this one can help provide the science needed for these efforts.

Flooding also creates "niches," which may contribute to the endemism, where flood tolerance of seedlings may be a key factor in determining the structure and function of these forests (Parolin et al. 2004) and vegetative propagation is important for some species. Flooding creates oxygen deficiency, reduced photosynthesis, and low water conductance so that flooding may be a greater source of mortality than desiccation. Fine root production increases with inundation, however, and fine roots do most of the nutrient uptake usually in a thick superficial root mat

(Junk 1997). Igapó forests also have low leaf litter fall, low leaf nutrient content, and a slow decomposition rate with reduced annual tree ring width and root production, compared to other flooded forests (Junk 1997).

Igapó forests are, at least, primary (1°) igapó forests which form as a result of a predictable and seasonal flood pulse, secondary (2°) igapó forests which form due to other more unpredictable and unseasonal black-water flooding (low amplitude, short duration; Junk et al. 2000; Neff et al. 2006; Junk 2011) common to low-order rivers (Junk et al. 2000), and successional igapó forests which form due to disturbances both natural (e.g., tree fall) and human (e.g., logging/harvesting, conversion to crop land and pasture; Myster 2007a) creating patches within 1° and 2° forests. River margins are common places of disturbance, and primary succession on sandy soil can take decades (Author, "unpublished data").

Igapó forests have a low density and low productivity of herbaceous plants. The trees grow slowly and can get up to 1000 years old (Junk et al. 2011) with adaptations such as a thick cuticle, sunken stomata, wax, and hairs to help reduce effects of drought on these sandy soils. Important tree families in igapó are Fabaceae, Euphorbiaceae, Sapotaceae, Lecythidaceae, Malpighiaceae, Combretaceae, and Ochnaceae (Ferreira and Prance 1998). Among the most common tree species are *Couepia paraensis*, *Hevea spruceana*, *Inga punctata*, *Macrolobium acaciifolium*, *Malouetia tamaquarina*, *Piranhea trifoliata*, *Pouteria elegans*, *Swartzia polyphylla*, and *Tabebuia barbata* (Wittmann et al. 2004).

Sampling of physical structural parameters of trees (see Myster 2016 for a review) at least 10 cm dbh in a 1 ha igapó forest plot found 546 stems and 119 species (Ayres 1993), 451 stems and 90 species (Ferreira 1991), and 44 tree species/796 stems/20–22 m^2 total basal area at 8.6 mean inundation height, 103/941/41 at 4.8, and 137/1130/34 at 2.1 (Ferreira 1997) as richness increased as height decreased. Inuma (2006) found 26 species at inundation levels between 5 and 7 m, 35 species at levels of 3–5 m, and 44 species at levels 1–3 m. In clear-water forests, Ferreira and Prance (1998) recorded 21, 24, and 30 tree species per ha (< 5 cm dbh), and Campbell et al. (1986) recorded 40 species (> = 10 cm dbh) for 1/2 ha.

Tree diversity decreases as flooding severity increases, and forests in black-water are less diverse (in plants and animals), with less litter production, smaller trees, low aboveground biomass, and less herbaceous growth than other flooded forests (e.g., várzea; Pinedo-Vasquez et al. 2011) and unflooded forests (e.g., *terra firme*; Myster 2017a) in the Amazon (Junk et al. 2010) and with few tree species in common. Igapó forests also have lower primary production (Ayres 1993), 30% less litter production, and at most 2/3 of the growth rate of trees in those other forests. Reserves of N, P, K, Ca, and Mg are higher in the biomass than in the soil (Junk et al. 2000) where leaves last longer and are smaller, vertically oriented, and scleromorphic (like white sand forest) with relatively more defense compounds.

Finally, the uniqueness and high degree of endemism of igapó forests may be due in part to the dynamism of the Amazon and its tributaries, often changing their routes within a time span of a few decades (Pires and Prance 1985; Junk 1989; Kalliola et al. 1991). It very well may be that forests that are unflooded today were flooded in the past and vice versa leading to species found in unflooded forest establishing ecotypes (Myster and Fetcher 2005) in the flooded forest (Wittmann et al. 2004, 2010a, b). This ecotropic dynamic and the ability of tree species from the surrounding unflooded *terra firme* forest to form ecotypes, combined with flooding and its associated environmental heterogeneity, may contribute to the unique biology and ecology found in igapó forests (Kalliola et al. 1991). Moreover the predictability of the flood pulse (Junk et al. 2010) – both past and present – facilitates adaptation, and this, along with differences in the surrounding biota and a variety of soil types (Junk 1989, Honorio 2006), may help create complex and diverse igapó forest associations throughout the Amazon Basin (Myster 2009).

1.2 Case Study: Primary (1°) Igapó Forests at Área de Conservación Regional Comunal de Tamshiyacu-Tahuayo

I conducted my studies of igapó forests at two different sites in the Peruvian Amazon, flying to Iquitos and then taking various boats and canoes from there. My first study site was the Área de Conservación Regional Comunal de Tamshiyacu-Tahuayo (ACRCTT; www.perujungle.com, Myster 2007b, 2009, 2010, 2015b; Fig. 1.2) located in Loreto Province, 80 miles southeast of Iquitos (~2° S, 75° W) with an elevation of 106 m. The reserve is part of one of the largest (420, 080 ha) protected areas in the Amazon (https://natureandculture.org/tag/procrel/) containing

Fig. 1.2 Photograph of the author and a scarlet macaw at ACRCTT

wet lowland tropical rainforest of high diversity (Daly and Prance 1989). ACRCTT is comprised of low, seasonally inundated river basins of the upper Amazon and named for two of the major rivers (the Tahuayo and the Tamshiyacu) which form boundaries to the north and west and create large fringing floodplains (Junk 1984). The substrate of these forests is composed of alluvial and fluvial Holocene sediments from the eastern slopes of the Andes. Annual precipitation ranges from 2.4 to 3.0 m per year, and the rainy season is between November and April (Kalliola et al. 1991). The average temperature is relatively steady at 26° C.

Within the ACRCTT are areas of regular, seasonal black-water runoff that create 1° igapó forests of differing flooding frequency, duration, and maximum water column height. Common tree species in those igapó forests include *Calycophyllum spruceanum*, *Ceiba samauma*, *Inga* spp., *Cedrela odorata*, *Copaifera reticulata*, and *Phytelephas macrocarpa* with understory palms such as *Guazuma rosea* and *Piptadenia pteroclada* which are also common (Daly and Prance 1989; Myster 2007b; Prance 1979; Puhakka et al. 1992).

1.2.1 1 ha Plot: Floristics and Physical Structure Sampling

In May 2011, my field assistant and I set up a 1 ha plot in a 1° igapó forest at ACRCTT which is underwater 3–5 months every year (Myster 2013, 2015a). My field assistant suggested a large area of this kind of forest, and we then choose the 1 ha plot randomly within it. We measured the diameter at breast height (dbh) of all trees at least 10 cm dbh in the plot. The dbh measurement was taken at the nearest lower point where the stem was cylindrical, and for buttressed trees, it was taken above the buttresses. We also identified each tree to species, or to genus in a few cases when species identification was not possible, using Romoleroux et al. (1997) and Gentry (1993) as taxonomic sources, and consulted both the Universidad Nacional de la Amazonía Peruana (www.unapiquitos.edu.pe) herbarium in Iquitos and the web site of the Missouri Botanical Garden (http://www.missouribotanicalgarden.org).

From the data, I compiled floristic tables of family, genus, and species and then generated these physical structure parameters:

1. The total number of stems, the mean dbh among those stems, and the number of stems in each of four size classes: 10 cm < 20 cm dbh, 20 cm < 30 cm dbh, 30 cm < 40 cm dbh, and those with a dbh ≥ 40 cm
2. The total basal area which is the sum of the basal areas of all individual stems ($\sum \pi r^2$, where r = the dbh of the individual stem/2)
3. The aboveground biomass (AGB) of all the trees (using the formula in Nascimento and Laurance (2002) suggested for tropical trees of these stem sizes)
4. The percent closure of the canopy (using the formula in Buchholz et al. (2004) for tropical trees)

5. The stem dispersion pattern (random, uniform, clumped) computed by comparing plot data to Poisson and negative binomial distributions using Chi-square analysis
6. If clumped, Green's index was computed toaccess degree of clumping (Ludwig and Reynolds 1988; Myster and Pickett 1992).

I found 16 families, 29 genera, and 31 species in the 1 ha plot (Table 1.1) where Fabaceae was the most abundant family with also the most genera and species. Arecaceae was also abundant, and Meliaceae was the only family with only one stem. *Astrocaryum*, *Tovomita*, *Parkia*, *Eschweilera*, and *Mouriri* were the genera with more than one species. The most abundant species were *Aldina latifolia* with nine stems and *Caraipa grandifolia*, *Mabea* sp., *Hydrochori* sp., *Virola elongata*, and *Psychotria lupulina* with eight stems. There were a total of 167 tree stems in the 1 ha plot: 84 of those were in the first size class, 58 in the second size class, 17 in the third size class, and 8 in the fourth size class, showing a steep decline in stem size with abundance. The average dbh was 22.3 cm. The total basal area was 6.52 m^2, and AGB was 202 Mg. The canopy had a closure of 12.31% with a weak clumped spatial pattern (Green's index, 0.17; Myster and Pickett 1992).

1.2.2 1 ha Plot: Seed Predation, Seed Pathogens, and Germination Experiment

To understand recruitment in these forests, I next set out seeds on transects for a week in the 1 ha plot sampled above and then scored them for loss due to predators, loss due to seed pathogens, and those that germinated. I found that the largest difference was among the three seed processes with seed predation taking two to three times (mean 51%) as many seeds as either pathogenic disease (16%) or germination (20%). Within each process, however, there were significant differences among the seed test species: *Attalea butyracea* seeds survived predation best, *Qualea paraensis* seeds survived pathogens best, and *Macrolobium acaciifolium* germinated the least (Author, "unpublished data"). *Drypetes amazonica* was the other tree species used in the experiment.

1.2.3 0.01 ha Flooding × Tree Fall Gap Plots: Soils, Floristics, and Physical Structure

I also set up smaller plots in a 1° igapó forest at ACRCTT which is underwater 1–2 months every year, in a 1° igapó forest at ACRCTT which is underwater 3–5 months every year, and in a 1° igapó forest at ACRCTT which is underwater at least 6 months every year (a flooding gradient in igapó; Junk et al. 2011). All plots were in 1° unlogged forests with fresh average-sized (100–300 m^2; Brokaw 1982)

Table 1.1 All tree stems and their density sampled in the 1 ha plot sorted first by family, then by genus, and finally by species

Family	Genus	Species	Stem density
Apocynaceae	*Microplum*	*anomala*	3
Arecaceae	*Astrocaryum*	*jauari*	7
Arecaceae	*Astrocaryum*	*murumuru*	1
Arecaceae	*Euterpe*	*precatoria*	5
Arecaceae	*Oenocarpus*	*mapora*	2
Arecaceae	*Socratea*	*exorrhiza*	3
Calophyllaceae	*Caraipa*	*grandifolia*	8
Chrysobalanaceae	*Hirtella*	*racemosa*	1
Chrysobalanaceae	*Licania*	sp.	4
Clusiaceae	*Tovomita*	*macrophylla*	3
Clusiaceae	*Tovomita*	sp.	4
Euphorbiaceae	*Mabea*	sp.	8
Fabaceae	*Aldina*	*latifolia*	9
Fabaceae	*Campsiandra*	*angustifolia*	5
Fabaceae	*Crudia*	*amazonica*	2
Fabaceae	*Cynometra*	*spruceana*	1
Fabaceae	*Hydrochori*	sp.	8
Fabaceae	*Macrosamanea*	*amplissima*	5
Fabaceae	*Parkia*	*auriculata*	1
Fabaceae	*Parkia*	*pectinata*	6
Fabaceae	*Pithecellobium*	sp.	2
Lauraceae	*Nectandra*	sp.	1
Lauraceae	*Ocotea*	*aciphylla*	4
Lecythidaceae	*Eschweilera*	*albiflora*	6
Lecythidaceae	*Eschweilera*	*parviflora*	7
Malpighiaceae	*Acmanther*	*latifolia*	4
Malvaceae	*Theobroma*	*cacao*	4
Melastomataceae	*Mouriri*	*apiranga*	2
Melastomataceae	*Mouriri*	*grandiflora*	1
Melastomataceae	*Mouriri*	*myrtifolia*	6
Meliaceae	*Guarea*	*macrophylla*	1
Moraceae	*Brosimum*	*lactescens*	5
Moraceae	*Maquira*	*coriacea*	5
Moraceae	*Trymatoco*	*amazonicus*	1
Myristicaceae	*Virola*	*elongata*	8
Rubiaceae	*Ferdinandusa*	*rudgeoides*	4
Rubiaceae	*Psychotria*	*lupulina*	9
Salicaceae	*Casearia*	sp.	2
Sapotaceae	*Manilkara*	*bidentata*	2
Sapotaceae	*Pouteria*	*elegans*	3
Vochysiaceae	*Qualea*	*paraensis*	5

Table 1.2 All tree stems and their density sampled in the five 100 m² plots put together were sorted first by family, then by genus, and finally by species. I1 refers to the 1° igapó forest at ACRCTT forest that was underwater 1–2 months per year, I2 refers to the 1° igapó forest at ACRCTT that was underwater 3–5 months per year, and I3 refers to the 1° igapó forest at ACRCTT that was underwater at least 6 months per year. F refers to the closed-canopy forest, and G refers to its tree fall gap

			I1	I2	I3
			F G	F G	F G
Araliaceae	*Dendropanax*	sp.	1 0	1 0	0 1
Chrysobalanaceae	*Licania*	sp.	3 2	1 2	1 1
Euphorbiaceae	*Alchornea*	sp.	1 0	0 2	1 0
Euphorbiaceae	*Hevea*	*brasiliensis*	1 1	2 0	0 0
Lecythidaceae	*Couratari*	*guianensis*	2 1	1 1	0 0
Melastomataceae	*Mouriri*	*myrtifolia*	4 2	0 2	0 1
Myristicaceae	*Virola*	*elongata*	1 2	1 0	2 0
Rubiaceae	*Genipa*	sp.	1 0	1 1	0 1
Total stems			14 8	7 8	4 4
Total species			8 5	6 5	3 4

tree fall gaps within them. Specifically, I set up five replicates and randomly placed 100 m² (10 m × 10 m) plots within each of the three 1° forest types and also in the center of each of their tree fall gaps, for a total of 30 plots (5 × 3 plus 5 × 3). This data is housed in the archives of the Long-Term Ecological Research (LTER) site in Puerto Rico (LTERDBAS #150; luq.lternet.edu).

Within those plots, I first took five soil samples without litter (Myster 2017b) in May 2009 at five different random locations in each of the three 1° forest plots. I found soil pH increased monotonically slightly from 3.65 to 3.82 in forests as months underwater increased as did organic matter (32.20–40.64%), phosphorus (25.50–39.00 ppm), and potassium (107–137 ppm), but nitrogen decreased monotonically from 148 to 101 ppm across the same flooding gradient. I next sampled the plots for tree stems using the same protocol as in the 1 ha plot (see Myster 2007b, 2010, 2015b) and found for floristics (Table 1.2) (1) common species between all three forest types and their gaps as well as among all three forest types, (2) tree stem density and richness decreased in forests as number of months underwater increased, and (3) dominance-diversity curves have more dominance by single species in the least flooded forest compared to other forests.

For physical structure (Table 1.3, Figs. 1.3 and 1.4), there was a significant effect of tree fall gap formation on canopy average height, canopy maximum height, basal area, density, AGB, turnover, and alpha diversity and a significant effect of forest type on species richness, genera richness, density, turnover, and alpha diversity. In general, there were fewer trees, but they were larger and more productive in the forest plots compared with the gap plots, and the most flooded plots had fewer trees, species, and genera compared with both the less flooded forest and non-flooded forest. In addition, the greatest amount of turnover was in the most flooded forests, and the intermediately flooded forest had the greatest richness and alpha diversity.

Table 1.3 F statistic and significant p-value summary table for all main and interactive effects. The main effect openness refers to the comparison between the closed-canopy forest and its gap regardless of forest type. The main effect forest type refers to the comparison between forest types regardless of whether the forest was closed-canopy or gap. A p-value between 0.05 and 0.01 is indicated by *, a p-value between 0.01 and 0.001 is indicated by **, and a p-value that is less than 0.001 is indicated by ***

Response variable	Main effect (openness)	Main effect (forest type)	Interaction effect
Species richness	3.66	8.21**	5.64**
Genera richness	2.62	7.84**	5.57**
Average height	30.21***	3.20	1.22
Maximum height	17.25**	0.62	0.09
Basal area	6.89*	0.20	0.10
Density	4.84*	13.68***	2.93
Aboveground biomass	21.26**	1.98	2.34
Turnover	7.02*	5.02*	3.02
Fisher's α diversity	4.98*	5.12*	6.43*
Family richness	2.22	3.33	1.98
No. of unique species	6.06*	2.01	3.33
Mean stem size	3.02	2.76	5.45*
No. of stems 1 < 2 cm dbh	5.55*	5.02*	7.63**
No. of stems 2 < 4 cm dbh	3.84	4.04	4.55
No. of stems 4 < 10 cm dbh	3.02	3.33	3.59
No. of stems At least 10 cm dbh	4.97	2.91	4.72
Green's index	5.43*	2.41	3.02
% canopy coverage	5.06*	4.97*	6.38*

Canopy structure was determined by traditional gap dynamics, but much of canopy diversity depended on the type of forest; tree density decreased as flooding increased, especially among the smallest stems; and there was evidence to suggest that the high biodiversity of the Amazon may be maintained in part by the existence of moderately flooded forest and gaps.

I also found (1) increased flooding decreased family richness in the closed-canopy forests but increased it in their gaps, with no trends for number of unique species, (2) flooding decreased stem size everywhere as did the number of stems as size increased especially for larger stems, (3) Green's index showed clumping only for least flooded forests with the closed-canopy forest showing more than its tree fall gap, and (4) both flooding and tree fall gap creation decreased canopy coverage perhaps as an additive effect. Further, among the stem size classes, only the smallest stems were significantly affected by openness and by the type of forest, with a significant interaction term where flooding significantly decreased the number of these smaller stems in all forests and their gaps, except those with the

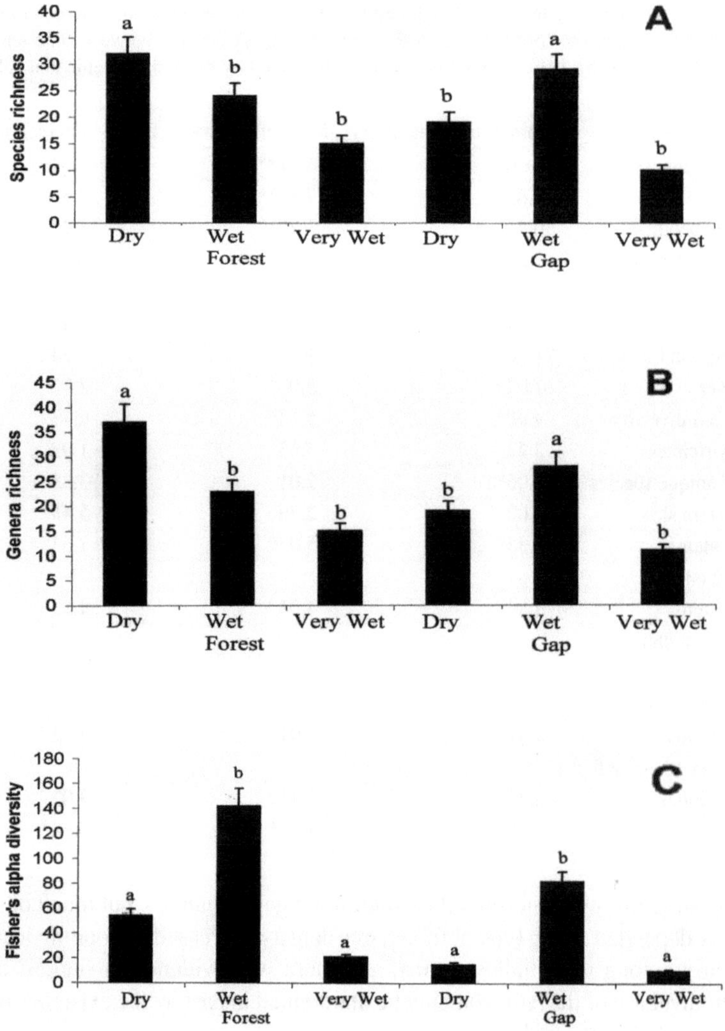

Fig. 1.3 Significant interactive effects from Table 1.3 are given as **means** and standard errors of (**a**) species richness, (**b**) genera richness, and (**c**) Fisher's α diversity. Means testing results are indicated by lowercase letters where significantly different groups have different letters. The 1° igapó forest at ACRCTT underwater for 1–2 months is labeled as dry, the 1° igapó forest at ACRCTT underwater 3–5 months is labeled as wet, and the 1° igapó forest at ACRCTT underwater at least 6 months is labeled very wet

highest level of flooding. Also tree fall gaps had significantly more, smaller stems than their forests in the two most flooded forests but not in the least flooded forest. I conclude that flooding is a greater stressor and influence on the structure of these forests than tree fall, and so, in these forests, gradients and disturbances overlap in their traditional roles..

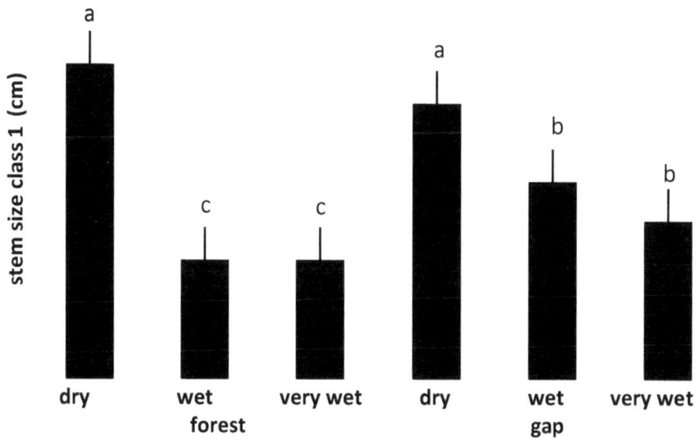

Fig. 1.4 The number of stems 1 < 2 cm dbh significant interactive effect from Table 1.3. Labelings are the same as in Fig. 1.3

1.2.4 0.01 ha Flooding × Tree Fall Gap Plots: Seed Predation, Seed Pathogens, and Germination

I set out seeds on transects for a week in these same plots and found (1) seed predation took more seeds then either seed pathogenic disease or germination for most seed species, but there were a few species that lost more seeds to pathogens than predators. Germination was lower than predation and pathogens for most species, but there were a few species where germination was higher than pathogens, and (2) predation decreased monotonically (and pathogens increased monotonically) as months underwater increased in black-water forests. I conclude that seed predation is the major post-dispersal filter for regeneration in these forests but pathogenic disease can play a major role, especially in forests that have water in them for long periods each year so that flooding may change those forests dramatically by altering the actions of seed mechanisms and tolerances (Myster 2017c).

1.3 Case Study: Secondary (2°) Igapó Forests at Sabalillo Forest Reserve

My second study site was Sabalillo Forest Reserve (SFR, 3° 20′ 3″ S, 72° 18′ 6″ W; Frederickson et al. 2005; Moreau 2008) established in 2000 and operated by the Project Amazonas (www.projectamazonas.org). SFR is located on both sides of the upper Rio Apayacuo, 172 km east of Iquitos, Peru. The reserve is part of 25,000 acres set aside over the last decade and is comprised of low, seasonally inundated river basins of the upper Amazon. The substrate of these forests is composed of alluvial and fluvial Holocene sediments from the eastern slopes of the Andes. Annual precipitation is 3.297 m per year (Choo et al. 2007). Within the SFR are areas of black-water runoff that create igapó forests of differing flooding duration and maximum water column height, where the rainy season is between November and April.

1.3.1 Small Plots: Floristics and Physical Structure

I investigated the floristics and physical structure of 2° igapó forests by setting up and sampling, in June 2013, five successional areas – island, oxbow lake, river margin, sandy beach, and side creek (Ferreira 1997) – close to a black-water river at SFR (Author, "unpublished data"). In each area we set up 10 5 m × 5 m continuous plots for a total sampling area of 250 m^2 per area. We measured the diameter at breast height (dbh) of each tree at least 1 cm dbh within each plot. The dbh measurement was again taken at the nearest lower point where the stem was cylindrical, and for buttressed trees it was taken above the buttresses. We identified its species, or to genus in a few cases, using Romoleroux et al. (1997) and Gentry (1993) as taxonomic sources. We also consulted the Universidad Nacional de la Amazonía Peruana herbarium and the web site of the Missouri Botanical Garden <www.mobot.org>. I found (1) 24 plant families where Urticaceae was the most common family, but Rubiaceae and Euphorbiaceae were also common. Most families had a monotonic decline in stem number as stems get thicker, and there were no stems with a dbh greater than 29 cm. The most common species were *Cecropia membranacea*, *Sapium glandulosum*, *Pourouma guianensis*, and *Byrsonima arthropoda*, and (2) the greatest number of stems was 47 in the island area and 18 in the forest underwater the longest. Mean stem size, species richness, Fisher's α, basal area, and AGB were lowest in the sandy beach and highest in the forest underwater the shortest. Results show that, as soil became less sandy and with more clay content, all structural parameters except stem number increased, suggesting an increase in forest community complexity as soil increases in water retention capacity and nutrients. I conclude that the severe and unpredictable flooding seen in these successional areas reduces forest structure more than the predictable and seasonal flooding that blackwater floodplain forests receive.

1.3.2 Small Plots: Soil Bulk Density and Its Predictive Ability

Finally, I collected three soil samples in each area by driving a 3-in-diameter ring into a depth of 10 cm and extracting the soil. Back in Iquitos, each soil sample was dried for three 4-min cycles in the microwave and then weighted. Soil bulk density is the weighted soil sample/volume of container, expressed as g/cm^3. I found that soil bulk density was highest in the sandy beach and lowest in the forest underwater the longest, and linear regression analysis showed that soil bulk density could best predict mean stem size, species richness, and Fisher's α (Author, "unpublished data").

1.4 Compilation Studies

I established 1 ha plots in all 5 of the most common forest types in Western Amazonia: *terra firme*, white sand and palm (unflooded), várzea, and igapó (flooded). I found in those plots that the most common families were Arecaceae, Fabaceae, and Clusiaceae which were also found in all five plots along with Euphorbiaceae; ordinations based on familial data showed that palm forest and white sand forest were the most different and that *terra firme* forest, várzea forest, and igapó forest were similar, but ordinations based on physical structure parameters showed a more spread out and individualistic pattern where várzea was most different, with palm/*terra firme* and white sand/igapó similar to each other. Further *terra firme* had the most significant species associations, and igapó had the least. I found also that *terra firme* and várzea had the most negative associations relative to positive associations and that igapó, palm, and white sand had the most positive associations relative to negative associations (Author, "unpublished data"). I concluded that soil characteristics seem to largely determine floristic composition and flooding largely determines physical structure. As forests become more stressed, either by flooding or loss of soil fertility, positive associations became more common suggesting less competition and more facilitation among the trees. Finally results that show a hierarchy of uniqueness among the forest types, with some variation depending on the parameters used, suggested a priority for conservation and management.

I also sampled soils and the seed rain in these five forest types and found (1) soil pH of the non-flooded forests was very similar to flooded forests but became more basic with increased flooding; (2) soil organic matter was lowest in the two non-flooded *terra firme* forests and also increased with flooding; (3) nitrogen (N) was lowest in the palm forest, phosphorus (P) was lowest in *terra firme*-low terrace forest, and potassium (K) was lowest in the *terra firme*-high terrace forest; (4) while N decreased sharply with flooding, both P and K increased with length of the flooding period; and (5) for some non-flooded forests, there was a correspondence between soil fertility and floristic similarity. I concluded that flooding has significant effects on nutrient availability of Amazonian forest soils by increasing the concentration of some nutrients but decreasing it for others.

For the seed rain, I found (1) all forests except black-water tahuampa contained seeds of tree species that have been sampled in other studies within a forest of the same type, but in all forests there were seeds of several tree species that have not yet sampled within their forest type; (2) total seed load peaked in the early part of the year – near the end of the rainy season – and then decreased monotonically over the remainder of the year for all forests; (3) species richness was greater in unflooded forests compared to flooded forests, and the largest number of species were found in *terra firme*; (4) seeds were more evenly distributed among species in the unflooded forests compared to the flooded forests; and (5) alpha diversity was much greater in *terra firme* compared to all other forests. I concluded that, for the unflooded forests, seed species number and richness increased with soil fertility, but for the flooded forests seed species number and richness decreased with months underwater. When taken together, results suggested that for forests across the Amazonian landscape, differences in flooding regime may have a greater effect on both seed rain load and seed species richness than differences in availability of soil nutrients.

Finally, field experiments in the five forest types have shown that seed predation took more seeds than either seed pathogenic disease or germination for most seed species but there were a few species that lost more seeds to pathogens than predators. Germination had the lowest percentage (%) for most species, but again there were a few species where germination % was higher than pathogens; within the unflooded forests, there was the most predation in *terra firme* forest, palm forest lost more seeds to pathogens than predators, and white sand forest predation levels were between the other two forest types; within the flooded forests, predation decreased as water went from white to black (at the same inundation levels), and predation decreased monotonically (and pathogens increased monotonically) as months underwater increased in black-water forests, and there was significantly more seed predation in the unflooded forests compared to the flooded forests but significantly more germination in the flooded forests compared to the unflooded forests. I concluded that seed predation is the major post-dispersal filter for regeneration in these forests but pathogenic disease can play a major role, especially in forests that have water in them for long periods each year so that flooding may change those forests dramatically by altering the actions of seed mechanisms and tolerances (Myster 2017c).

1.5 About This Book

Here I take advantage of my many years working in the Peruvian Amazon, and my award from the Fulbright Foundation to work at UNAP, to edit this book based solely on igapó forests. Although this will be the first book focused exclusively on igapó forests of the Amazon Basin – and thus its chapter scope and detail are unique – there have been many books written on parts, or all, of the Amazon in the past. And among these books, perhaps the most recent books that have been published which are closest to this one were on the central Amazon (Junk 1997; Junk et al. 2000, 2010). I look upon these books, however, as complementary to this book – especially the book on white water (várzea) forests (Padoch and Pinedo-Vasquez 2010) also published by Springer-Verlag – not as competition.

My organizing theme will be the structure, function, and dynamics of igapó forests in the Amazon Basin: how they were in the past, how they are changing today, and how they are likely to change in the future. I focus on their uniqueness due to their high level of complexity defined as the many ways that different components of igapó forests in the Amazon Basin ecosystem interact and also on how those interactions are on a higher order compared to other tropical forests (Myster 2017a). Chapter authors will discuss interactions within their own area of interest and will ask if the drivers for changes (e.g., climate change, human disturbances, tree recruitment mechanisms, stress) differ now compared to the past.

The book will contain sections devoted to (1) igapó over space and time; (2) water and light; (3) soil and the carbon cycle; (4) litter, fungi, and invertebrates; (5) vertebrates; (6) plants; and (7) human impacts and management. Authors will prepare chapters that consist of reviews of what is known about their topic, of the research they have done, and of what research needs to be done in the future often with a new conceptual and/or mathematical model. Authors will be concerned about the structure, function, and dynamics of igapó forests: how they were in the past, how they are changing today, and how they are likely to change in the future.

My concluding chapter will first summarize the results from each of the preceding chapters, then synthesize those results adding to conceptual models, and finally suggest future avenues for research. Conceptual models will make use of tree replacement, or barriers to replacement, by individual trees within these forests (Myster 2012).

Acknowledgments I thank Dr. Paul Beaver and the staff at the Área de Conservación Regional Comunal de Tamshiyacu-Tahuayo and Dr. Devon Michaels and the staff at the Sabalillo Forest Reserve for their help in facilitating my research in the Peruvian Amazon. I also thank the Fulbright Foundation for a teaching/research award at the Universidad Nacional de la Amazonía Peruana in Iquitos, Peru.

References

Ayres JMC (1993) As matas de varzea do Mamiraua. MCT-CNPq-Programa do tropic umido, Sociedade civil de Mamiraua, Brasil

Brokaw NVL (1982) The definition of treefall gap and its effect on measures of forest dynamics. Biotropica 11:158–160

Buchholz T, Tennigkeit T, Weinreich A (2004) Maesopsis eminii – a challenging timber tree species in Uganda – a production model for commercial forestry and smallholders. Proceedings of the international union of forestry research organizations (IUFRO) conference on the economics and management of high productivity plantations, Lugo, Spain

Burke MK, King SL, Gartner D, Eisenbies MH (2003) Vegetation, soil, and flooding relationships in a Blackwater floodplain forest. Wetlands 23:9980–1002

Campbell DG, Daly DC, Prance GT, Maciel UN (1986) Quantitative ecological inventory of *terra firme* and varzea tropical forest on the Rio Xingu, Brazilian Amazon. Brittonia 38:369–393

Choo JPS, Martinez RV, Stiles EW (2007) Diversity and abundance of plants with flowers and fruits from October 2001 to September 2002 in Paucarillo Reserve, Northeastern Amazon. Peru Revisita Peru Biology 14:25–31

Daly DG, Prance GT (1989) Brazilian Amazon. In: Campbell DG, Hammond HD (eds) Floristic inventory of tropical countries. New York Botanical Garden, Bronx, pp 401–426

Ferreira LV (1991) Oefeito do period de inundacao, na distribuicao, fenologia e regeneracao de plantas emuma floresta de igapo na Amazonia Central. Ms thesis. Instituto Nacional de Pesquisas da Amazonia, INPA, Brasil

Ferreira LV (1997) Effects of the duration of flooding on species richness and floristic composition in three hectares in the Jau National Park in floodplain forests in central Amazonia. Biodivers Conserv 6:1353–1363

Ferreira LV, Prance GT (1998) Species richness and floristic composition in four hectares in the Jau national park in upland forest in Central Amazonia. Biodivers Conserv 7:1349–1361

Ferreira LV, Stohlgren TJ (1999) Effects of river level fluctuation on plant species richness, diversity, and distribution in a floodplain forest in Central Amazonia. Oecologia 120:582–587
Frederickson ME, Greene MJ, Gordon DM (2005) 'Devil's garden' bedeviled by ants. Nature 437:495–496
Gentry AH (1993) A field guide to the families and genera of woody plants of Northwest South America (Colombia, Ecuador, Peru) with supplementary notes on herbaceous taxa. Conservation International, Washington, DC
Honorio EN (2006) Floristic relationships of the tree flora of Jenaro Herrera, an unusual area of the Peruvian Amazon. M.S. thesis, University of Edinburgh, Edinburgh, UK
Inuma JJ (2006) Comparação na diversidade e estrutura das comunidades de plantas lenhosas da terra firme, várzea e igapó do Amaná, Amazônia Central. PhD thesis, Instituto Nacional de Pesquisas da Amazônia
Junk WJ (1984) Ecology of the Varzea, floodplains of Amazonian white-water rivers. In: Junk WJ (ed) The Amazon: limnology and landscape ecology of a mighty tropical river and its basin. Kluwer, Dordrecht, pp 215–243
Junk WJ (1989) Flood tolerance and tree distribution in central Amazonian floodplains. In: Holm-Nielsen LB, Nielsen IC, Balslev H (eds) Tropical forests: botanical dynamics, speciation and diversity. Academic, New York, pp 47–64
Junk WJ (1997) The central Amazon floodplain: ecology of a pulsing system. Springer, Berlin
Junk WJ, Furch K (1985) The physical and chemical properties of Amazonian waters and their relationship with the Biota. In: Prance GT, Lovejoy TE (eds) Amazonia. Pergamon Press, Ltd., Oxford, pp 3–17
Junk WJ, Ohly JJ, Piedade MTF, Soares MGM (2000) The Central Amazon floodplain: actual use and options for a sustainable management. Backhuys Publishers, Leiden
Junk WJ, Piedade MTF, Parolin P, Wittman F, Schongart J (2010) Amazonian floodplain forests: ecophysiology, biodiversity and sustainable management. Ecological Studies. Springer, Berlin
Junk WJ, Piedade MTF, Schongart J, Cohn-Haft M, Adency JM, Wittmann F (2011) A classification of major naturally-occurring Amazonian lowland wetlands. Wetlands 31:623. https://doi.org/10.1007/s13157-011-0190-7
Kalliola RS, Jukka M, Puhakka M, Rajasilta M (1991) New site formation and colonizing vegetation in primary succession on the western Amazon floodplains. J Ecol 79:877–901
Lamotte S (1990) Fluvial dynamics and succession in the Lower Ucayali River basin, Peruvian Amazonia. For Ecol Manage 33:141–156
Ludwig JA, Reynolds JF (1988) Statistical ecology. Wiley, New York
Moreau CS (2008) Unraveling the evolutionary history of the hyperdiverse ant genus *Pheidole* (Hymenoptera: Formicidae). Mol Physiol Evol 48:224–239
Myster RW (2001) Mechanisms of plant response to gradients and after disturbances. Bot Rev 67:441–452
Myster RW (2007a) Post-agricultural succession in the Neotropics. Springer, Berlin
Myster RW (2007b) Interactive effects of flooding and forest gap formation on composition and abundance in the Peruvian Amazon. Folia Geobot 42:1–9
Myster RW (2009) Plant communities of Western Amazonia. Bot Rev 75:271–291
Myster RW (2010) Flooding duration and treefall interactive effects on plant community richness, structure and alpha diversity in the Peruvian Amazon. Ecotropica 16:43–49
Myster RW (2012) Plants replacing plants: the future of community modeling and research. Bot Rev 78:2–9
Myster RW (2013) The effects of flooding on forest floristics and physical structure in the Amazon: results from two permanent plots. For Res 2:112. https://doi.org/10.4172/2168-9776.1000112
Myster RW (2015a) Black-water forests (igapó) vs. white-water forests (várzea) in the Amazon: floristics and physical structure. The Biologist (Lima) 13:391–406
Myster RW (2015b) Flooding x tree fall gap interactive effects on black-water forest floristics and physical structure in the Peruvian Amazon. J Plant Inter 10:126–131
Myster RW (2016) The physical structure of Amazon forests: a review. Bot Rev 82:407–427

Myster RW (2017a) Forest structure, function and dynamics in Western Amazonia. Wiley, Oxford
Myster RW (2017b) A comparison of the forest soils in the Peruvian Amazon: *Terra firme*, palm, white sand and igapó. J Soil Sci Envir Manag 8:130–134
Myster RW (2017c) Comparing and contrasting flooded and unflooded forests in Western Amazonia: seed predation, seed pathogens, germination. Community Ecol 18:169–174
Myster RW, Fetcher N (2005) Ecotypic differentiation and plant growth in the Luquillo Mountains of Puerto Rico. J Trop For Sci 17:163–169
Myster RW, Pickett STA (1992) Effects of palatability and dispersal mode on spatial patterns of tree seedlings in old fields. Bull Torrey Bot Club 119:145–151
Nascimento HEM, Lawrance WF (2002) Total aboveground biomass in central Amazonian rainforest: a landscape-scale study. For Ecol Manage 68:311–321
Neff T, Lucas RM, Dos-Santos JR, Brondizio ES, Freitas CC (2006) Area and age of secondary forests in Brazilian Amazonia 1978–2002: an empirical estimate. Ecosystems 9:609–623
Padoch C, Pinedo-Vasquez M (2010) The Amazon várzea: the decade past and the decade ahead. Springer, Berlin
Pallardy S (2008) Physiology of Woody plants. Academic press, New York
Parolin P, DeSimone O, Haase K, Waldhoff D, Rottenberger S, Kuhn U, Kesselmeier J, Kleiss B, Schmidt W, Piedade MTF, Junk WJ (2004) Central Amazonian floodplain forests: tree adaptations in a pulsing system. Bot Rev 70:357–380
Pinedo-Vasquez M, Ruffino ML, Padoch C, Brondizio ES (2011) The Amazon Varzea: the decade past and the decade ahead. Springer, Berlin
Pires JM, Prance GT (1985) The vegetation types of the Brazilian Amazon. In: Prance GT, Lovejoy TE (eds) Amazonia. Pergamon press, Oxford, pp 109–145
Prance GT (1979) Notes on the vegetation of Amazonia III. The terminology of Amazonian forest types subject to inundation. Brittonia 31:26–38
Puhakka M, Kalliola R, Rajasilta M, Salo J (1992) River types, site evolution and successional vegeation patterns in Peruvian Amazonia. J Biogeogr 19:651–665
Romoleroux K, Foster R, Valencia R, Condit R, Balslev H, Losos E (1997) Especies lenosas (dap => 1 cm) encontradas en dos hectareas de un bosque de la Amazonia ecuatoriana. In: Valencia R, Balslev H (eds) Estudios Sobre Diversidad y Ecologia de Plantas. Pontificia Universidad Catolica del Ecuador, Quito, pp 189–215
Salati E (1985) The climatology and hydrology of Amazonia. In: Prance GT, Lovejoy TE (eds) Amazonia. Pergamon Press, Ltd., Oxford, pp 18–48
Tukey HB (1970) The leaching of substances from plants. Annu Rev Plant Physiol 21:305–324
Wittmann FW, Junk J, Piedade MTF (2004) The varzea forests in Amazonia: flooding and the highly dynamic geomorphology interact with natural forest succession. For Ecol Manage 196:199–212
Wittmann F, Junk WJ, Schongart J (2010a) Phytogeography, species diversity, community structure and dynamics of central Amazonian floodplain forest. In: Junk WJ, Piedade MTF, Parolin P, Wittman F, Schongart J (eds) Amazonian floodplain forests: ecophysiology, biodiversity and sustainable management. Ecological Studies. Springer, Berlin
Wittmann A, Lopes A, Conserva A, Wittmann F, Piedade M (2010b) Seed germination and seedling establishment on Amazonian floodplain trees. In: Junk WJ, Piedade MTF, Parolin P, Wittman F, Schongart J (eds) Amazonian floodplain forests: ecophysiology, biodiversity and sustainable management. Ecological Studies. Springer, Berlin, pp 259–280

Part I
Igapó Over Space and Time

Chapter 2
Diversity of Dispersal Systems in Igapó Forests: An Analysis of Local Tree Diversity, Species Turnover, and Dispersal Systems

María Natalia Umaña, Diego F. Correa, Ángela Cano, Luisa F. Casas, Sasha Cárdenas, Boris Villanueva, and Pablo Stevenson

2.1 Introduction

Species are not uniformly distributed across the landscape, with most of them exhibiting a patchy distribution. These patchy distributions are often attributed to environmental heterogeneity, but dispersal limitation likely also plays a key role (Nathan and Muller-Landau 2000; Muller-Landau et al. 2008; Tilman and Pacala 1993; Chust et al. 2006; Hubbell 2001; Vormisto et al. 2004). Indeed, species with different dispersal abilities play a critical role in determining the spatial patterns in trees (Seidler and Plotkin 2006) as well as patterns of species diversity (Tilman and Pacala 1993). Thus, a key to understanding the spatial distributions of individual species, and the spatial turnover in species composition is the investigation of dispersal abilities and mechanisms across species in a region. Further insights can be gained by comparative investigations of dispersal systems within and between communities that differ in their environments, thereby allowing for the quantification of the relative importance of dispersal system, space, and environmental heterogeneity for determining the turnover of species composition.

M. N. Umaña (✉)
School of Forestry and Environmental Studies, Yale University, New Haven, CT, USA
e-mail: maumana@umich.edu

D. F. Correa
School of Agriculture and Food Sciences, The University of Queensland, Brisbane, Queensland, Australia

Á. Cano · L. F. Casas · S. Cárdenas · P. Stevenson
Departamento de Ciencias Biológicas, Laboratorio de Ecología de Bosques Tropicales y Primatología, Universidad de los Andes, Bogota, Colombia
e-mail: pstevens@uniandes.edu.co

B. Villanueva
Ciencias Forestales- Grupo de Investigación en Biodiversidad y Dinámica de Ecosistemas Tropicales, Universidad del Tolima, Ibagué, Tolima, Colombia
e-mail: bsvillanuevat@ut.edu.co

R. W. Myster (ed.), *Igapó (Black-water flooded forests) of the Amazon Basin*,
https://doi.org/10.1007/978-3-319-90122-0_2

The diversity of dispersal systems in plants is vast and includes the development of distinct morphological traits. The most common traits are fleshy structures to be dispersed by animals, wings for dispersal by wind, floating mechanisms to allow water dispersal, and explosive dehiscence (van der Pijl 1982; Correa et al. 2013; Correa, Álvarez and Stevenson 2015). Examinations of the spatial distributions of dispersed seeds have demonstrated that there is considerable variation in the shapes and sizes of seed shadows of plants with different dispersal systems suggesting their potentially great importance for determining the spatial structure of plant communities. For instance, it has been shown that, on average, animal-dispersed species have longer dispersal distances than wind-dispersed species (Clark et al. 2005; Correa et al. 2015). Such long-distance dispersal is likely to be important in tropical forests where a large proportion of trees have fruits or seeds attractive to mammals and birds. Indeed, van Roosmalen (1985) estimated that 87–90% of woody species in tropical forest are animal dispersed. Most of these species are transported long distances, thereby facilitating the avoidance of negative distance and density dependent processes and maintaining species diversity (Janzen 1970; Connel 1971). However many other species have diaspores easily dispersed by air or water. For example, many Fabaceae species – the most species-rich plant family in the tropics – are adapted to abiotic dispersal (Richards 1996).

Although plant communities in tropical rain forests have been described as mainly animal dispersed, there is likely a variation in the importance of different dispersal systems among rain forests due to environmental heterogeneity. For example, South American rain forests can be broadly classified as *terra firme* and flooded forests. In the floodplains, species are submerged during 3–4 months every year and can be classified in black-water flooded forest (called *igapó*) or white-water flooded forest (*várzea*) (Ayres 1993; Irion et al. 1997). Flooded forests exhibit particular conditions that influence the community assembly and structure (Myster 2016). Excess of water results in complex conditions such as impeded gas exchange, hypoxia, anoxia, and high CO2 concentrations in the root systems, among others (Bailey-Serres and Colmer 2014). Because of these stressful conditions, flooded forests are mainly structured by environmental filtering (Umaña et al. 2012). Conversely, *terra firme* forests are less physiologically stressful, and this likely permits a higher number of species to coexist. The important abiotic differences between these two landscape units may result in different adaptations allowing for the successful colonization and survival of plant species. In particular, we propose that one mechanism by which plants may be sorted into flooded and non-flooded conditions is to exhibit seed dispersal strategies that take advantage of the variation in the available dispersal agents between the two forest types. For example, flooded forest should present a high relative importance of dispersal facilitated by water. In contrast, *terra firme* forest should exhibit a dominance of terrestrial animal-dispersed species. Thus, our main hypothesis is that the relative proportion of each dispersal system in plant communities should vary depending on the availability of dispersal agents. In particular, we expect that floodplains should have a more diverse dispersal system composition, since dispersal by common forest dwellers (such as birds, primates, tapirs, and bats) occurs along with dispersal via water and fish.

Our predictions on the types of dispersal systems and their diversity within forest types can be expanded to understand the turnover of species and the relationships between plant communities and dispersal systems within and across forest types. In particular, forests dominated by species with long-distance dispersal systems should have low turnover in species through space. We expect a lower species turnover among flooded forests than among *terra firme* forests because water-dispersed seeds travel longer distances by water and fish (Gottsberger 1978; Horn 1997) than seeds dispersed by birds and mammals. In addition to species turnover, the functional composition (i.e., the dispersal system composition) is also expected to change among forests. While both, the species composition and dispersal system composition are likely to turnover through space, we predict that, if dispersal systems are related to forest types, the turnover of species between assemblages in the same forest type will be higher than the turnover of dispersal systems. Lastly, we predict that the species and dispersal system compositions will have high levels of turnover when comparing assemblages in different forest types.

The main objective of this study is to evaluate the relationship between species diversity and dispersal systems as well as to quantify the species and dispersal systems turnover between flooded and non-flooded tree assemblages in Colombian tropical forests. In particular, we ask: (1) How does the proportion of dispersal systems vary depending on the availability of dispersal agents across the three different types of forests, *terra firme* and two in floodplains (*igapó* and *várzea*)? (2) Are there higher levels of beta diversity in forests dominated by animal-dispersed species (i.e., endozoochory)? And (3) is the spatial turnover in the dispersal system composition of tree communities constrained relative to the species turnover, and how does this vary within and between forest types?

2.2 Methods

2.2.1 Study Site

We established 22 1-ha plots in 5 sites in lowland tropical mature forests in Colombia. We sampled three different landscape units: two types of floodplains (10 plots) and *terra firme* forests. Seven plots (two in *várzea* floodplains and five in *terra firme*) were established in the Tinigua National Park, Meta (2°32′ N, 74°3′ W). Six plots (two in *igapó* floodplains and four in *terra firme*) were established in the Caparú Biological Station, Vaupés, in the basin of the Apaporis River (1°4′ S, 69°31′ W) (Cano and Stevenson 2009; Umaña et al. 2012). Five plots (two in *igapó* and three in *terra firme)* were established in the Tomo Grande Reserve, Vichada (4°50′ N, 72°16′ W) (Correa and Stevenson 2010). Two plots in *várzea* floodplains were established in the Hacienda San Juan de Carare, Santander (6°1′ N, 74°11′ W). Finally two additional flooded plots (one in *várzea* and one in *igapó*) were established in the Casanare Department (5°40′ N, 70° 08′ W). In each plot, all woody stems ≥10 cm diameter at breast height (DBH), including trees, palms, and lianas, were tagged, measured for DBH, and identified to species or morphospecies. These plots have remained with low levels of anthropic disturbances, although the plots in Tomo Grande and some in San Juan are located within forest fragments.

To obtain a list of dispersal systems for all the species ($n = 1079$), we assembled a database containing information reported in the scientific literature (van Roosmalen 1985; Stevenson et al. 2000; Pennington et al. 2004), herbaria (MOBOT and Universidad de los Andes), and online herbaria (Field Museum Chicago Kew, New York Botanical Garden, Herbario Nacional Colombiano (COL), and Herbario Amazónico Colombiano (COAH)). In this database, anemochory, hydrochory, endozoochory, synzoochory, myrmecochory, and explosive dehiscence were assigned to each species according to diaspore morphology (including fleshiness, seed size, and the presence of additional structures) (Rodley 1930; van der Pijl 1972; Correa et al. 2013, 2015). Species that could not be assigned into any of the dispersal systems were excluded from the analysis (however, the proportion of these species was low, 6.33%).

2.2.2 *Statistical Analyses*

We quantified species diversity by calculating the Shannon index, and we used the same index to calculate the evenness of dispersal system for each plot. The Shannon index for dispersal system was calculated in two ways: by using the number of species per system and by using the number of individuals per species. Then, we correlated species diversity with dispersal system diversity.

To evaluate the similarity in tree and palm species composition among the three forest types, we performed a non-metric multidimensional scaling (NMDS) by using the function "meta-MDS" of the statistical package "Vegan" (Oksanen et al. 2013). For this analysis, we used the Chao-Jaccard abundance-based estimator (Chao et al. 2005) because it is more appropriate in diverse forests where rare species are abundant.

To quantify the variation in diversity of dispersal systems across each type of forest, we calculated the proportion of species per dispersal system. To evaluate the species and system turnover among regions, we calculated the Raup-Crick index (Chase et al. 2011). This metric estimates compositional and trait dissimilarity between two communities, accounting by differences in local diversity (α-diversity) and species occupancy rates in the study system. To calculate this metric, species were weighted by their frequency of occupancy, and the null distribution was generated using 999 iterations. The null model utilized was an independent swap where the observed occupancy rates of all species and the α-diversity of all plots were retained in all null communities. The Raup-Crick metric ranges from -1 to 1, where the value approaches 1 if local communities are more dissimilar than expected by chance; if it is close to zero, the observed dissimilarity is indistinguishable from random; and if it is near -1, the two assemblages are more similar than expected by chance (Chase et al. 2011). Thus, from this analysis, we can address whether the compositional change in dispersal systems from one plot to another is higher, lower, or no different than expected given the observed species community data matrix. All the statistical analyses were performed in R statistical software (v. 2.14.0; R Core Team Development 2011) using the R code for Raup-Crick coming from the supplemental material in Chase et al. (2011).

2.3 Results

As expected the *terra firme* forests exhibited the highest number of species, although some igapó plots in Central Amazonia showed similar levels of diversity than *terra firme* forest plots. *Várzea* plots showed the lowest levels of diversity (Fig. 2.1). Plots established in Caparú Biological Station and Tinigua National Park exhibited the highest species diversity followed by Tomo Grande, San Juan, and Casanare. Within each region, flooded forest also showed lower species diversity than *terra firme* forest.

The NMDS analysis showed that each landscape unit forms an independent floristic unit (Fig. 2.2). In particular, the flooded forests exhibited a more widespread distribution than *terra firme* forest, which indicates higher divergence in species composition within *igapó* and *várzea* compared with non-flooded forest. In addition, plots in the same site and in the same type of forest showed highly similarity in species composition. Indeed, the most abundant species within each region were different, for example, for Caparú Biological Station in *terra firme* forest, the palm *Oenocarpus bataua* (Arecaceae) was the most abundant species, a result that is not surprising for the Amazon forest. However, for Tinigua National Park in *terra firme* the most abundant species was *Bambusa guadua* (Poaceae). We also found important differences in species abundance within the same region depending on the forest type. For example, the most abundant species in Casanare in *várzea* forest was *Phyllanthus elsiae* (Phyllanthaceae), while for Casanare *igapó* plot, *Tachigali* cf. *chrysophylla* (Fabaceae) was the most common species. Furthermore, our results showed that the dominance was highly variable among the different plots. In Casanare *igapó*, *Tachigali* cf. *chrysophylla* showed 257 individuals in 1 ha plot, while in Caparú Biological Station *igapó* forest, *Caraipa densifolia* was the most abundant species, represented by 63 individuals.

In terms of dispersal system diversity, *igapó* exhibited the highest level of diversity than várzea and *terra firme* forests (Figs. 2.1 and 2.3). Because the diversity of dispersal systems had the same trend when calculating the index by species (number of species per dispersal system) and by individuals (number of individuals per dispersal system), we decided to show only the results of species per system.

By comparing the species similarity between plots within and across different types of forests, we found that species turnover was higher between different types of forests than within same type of forest (Fig. 2.4a, b). In terms of species turnover within type of forests, we found lower beta diversity within the *várzea* and within *igapó* than within *terra firme* plots (Fig. 2.4a).

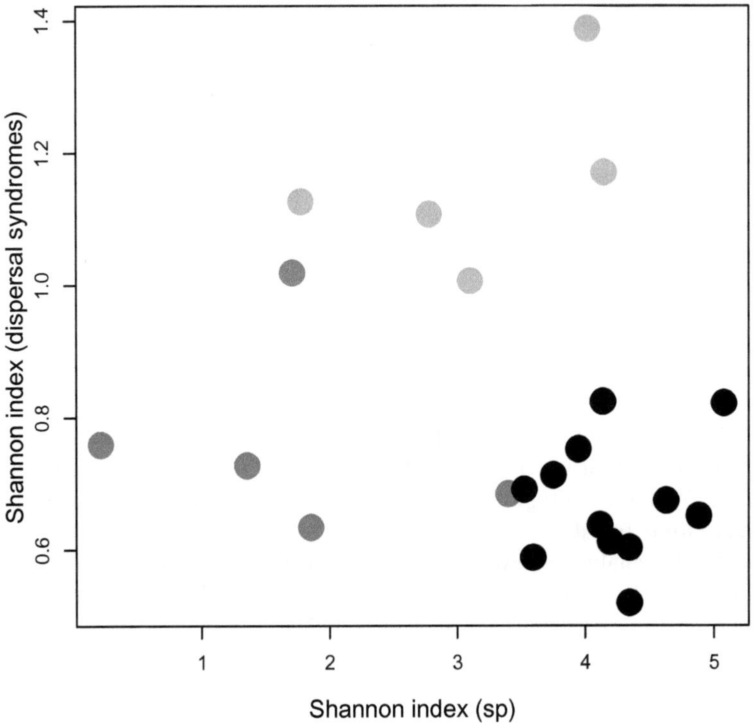

Fig. 2.1 Relationship between Shannon index for species and Shannon index for dispersal systems by species. Light gray dots represent *igapó* plots, dark gray dots represent *várzea* plots, and black dots represent *terra firme* plots

A central aim of this work was to determine if the observed turnover in dispersal systems between landscape units was any different from that expected given the within site species and functional diversity. To accomplish this, we utilized a null modeling approach. Based on these analyses, we found that, given the observed plot species and functional diversity and species occupancy rates, the dispersal system turnover was generally lower than expected by chance. In particular, we generally found negative Raup-Crick values when comparing forest plots from the three different landscape units (Fig. 2.4a, b). This indicates that, while there is dispersal system turnover, it is lower than expected given the observed local species and functional diversities.

2.4 Discussion

In this study, we asked whether species turnover reflected the variation in dispersal systems among three different habitats in a tropical region. To date, the diversity and turnover of dispersal systems in tropical forests have been poorly explored at

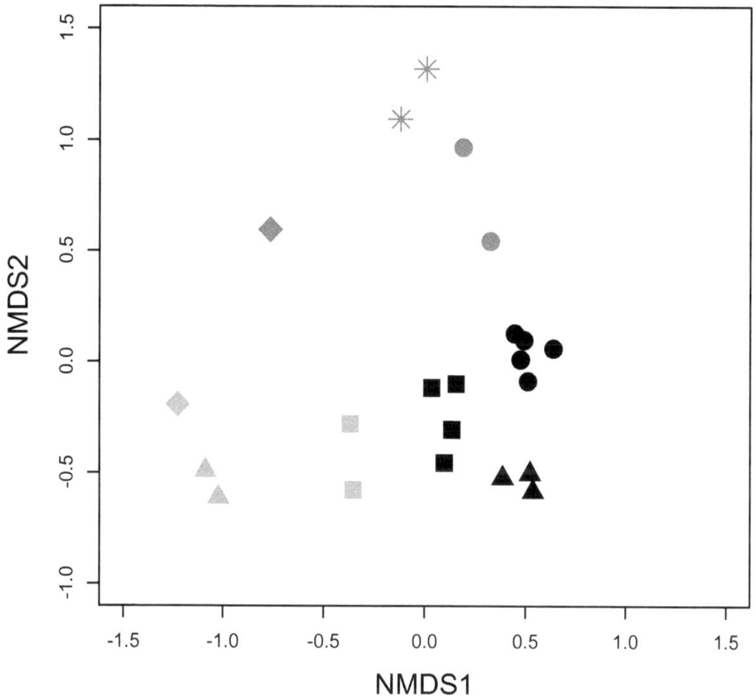

Fig. 2.2 NMDS of 22 1-ha plots of 3 forests communities. Colors represent different forest types: light gray symbols represent coefficients calculated *igapó* plots, dark gray symbols represent coefficients calculated for *várzea* plots, and black symbols represent *terra firme* plots. Symbols represent different sites: circles represent plots in Tinigua National Park, squares represent plots in Caparú Biological Station, triangles represent plots in Tomo Grande, stars represent plots in Hacienda San Juan, and diamonds represent plots in Casanare

a community level. In this study, we found that the diversity of dispersal systems is highly dependent on the type of forest. In terms of species diversity, our results showed that *terra firme* forests harbor the highest levels of species diversity (Gentry 1988; Phillips et al. 1994; Pitman et al. 1999; Cano and Stevenson 2009). However, flooded forest did not always have low levels of species diversity. This may be due to different biogeographical and historical legacies of different plots in our study region. That said, our general finding is that, within regions that have similar biogeographic histories, *terra firme* forests have more species than flooded forests. These results are not surprising since periods of inundation in flooded forests impose some restrictions to the establishment of species (Ferreira 1997; Junk 1989) generating a strong filter resulting in compositional differences in floras. Contrary to what we expected based on Ayres (1993), *várzea* and not *igapó* forests had the lowest species diversity; these findings suggest that there are other important mechanisms affecting the levels of diversity besides the flooding regime in these plains. We think that this difference could be explained by the fact that the vársea forests used in this study were relatively young and in early stages of succession.

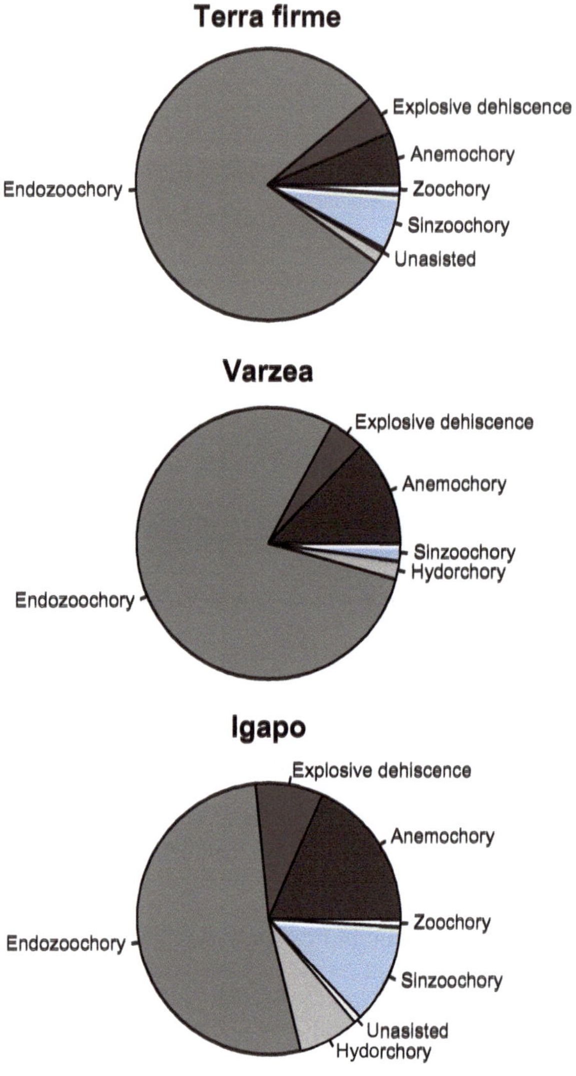

Fig. 2.3 Proportion of species with certain dispersal syndrome in each of the three forest types. *Terra firme* plot from Hilly forest in Caparú Biological Station; *várzea* plot from San Juan; *igapó* plot from Caparú Biological Station

Flooded environments facilitate the establishment of species with abiotic dispersal systems as anemochory and hydrochory (Gentry 1983; Correa et al. 2013, 2015). Interestingly, *igapó* forest, which has been described as a more stressful environment and more challenging for species establishment (Junk et al. 1989; Ayres 1993; Irion et al. 1997), had the highest diversity of dispersal systems. This finding supports the expectation that seed dispersal systems are more diverse as the variation of potential vectors increases in this forest (Umaña et al. 2011). Thus, flooding events permit the success of additional dispersal systems not seen in non-flooded, though generally less stressful, forests. Thus, in terms of the diversity of dispersal systems, flooded *igapó* does not represent an environment that strongly filters dispersal system. On the other hand, *várzea* showed less dispersal system diversity, due

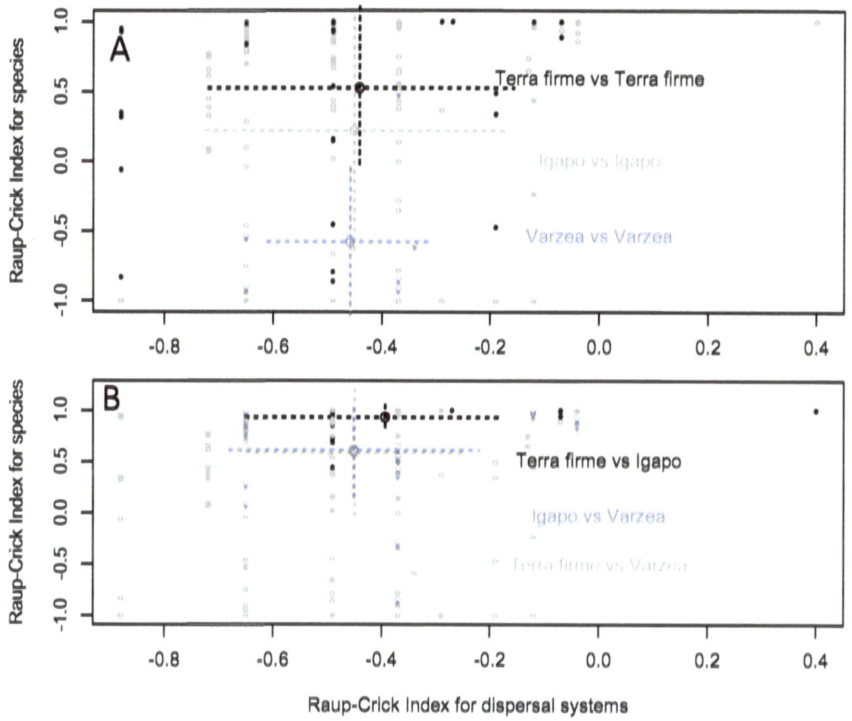

Fig. 2.4 Relationship between Raup-Crick species beta diversity values and Raup-Crick trait beta diversity values. (**a**) Dashed lines represent the mean and standard deviation of coefficients calculated within each type of forest, dark gray dots represent *várzea* forest, light gray dots represent *igapó* forest, and black dots represent *terra firme* plots. (**b**) Dashed lines represent the mean and standard deviation of coefficients calculated between plots from different forests, black dots represent comparison between *terra firme* and *igapó* plots, light gray dots represent comparison between *várzea* and *terra firme* plots, and dark gray dots represent the comparison between *igapó* and *váreza* plots. Open dots represent coefficient calculated between plots from the same site, dots represent Tomo Grande plot, triangles represent Caparú plots, squares represent Macarena plots, and diamonds represent San Juan plots

to the low number of species found in the plots and the strong dominance of few species.

In contrast to flooded forests, *terra firme* is an environment where animal dispersal is much more common and for this reason is not surprising to find a dominance of animal dispersal systems (Howe and Smallwood 1982; Jordano 2000; Link and Stevenson 2004; Arbeláez and Parrado-Roselli 2005; Correa et al. 2015). Given that *terra firme* environments are older and less disturbed than floodplains, these forests show less establishment opportunities for wind-dispersed species that tend to have small-seeded species and high light demands (Foster and Janson 1985). Thus, knowing that dispersal is relevant to plant success (Terborgh et al. 2002), large-seeded plants are well suited for shaded conditions (Foster and Janson 1985). In addition, large-seeded plants are dispersed by large frugivores (Fragoso and Huffman 2000; Peres and van Roosmalen 2002; Stevenson et al. 2005; Defler and

Defler 1996) and scatter-hoarding rodents (Jansen et al. 2012; Jansen et al. 2002), suggesting that animal dispersion represents the best strategy for plants in terra firme forests.

The importance of dispersal limitation can be variable among different landscape units; hence, our second goal was to evaluate if communities that experienced a strong environmental filter would be more similar in species composition. In terms of species turnover, we found that plots in the same landscape unit exhibited high values of floristic similarity while plots far away from one another in different landscape units had higher levels of species turnover. On local scales, species turnover was higher within plots from the same forest than from plots of different forests. These findings indicate that there is a filter within each type of forest, which constrains the successful establishment of particular species and maintains the floristic identity of each forest type.

In terms of turnover of dispersal systems, all the forests showed high levels of similarity in dispersal systems. However, we should consider that the analysis considered seven different types of dispersal systems, and all animal-dispersed species were grouped in one single category. It is possible that considering a more detailed classification, separating seed dispersal by birds, rodents, insects, fish, etc., the patterns show a higher turnover. Thus, the observed patterns hide finer levels of turnover between forest types that cannot be detected with our coarse delineations of dispersal types. That said, this result supports the analyses that quantified the local diversity of dispersal systems. Specifically, we found that dispersal system diversity increases in flooded forests relative to non-flooded forests primary by the addition of abiotic dispersal to biotic dispersal. Thus, there is no complete turnover of dispersal systems between the two types of forest. Rather, the non-flooded forests are a nested subset of the flooded forests, and the compositional difference is not a matter of turnover of systems but a loss of a few systems. Our null model controlled for differences in local diversity between the plots being compared. Because this difference in system diversity drives the compositional difference between plots, rather than turnover per se, and it was controlled in the null model, we found lower-than-expected dispersal system turnover.

Although we expected that most of the large differences in species composition would occur between flooded and non-flooded forests, our results showed that *igapó* and *terra firme* forests exhibited higher levels of species turnover than *várzea* forests (Fig. 2.3a). A possible explanation for this pattern is that *terra firme* and *igapó* are older forests than *várzea*. Therefore, geologically younger *várzea* forests may not have been stabilized enough in order to reliably detect differences in species composition, while *terra firme* and *igapó* forests have been present since longer time. Thus, this finding would reflect the footprint of biogeographical process along time and suggests an important role of historical processes in determining plant community structure in tropical forests (Chave et al. 2008; Fine and Kembel 2011; Tuomisto et al. 2003).

In summary this study concludes that that floodplains and *terra firme* exhibit important differences of dispersal systems. Animal dispersion dominates *terra firme* forest, while flooded forest offers more opportunities for successful wind and water dispersion. Species and trait turnover suggest that the traditional view of lowland tropical forests as flooded and non-flooded systems is not accurate, as tropical forests have been shaped by historical events occurring at different moments in space and time which influence the species overlapping among and within the tree forests. Despite the variety of plants that inhabit *terra firme* forests, the diversity of dispersal systems is not correspondingly high. In fact, in these forests, endozoochory is the dominant trait, strongly suggesting that it is the best strategy of seed dispersal in mature, closed-canopy conditions, where abiotic dispersal is not as effective. In floodplains, on the other hand, abiotic dispersion is more important. In particular, *igapó* exhibits a much more diverse forest in terms of dispersal systems.

Acknowledgments We would like to thank the curators and researchers from MO who allowed us to access to the collections and assisted us with the data analyses, in particular to Peter Jørgensen, Iván Jiménez, and Sebastían Tello. We thank also to the students, researchers, and volunteers that collected the data and make this work possible, especially to Ana Belén Hurtado and Natalia Norden. The financial and logistic support to gather the data and to undertake these analyses was provided by Universidad de los Andes, Ecopetrol, Colciencias, Banco de la República, Primate Conservation Inc., Margot Marsh Foundation, Conservation International, Missouri Botanical Garden (Bascom Fellowship), and Lincoln Park Zoo.

References

Arbeláez MV, Parrado-Roselli A (2005) Seed dispersal modes of the sandstone plateau vegetation of the middle Caquetá River. Biotropica 37:64–72

Ayres JM (1993) As matas de várzea do Mamirauá, Medio rio Solimøes. CNPq, Brasília

Bailey-Serres J, Colmer TD (2014) Plant tolerance of flooding stress – recent advances. Plant, Cell and Environment 37:2211–2215

Cano A, Stevenson PR (2009) Diversidad y composisción florística de tres tipos de bosque en la Estación Biológica Caparú, Vaupés. Revista Colombiana Forestal 12:63–80

Cano A and Stevenson PR (2009) Diversidad y composición florística de tres tipos de bosque en la Estación Biológica Caparú, Vaupés. Colombia Forestal 12: 63-80.

Chase JM, Kraft NJB, Smith KG, Vallend M, Inouye D (2011) Using null models to disentangle variation in community from variation in α- diversity. Ecosphere 2:1–11

Chave J, Condit R, Muller-Landau HC, Thomas SC, Ashtin PS, Bunyavejchewin S, Co LL, Dattaraja H, Davies S, Esufali S, Ewango CEN, Feeley KJ, Foster RB, Gunatilleke N, Gunatilleke S, Hall P, Hart TB, Hernández C, Hubbell SP, Itoh A, Kiratiprayoon S, LaFrankie JV, de Lao SL, Makana J-R, Noor MNS, Kassim AR, Samper C, Sukumar R, Suresh HS, Tan S, Thompson J, Tongco MDC, Valencia R, Vallejo M, Villa G, Yamakura T, Zimmerman JK, Elizabeth L (2008) Assessing evidence for a pervasive alteration in tropical tree communities. PLoS Biol 6:e45. https://doi.org/10.1371/journal.pbio.00645

Chao A, Chazdon R, Colwell RK (2005) new statistical approach for assessing compositional similarity based on incidence and abundance data. Ecology Letters 8:148–159

Chust G, Chave J, Condit R, Salomon A, Lao S, Rolando P (2006) Determinants and spatial modeling of tree β-diversity in a tropical forest landscape in Panama. J Veg Sci 17:83–92

Clark C, Poulsen JR, Bolker BM, Connor EF, Parker VT (2005) Comparative seed shadows of bird-, monkey , and wind-dispersed trees. Ecology 86:2684–2694

Connel JH (1971) On the role of natural enemies in preventing exclusion in some marine animals and in rain forest trees. In: Deb Boer PJ, Gradwell GR (eds) Dynamics of populations. Pudoc, Wageningen, pp 298–312

Correa DF, Stevenson PR (2010) Estructura y diversidad de bosques de galería en una sabana estacional de los Llanos Orientales colombianos (Reserva Tomo Grande, Vichada). Orinoquía 14(Suppl. 1):13–48

Correa DF, Stevenson PR, Álvarez E, von Hildebrand P (2013) Frequency and abundance patterns of plant dispersal systems in Colombian forests and their relationships with geographic regions of the country. Colombia Forestal 16:33–51

Correa DF, Álvarez E, Stevenson PR (2015) Plant dispersal systems in Neotropical forests: availability of dispersal agents or availability of resources for constructing zoochorus fruits? Glob Ecol Biogeogr 24:203–214

Defler TR, Defler SB (1996) Diet of a group of *Lagothrix lagotrhicha lagothricha* in southeastern Colombia. Int J Primatol 17:161–190

Ferreira LV (1997) Effects of the duration of flooding on species richness and floristic composition in three hectares in the Jaú National Park in floodplain forests in central Amazonia. Biodivers Conserv 6:1353–1363

Fine PVA, Kembel SW (2011) Phylogenetic community structure and phylogenetic turnover across the space and edaphic gradients in Western Amazonian tree communities. Ecography 34:552–565

Foster SA, Janson CH (1985) The relationship between seed seize and establishment conditions in tropical woody plants. Ecology 66:773–780

Fragoso MV, Huffman JM (2000) Seed-dispersal and seedling recruitment patterns by the last Neotropical megafauna element in Amazonia, the tapir. J Trop Ecol 16:369–385

Gentry AH (1983) Dispersal ecology and diversity in Neotropical forest communities. In: Kubitzki K (ed) Dispersal and distribution. Velag Paul Parey, Berlín, pp 303–314

Gentry AH (1988) Tropical forest biodiversity: distributional patterns and their conservational significance. Oikos 63:19–28

Gottsberger G (1978) Seed dispersal by fish in the inundated regions of Humaita, Amazonia. Biotropica 10:170–183

Horn MH (1997) Evidence for dispersal of fig seeds by the fruit-eating characid fish *Brycon guatemalensis* Regan in a Costa Rican tropical rain forets. Oecologia 109:259–264

Howe HF, Smallwood J (1982) Ecology of seed dispersal. Annu Rev Ecol Syst 13:201–228

Hubbell SP (2001) The unified theory of biodiversity and biogeography. Princeton Univ. Press, Princeton

Irion G, Junk WJ, de Mello JASN (1997) The large Central Amazonian river floodplains near Manaus: geological, climatological, hydrological and geomorphological aspects. In: Junk WJ (ed) The Central Amazon floodplain: ecology of a pulsing system. Springer, Berlin Heidelberg New York, pp 23–46

Jansen PA, Bartholomeous BF, Ouden J, Wieren SE (2002) The role of seed size in dispersal by a scatter-hoarding rodent. In: Levey DJ, Silva W, Galetti M (eds) Seed dispersal and frugivory: ecology, evolution, and conservation. CABI Pub, New York, pp 209–225

Jansen PA, Hirsch BT, Emsens W-J, Zamora-Gutierrez V, Wikelski M, Kays R (2012) Thieving rodents as substitute dispersers of megafaunal seeds. PNAS 109:12610–12615

Janzen DH (1970) Herbivores and the number of tree species in tropical forest. Am Nat 104:501–528

Jordano P (2000) Fruits and frugivory. In: Fenner M (ed) Seeds: the ecology of regeneration in plant communities, 2nd edn. CABI Pub. Wallingford, UK, pp 125–166

Junk WJ (1989) The flood tolerance and tree distribution in central Amazonia. In: Holm-Nielsen LB, Nielsen IC, Balsev H (eds) Tropical forest botanical dynamics: speciation and diversity. Academic Press, London, UK pp 47–64

Junk WJ, Bayley PB, Sparks RE (1989) The flood pulse concept in river-floodplain system. In: Dodge DP (ed) Proceedings of the international large river symposium, vol 106. Can. Spec. publ. fish. Aquat. Sci, London, UK pp 110–127

Link A, Stevenson PR (2004) Fruit dispersal syndromes in animal disseminated plants at Tinigua National Park, Colombia. Revista Chilena Historia Tropical 77:319–334

MullerLandau HC, Wright SJ, Calderón O, Condit R, Hubbell S (2008) Interspecific variation in primary seed dispersal in a tropical forest. J Ecol 96:653–667

Myster RW (2016) The physical structure of Amazon forests: a review. Bot Rev 82:407–427
Nathan R, Muller-Landau HC (2000) Spatial patterns of seed dispersal, their determinants and consequences for recruitment. TREE 15:278–285
Oksanen J, Blanchet FG, Kindt R, Wagner HH (2013) Vegan: community ecology package. R package vegan, vers. 2.2-1. Community ecology package.
Pennington TD, Reynel C, Daza A, Wise R (2004) Illustrated guide to the trees of Peru. David Hunk Books Pub, Sherbourne, UK
Peres CA, Van Roosmalen M (2002) Primate frugivory in two species-rich neotropical forests: implications for the demography of large-seeded plants in over-hunted areas. In: Levey DJ, Silva WR, Galetti M (eds) Seed dispersal and frugivory: ecology, evolution and conservation. CABI Pub. New York, USA, pp 407–421
Phillips OL, Gentry AH, Sawyer SA, Vásquez R (1994) Dynamics and species richness of tropical rain forests. PNAS 91:2805–2809
Pitman NCA, Terborgh J, Silman M, Nuñez P (1999) Tree species distributions in an upper Amazonian forest. Ecology 80:2651–2661
Rodley HN (1930) The dispersal of plants throughout the world. L. Reeve and Co. Limited. Ashford, Kent, UK. 744pp.
Richards PW (1996) Seed dispersal. In: The Tropical rain forest an ecological study, 2nd edn. Cambrigde Univ. Press, Cambridge, pp 109–115
Seidler TG, Plotkin JB (2006) Seed dispersal and spatial patterns in tropical trees. PLoS Biol 4:e344. https://doi.org/10.1371/journal.pbio.0040344
Stevenson PR, Quiñones Fernandez MJ, Castellanos Torradio MC (2000) Guía de frutos de los Bosques del Río Duda, Macarena, Colombia. Asociación para la defense de La Macarena-IUCN (the Netherlands)
Stevenson PR, Link A, Ramírez BH (2005) Frugivory and seed fate in *Bursera inversa* (Burseraceae) at Tinigua Park, Colombia: implications for primate conservation. Biotropica 37:431–438
Swamy V, Terborgh J, Dexter KG, Best BD, Alvarez P, Cornejo F (2010) Are all seeds equal? Spatially explicit comparisons of seed fall and sapling recruitment in a tropical forest. Ecol Lett 14:195–201
Terborgh J, Pitman NC, Silman M, Núñez P (2002) Maintenance of tree diversity in tropical forests. In: Levey DJ, Silva WR, Galetti WR (eds) Seed dispersal and frugivory: ecology, evolution, and conservation. CABI Pub. New York, USA, pp 1–17
Tilman D, Pacala S (1993) The maintenance of species richness in plant communities. In: Ricklefs RE, Schluter D (eds) Species diversity in ecological communities. Univ. of Chicago Press, Chicago, pp 13–25
Tuomisto H, Ruokolainen K, Yli-Halla M (2003) Dispersal, environment and floristic variation of western Amazonian forest. Science 299:241–244
Umaña MN, Stevenson PR, Hurtado AB, Medina I (2011) Dispersal syndromes among three landscape units in Colombian lowland Amazonia. Int J Plant Repro Biol 3:155–159
Umaña MN, Norden N, Stevenson PR (2012) Determinants of plant community assembly in a mosaic of landscape units in Central Amazonia: ecological and phylogenetic perspectives. Plos One 7:e45199. https://doi.org/10.1371/journal.pone.0045199
Van der Pijl L (1972) Principles of dispersal in higher plants, 2nd edn. Springer, Nueva York 162 p
Van der Pijl L (1982) Dispersal in higher plants, 3rd edn. Springer, Berlin
van Roosmalen MGM (1985) Fruits of the Guiana flora. Institute of Systematic Botany. Utrecht Univ, Utrecht, p 483
Vormisto J, Svenning J-C, Hall P, Balslev H (2004) Diversity and dominance in palm Arecaceae: communities in *terra firme* forests in the western Amazon basin. J Ecol 92:577–588

Part II
Water and Light

Chapter 3
Mercury in Black-Waters of the Amazon

Daniele Kasper, Bruce Rider Forsberg, Helena do Amaral Kehrig, João Henrique Fernandes Amaral, Wanderley Rodrigues Bastos, and Olaf Malm

3.1 Mercury in the Amazon Black-Waters

Mercury (Hg) in the Amazon region was related to the goldmining for many decades. In fact, the use and release of mercury by goldminers were intense, especially between 1970s and 1990s. It is estimated that, at this time, more than 1,600,000 goldminers worked in Brazil (Pfeiffer and Lacerda 1988). The Hg emissions from goldmining were estimated in 31 t.year^{-1} in Brazil (Lacerda 2003). Several studies were done in the Amazon region to assess the impact of goldmining activities on biota, environment, and riverine people (e.g., Akagi et al. 1995; Malm et al. 1990; Malm et al. 1995; Padovani et al. 1995). However, studies done in the last 20 years have shown that the Amazon region has mean mercury concentrations naturally higher than the global averages, and the impact of goldmining on these concentrations is relatively local to the region near the impact (Lechler et al. 2000). The Hg emitted by the goldmining in the Amazon, for example, corresponds to <3% of the total Hg load present in the superficial soils of the Negro (Fadini and Jardim 2001) or Tapajós River basins (Roulet et al. 1998b).

Riverine people that consume fish from Negro River showed higher Hg concentrations in the hair than those that consume fish from Tapajós and Madeira basins (Silva-Forsberg et al. 1999). The last two basins were important goldmining regions, while there is little documented history of goldmining in the Negro, with the

D. Kasper (✉) · O. Malm
Universidade Federal do Rio de Janeiro, Rio de Janeiro, RJ, Brazil

B. R. Forsberg · J. H. F. Amaral
Instituto Nacional de Pesquisas de Amazonia, Manaus, AM, Brazil

H. do Amaral Kehrig
Universidade Estadual do Norte Fluminense, Campos dos Goytacazes, RJ, Brazil

W. R. Bastos
Universidade Federal do Rondonia, Porto Velho, RO, Brazil

R. W. Myster (ed.), *Igapó (Black-water flooded forests) of the Amazon Basin*,
https://doi.org/10.1007/978-3-319-90122-0_3

exception of the Branco sub-basin, where significant mining occurred during 1980s and 1990s. Those results were associated with the naturally elevated mercury concentrations of Negro River basin (Silva-Forsberg et al. 1999). In this basin, the largest black-water tributary of the Amazon, the soils are naturally rich in Hg (Fadini and Jardim 2001) and supply the element to the aquatic system by pedologic processes (Fadini and Jardim 2001; Sousa 2015). Hydromorphic spodosols, abundant in the most Amazon black-water basins, are conducive to transport of Hg to aquatic system as well as organic matter (Sousa 2015). Therefore, this results in a strict positive relationship between dissolved organic carbon (DOC) and the total mercury content dissolved in Amazonian black-waters (Sousa 2015) and highest total mercury concentrations in black-water of Amazon streams (e.g., Sousa 2015) or lakes (e.g., Brito et al. 2017).

Hg in the aquatic system is more efficiently transferred to biota in its methylated form, methylmercury (MeHg, CH_3Hg^+). Therefore, an important stage of mercury cycle is it methylation. The product, MeHg, is highly toxic to biota, has high absorption capacity by organisms, and is the main responsible for mercury bioaccumulation and trophic transfer (Watras et al. 1998). The methylation can occur by abiotic or biotic processes. In abiotic methylation, a methyl radical is transferred to inorganic Hg (e.g., Hg^{2+}) by methylcobalamin, humic, or fulvic acids (Mauro et al. 1999; Mason and Benoit 1986). However, the biotic methylation is more common, and the transference of methyl radical is intermediated by organisms, mainly methanogens and iron- and sulfate-reducing bacteria (Compeau and Bartha 1985; Correia et al. 2012; Gilmour et al. 2013; Kerin et al. 2006).

Environmental conditions that raise the microbial activity, in general, increase also the mercury methylation. Therefore, the methylation is influenced by a diverse array of physical and chemical environmental conditions, for example, temperature, dissolved oxygen, pH, and redox potential. Neutral to moderately acid conditions (pH between 5 and 7) favor MeHg formation, and alkaline conditions favor dimethylmercury formation (CH_3HgCH_3), a volatile compound that can volatilize from water to air (Winfrey and Rudd 1990). Therefore, many studies have shown an inverse correlation between pH values and Hg concentrations in biota as an indicative of MeHg availability on environment (e.g., Ikingura and Akagi 2003; Svobodová et al. 1999). Anoxic conditions also favor the process, and negative correlation between aqueous MeHg concentrations and dissolved oxygen levels has been observed in Amazonian aquatic habitats (Brito et al. 2017; Kasper et al. 2014, 2017).

The flooded forests, macrophyte beds (Guimarães et al. 2000), poorly drained soils (Branfireun et al. 1996), and anoxic hypolimnion of lakes and reservoirs (Brito et al. 2017; Kasper et al. 2014) have some environmental conditions favorable to methylation. Therefore, they are known sites favorable to methylation, and, consequently, high MeHg concentrations are commonly observed on them. However, these habitats are not homogenous, and the methylation rate can vary within them. For example, in the floodplain of the Tapajós basin, a clear-water river, Guimarães et al. (2000) observed highest methylation in the sediment below aquatic macrophyte beds (2.3–8.9%), intermediate rates in the sediment of flooded forests (3.2–4.5%), and lowest in the sediment of center of lake (0.2–0.56%). The MeHg production in surface sediments of macrophytes and flooded forests is favored by the intense incorporation and degradation of fresh and labile organic matter originated from deposits of gross materials such as leaves, fruits, and flowers (Roulet et al. 2001).

The potential of Hg methylation also can vary temporally if the environmental conditions conducive to methylation are temporally variable. In the Amazon basin, the inundation of extensive alluvial floodplains occurs in response to the flood-pulse that seasonally provides sites favoring methylation (Guimarães et al. 2000; Junk 1993; Roulet et al. 2000). Higher MeHg concentrations in water (Kasper et al. 2017), in the sediments (Roulet et al. 2001) and in the plankton (Kasper et al. 2014) of Amazon Rivers during high-water season, and in carnivorous fish (Belger and Forsberg 2006; Kasper et al. 2014) were correlated to the extent of floodable areas occurring during this hydrological period. When these floodplains and soils are dry, or the water column of lakes and reservoirs is shallow or destratified, methylation and, consequently, MeHg concentrations decrease (Brito et al. 2017; Guimarães et al. 2000; Kasper et al. 2014; Roulet et al. 2000).

The high DOC and low pH of the black-water aquatic systems, when associated with anoxic environments (mainly during high-water season), make these aquatic systems conducive to high mercury concentration and its methylation. Consequently, their biota normally presents higher mercury levels than those from clear- or white waters, including the riverine people that feed local fish (Belger and Forsberg 2006, Silva-Forsberg et al. 1999, Table 3.1). When comparing Hg concentrations in fish from different water types of the Amazon (Table 3.1), it is noticeable that fish from black-water systems, without anthropogenic impacts, have Hg concentrations in the range of impacted no-black-water systems (e.g., deforestation, mining, or damming in clear or white waters). Although black-water basins have high Hg and their floodable areas (igapós) are known as important sites to Hg methylation, few studies assessed those environments.

3.2 Case Study: Mercury in the Negro Basin, a Black-Water River

The Negro River is the largest black-water tributary of the Amazon River. This basin is absent of significant roads and hard to navigate in its upper portion, and also its sandy soils are nutrient poor and not favorable for agriculture activity. It also has many protected areas, like natural and indigenous reserves, which avoid the anthropogenic interference and human occupation. Therefore, it is one of the rare basins of the world with river and its floodplain still in almost pristine conditions, allowing studies of natural biogeochemical cycles. To subsidize further studies in Negro River basin, this section brings an integrated analysis where we combined unpublished and published data of total mercury and methylmercury with environmental variables sampled along the Negro River, in its main tributaries, and in some lakes within its floodplain.

Table 3.1 Total mercury concentrations ([THg]) in muscle of fish from different sites of the Amazon. [THg] = mean ± standard deviation (minimum-maximum) wet weight (w.w.). The black-water aquatic systems are in bold

Site	Site characteristic	[THg] μg.kg^{-1} w.w.	Reference
***Leporinus* spp. (herbivore/omnivore)**			
Negro basin (Brazil)	**Natural, black-water**	**(10–516)**	Barbosa et al. (2003)
Negro basin (Brazil)	**Natural, black-water**	**(17–219)**	Dorea et al. (2006)
Natural reserve (Suriname)	Natural	90	Mol et al. (2001)
Las Marías river (Venezuela)	Deforestation, clear-water	200 ± 190	Kwon et al. (2012)
Tapajós basin (Itaituba city, Brazil)	Goldmining, clear-water	65 ± 27 (27–126)	Santos et al. (2000)
Maroni river (French Guiana)	Goldmining	329 ± 101[ab]	Maury-Brachet et al. (2006)
Petit-Saut reservoir (French Guiana)	**Goldmining, black-water**	**403 ± 40[ab]**	Durrieu et al. (2005)
Samuel reservoir (Brazil)	Cassiterite mining, clear-water	132 ± 63	Kasper et al. (2012)
***Cichla* spp. (carnivore)**			
Negro basin (Brazil)	**Natural, black-water**	**(39–2441)**	Barbosa et al. (2003)
Negro basin (Brazil)	**Natural, black-water**	**337 ± 244 (31–1469)**	Belger and Forsberg (2006)
Puruzinho lake (Brazil)	**Natural, black-water**	**676 (418–836)**	Azevedo-Silva et al. (2016)
Natural reserve (Suriname)	Natural	390 ± 160 (250–570)	Mol et al. (2001)
Tapajós basin (Itaituba city, Brazil)	Goldmining, clear-water	376 ± 155 (214–611)	Santos et al. (2000)
Tapajós basin (Itaituba city, Brazil)	Goldmining, clear-water	917	Uryu et al. (2001)
Tapajós basin (Itaituba city, Brazil)	Goldmining, clear-water	990 ± 330	Kehrig and Malm (1999)
Tapajós basin (Santarém city, Brazil)	Deforestation, clear-water	175	Uryu et al. (2001)
Tapajós basin (Santarém city, Brazil)	Deforestation, clear-water	180 ± 110	Kehrig and Malm (1999)
Madeira basin (Brazil)	Sampling after 15–20 gold rush, white water	414 ± 228	Bastos et al. (2006)
Upper Madeira basin (Brazil)	Goldmining, white water	660 (120–2210)	Boischio and Henshel (2000)

(continued)

Table 3.1 (continued)

Site	Site characteristic	[THg] $\mu g.kg^{-1}$ w.w.	Reference
Coppename river (Suriname)	Goldmining, clear-water	240	Mol et al. (2001)
Saramacca river (Suriname)	Goldmining, clear-water	730	Mol et al. (2001)
Balbina reservoir (Brazil)	**Black-water**	**383 (193–727)**	Kasper et al. (2014)
Balbina reservoir (Brazil)	**Black-water**	**252 ± 158 (70–723)**	Kehrig et al. (2009)
Samuel reservoir (Brazil)	Cassiterite mining, clear-water	490 ± 277 (286–1472)	Kasper et al. (2012)
Brokopondo reservoir (Suriname)	Goldmining	740 ± 540 (160–2520)	Mol et al. (2001)
Tucuruí reservoir (Brazil)	Goldmining	547 ± 555 (110–3350)	Kehrig et al. (2009)
Tucuruí reservoir (Brazil)	Goldmining	1100 ± 810	Porvari (1995)
***Hoplias malabaricus* (carnivore)**			
Negro basin (Brazil)	**Natural, black-water**	**(120–1592)**	Barbosa et al. (2003)
Negro basin (Brazil)	**Natural, black-water**	**350 ± 250 (55–1008)**	Belger and Forsberg (2006)
Puruzinho lake (Brazil)	**Natural, black-water**	**407 (256–915)**	Azevedo-Silva et al. (2016)
Blanco river (Bolivia)	Natural, white water	89 ± 39 (41–196)	Pouilly et al. (2012)
San Matín river (Bolivia)	Natural, clear-water	74 ± 27 (29–133)	Pouilly et al. (2012)
San Jorge basin (Colombia)	Natural	(210–430)	Olivero-Verbel et al. (2004)
Las Marías river (Venezuela)	Deforestation, clear-water	1130 ± 180	Kwon et al. (2012)
Cupari basin (Brazil)	Goldmining	400 ± 94	Sampaio da Silva et al. (2009)
Itapacurazinho basin (Brazil)	Goldmining	1398 ± 691	Sampaio da Silva et al. (2009)
Jacaré basin (Brazil)	Goldmining	810 ± 261	Sampaio da Silva et al. (2009)
Paraná basin (Brazil)	Goldmining	295 ± 79	Sampaio da Silva et al. (2009)
Restinga basin (Brazil)	Goldmining	501 ± 229	Sampaio da Silva et al. (2009)
Upper Madeira basin (Brazil)	Goldmining, white water	380 (80–1060)	Boischio and Henshel (2000)

(continued)

Table 3.1 (continued)

Site	Site characteristic	[THg] $\mu g.kg^{-1}$ w.w.	Reference
Itênez river (Bolivia)	Goldmining and deforestation, clear-water	128 ± 69 (51–319)	Pouilly et al. (2012)
Saramacca river (Suriname)	Goldmining, clear-water	560 ± 170 (240–960)	Mol et al. (2001)
Coesewijne river (Suriname)	Goldmining, clear-water	360 ± 90 (260–460)	Mol et al. (2001)
Mina Santa Cruz marsh (Colombia)	Goldmining	322	Olivero and Solano (1998)
Brokopondo reservoir (Suriname)	Goldmining	410	Mol et al. (2001)
Serrasalmus rhombeus **(carnivore)**			
Negro basin (Brazil)	**Natural, black-water**	**(63–1085)**	Dorea et al. (2006)
Natural reserve (Suriname)	Natural	350 ± 140 (180–590)	Mol et al. (2001)
Las Marías river (Venezuela)	Deforestation, clear-water	590 ± 350	Kwon et al. (2012)
Coppename river (Suriname)	Goldmining, clear-water	600 ± 130 (480–810)	Mol et al. (2001)
Saramacca river (Suriname)	Goldmining, clear-water	590 ± 400 (310–870)	Mol et al. (2001)
Coesewijne river (Suriname)	Goldmining, clear-water	1750 ± 160 (1640–1860)	Mol et al. (2001)
Commewijne river (Suriname)	Goldmining, clear-water	590 ± 80 (500–630)	Mol et al. (2001)
Maroni river (French Guiana)	Goldmining	411 ± 220[b]	Fréry et al. (2001)
Balbina reservoir (Brazil)	**Black-water**	**600 ± 400 (50–900)[c]**	Kehrig et al. (1998)
Samuel reservoir (Brazil)	Cassiterite mining, clear-water	696 ± 178 (504–979)	Kasper et al. (2012)
Brokopondo reservoir (Suriname)	Goldmining	1180 ± 620 (420–4620)	Mol et al. (2001)

[a]mean ± standard error
[b]concentrations in dry weight transformed to wet weight: [THg] dry weight * 0.22
[c]methylmercury concentration

Samples were collected during expeditions between 1990s and 2012. Water samples were collected at 25 sites (4 sites along Negro main stem and at 1 site in each of its 21 main tributaries) during July 2011 (high-water season) and December 2012 (low-water season). Sediment and plankton samples were collected at five lakes at Negro floodplain during January 2004. Limnological measurements were made in situ at the same sites and time of water and plankton sampling. Fish samples were collected in Padauiri River, a tributary of Negro River during 1990s. Dolphin was collected in Negro River during January 2004 (Fig. 3.1).

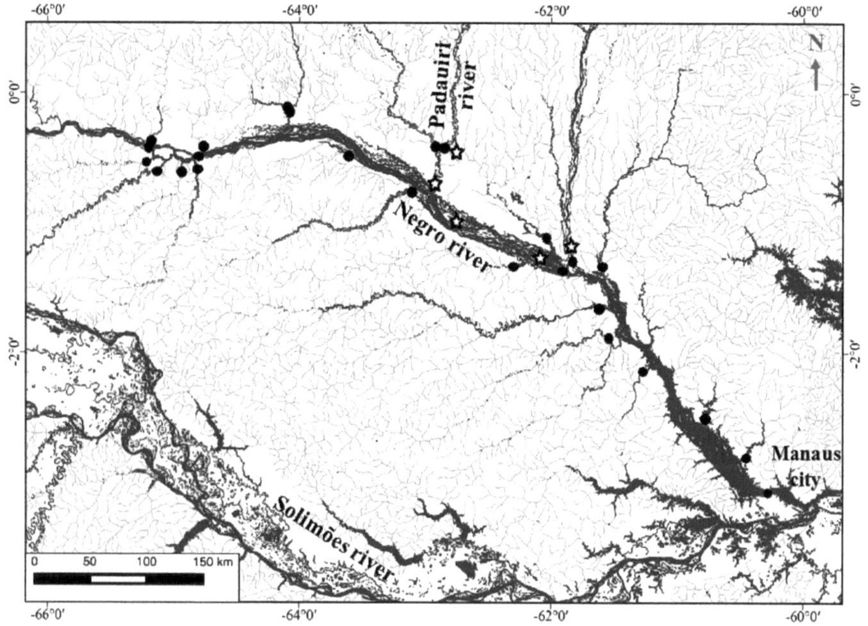

Fig. 3.1 Sampling sites in the Negro basin. Filled circles are sampling sites of water in the Negro main channel and its tributaries, and stars are sampling sites of plankton and sediment in lakes. Fishes were collected in the Padauiri River and dolphin in the Negro main channel

Unfiltered water samples were assessed to total mercury (THg) and MeHg concentrations following EPA (2002) and EPA (2001), respectively. THg concentrations in suspended matter (material retained in GF-5 filters after water subsamples filtration) and in sediment were analyzed according to Bastos et al. (1998). Plankton were sampled with conical nets of 25 and 70 μm that collect mainly phytoplankton and zooplankton, respectively. Here and after we will refer to the sample collected in 25 μm net as phytoplankton and in 70 μm as zooplankton. In order to obtain more pure phytoplankton and zooplankton samples, the filtered material was separated according to the methodology described in Palermo (2008). Plankton, skinless dorsal muscles of fish, and dolphin muscle were analyzed to THg and MeHg concentrations following Bastos et al. (1998) and Kehrig and Malm (1999), respectively. Details of the sample sites, collect procedures, and laboratory analyses are described in Kasper et al. (2014, 2015, 2017) and Kehrig and Malm (1999).

THg concentrations in water did not vary between two hydrological seasons, while MeHg concentrations were highest during high-water season (Table 3.2). Since THg was similar between seasons, probably the highest MeHg concentrations observed occur due to increase of MeHg formation during high water. At this time, the extensive areas of flooded forests, flooded macrophyte beds, and poorly drained soils besides the hypolimnion of the lakes present on Negro floodplain are the most likely source of MeHg to the river (Kasper et al. 2017). Negro basin was more acid

Table 3.2 Mean±standard deviation of variables measured in high-water and low-water seasons in the Negro main stem and its tributaries ($n = 25$). Variables taken in the superficial water column

Variable	Low-water	High-water
[THg] – water ($ng.L^{-1}$)	5.05 ± 4.17	4.14 ± 1.13
[THg] – suspended matter ($ng.g^{-1}$ dw)[a]	324 ± 106	583 ± 298
[MeHg] – water ($ng.L^{-1}$)[a]	0.05 ± 0.02	0.08 ± 0.04
Total suspended matter ($mg.L^{-1}$)	8.1 ± 7.0	6.6 ± 4.2
Dissolved oxygen ($mg.L^{-1}$)[a]	6.5 ± 0.5	2.6 ± 1.2
pH[a]	5.03 ± 0.79	4.81 ± 0.57
Dissolved organic carbon ($mg.L^{-1}$)[a]	11.1 ± 6.5	8.7 ± 3.4

[a]Indicates statistical difference (Paired t or Wilcoxon test)
[THg] = Total mercury concentration, [MeHg] = methylmercury concentration

and less oxygenated during high water (Table 3.2), showing that the basin upstream from sampling site has water more conducive to mercury methylation than during low-water season.

THg concentrations in unfiltered water were not correlated with limnological variables ($p > 005$, dissolved oxygen, pH, and DOC), while MeHg concentrations in water were negatively correlated with dissolved oxygen ($p < 0.05$) but unrelated to pH and DOC ($p > 0.05$). The strict relationship between DOC and THg concentrations in water probably occurs only with dissolved mercury, since the pedological processes that link Hg to DOC in their transport from soil to water occur with mercury in the dissolved fraction (Sousa 2015), and here the whole water were assessed. Fadini and Jardim (2001) also found a poor correlation between total organic carbon (that were very close to the DOC) and total mercury content in the same river system. However, MeHg concentrations were correlated to dissolved oxygen, highlighting the importance of flooded area and thermal structure mainly of the floodplain where the oxygen is depleted and methylation probably is more intense.

THg concentration of suspended matter was highest during high-water (Table 3.2). Since THg concentrations in water and total suspended matter concentrations were similar between seasons (Table 3.2), we conclude that highest THg concentration of suspended matter during high-water season that occurs due to Hg was more associated to particulate form during high-water than during low-water season. Some mechanisms could influence on this, such as (i) water conditions can prevail the Hg in the particulate form than in the dissolved form during high-water, or (ii) particles that compound the suspended matter can be different and more conducive to absorption and/or absorption of Hg during high-water. These are hypotheses that need to be tested.

THg concentration in plankton from lakes varied from 66 to 158 $ng.g^{-1}$ dw (Table 3.3). The %MeHg observed in these plankton samples (Table 3.3) were in the same range or higher than the values obtained previously in plankton from reservoirs, such as in Canada (Ontario, 48–62%, Paterson et al. 1998; La Grande-Quebec, 50–80%, Brouard and Doyon 1991; Downstream to Caniapiscau-Quebec, 32–65%, Schetagne et al. 2000); the USA (California, 20–67%, Stewart et al. 2008), Brazil

Table 3.3 Variables measured in low-water season in five floodplain lakes of the Negro basin. Values represent the only measurement done in each lake. Plankton samples and limnological variables taken in the superficial water column

	Lake				
Variable	Iara	Nazaré	Fernandola	Ramada	Araça
[THg] – phytoplankton (μg.kg^{-1} dw)	85	147	86	158	71
%MeHg – phytoplankton	n.a.	56	71	60	48
[THg] – zooplankton (μg.kg^{-1} dw)	77	122	108	155	66
%MeHg – zooplankton	57	60	53	67	51
[THg] – sediment (μg.kg^{-1} dw)	757	193	261	434	213
Dissolved oxygen (mg.L^{-1})	4.2	4.7	4.2	4.2	3.9
pH	4.6	4.4	4.2	4.7	5.8
Dissolved organic carbon (mg.L^{-1})	15.3	20.9	6.5	13.3	11.2

[THg] = Total mercury concentration, %MeHg = Percentage of the total mercury as methylmercury, n.a. = not assessed, dw = dry weight

(Vigário, Lajes, and Santana reservoirs in Rio de Janeiro, 2–32%, Palermo 2008; Balbina-Amazon, 1–34%, Kasper unpub. Data; Tucuruí-Amazon, 31–36%, Palermo 2008), and China (Baihua, 0.2–70%, Wang et al. 2011). Comparing to natural aquatic systems, the %MeHg of lakes in Negro basin were higher than many natural lakes, for example, in Argentina (Nahuel Huapi, 0.1–7%, Arcagni et al. 2018) and Canada (Ontario, 16–39%, Paterson et al. 1998; Quebec, 15–40%, Plourde et al. 1997), or rivers (Uatumã-Amazon, 1–34%, Kasper unpub. data). This high %MeHg observed in plankton from lakes in Negro basin demonstrates that the plankton is accumulating mainly MeHg (48–71%, Table 3.3) instead of inorganic mercury. The MeHg in the Negro basin waters (Table 3.2) is in the range of natural aquatic systems and probably are not the reason for the high %MeHg of plankton. Therefore, we believe that some processes, not evaluated in the present study, occur specially in the environmental conditions of Negro basin such as highest and fastest MeHg absorption by plankton and greater availability of MeHg to bioaccumulation contributing to the observed highest %MeHg in plankton. These mechanisms have never been studied in the Negro basin. The high %MeHg in the base of the food chain can result in elevated %MeHg due to biomagnification in organisms of higher trophic levels that are part of the plankton food web (such as some piscivorous fish).

Another hypothesis to explain this elevated %MeHg in plankton is that the Amazon lakes are highly conducive to Hg methylation. Brito et al. (2017) suggest that lakes of the Amazon are important contributors of MeHg to the basin, because they have favored conditions for methylation related to their thermal structure and aquatic habitat complexity. Therefore, the MeHg of plankton observed here can be a pattern observed in other Amazonian lakes, with higher %MeHg than the plankton encountered in rivers. Brito et al. (2017) did not assess the %MeHg but observed higher MeHg concentrations in plankton from Janauacá lake (Solimões basin) than those from Solimões River.

Table 3.4 Mean±standard deviation of total mercury ([THg]) and methylmercury ([MeHg]) concentration and percentage of the total mercury as methylmercury (%MeHg) in muscle of fish and dolphin from Negro River basin

Species	n	Trophic level	[THg] μg.kg^{-1} w.w.	[MeHg] μg.kg^{-1} w.w.	%MeHg
Fish					
Laemolyta taeniata	8	Omnivore	171±32 ^{a}	106±35 ^{a}	63±23 ^{a}
Satanoperca acuticeps	10	Omnivore	125±48 ^{a}	95±37 ^{a}	77±13 ^{a}
Cyphocharax abramoides	8	Detritivore	78±26 ^{a}	66±20 ^{a}	82±4 ^{ab}
Osteoglossum ferreirai	12	Carnivore (insectivore)	643±262 ^{bc}	631±231 ^{b}	100±6 ^{c}
Cichla monoculus	10	Carnivore	433±103 ^{b}	389±58 ^{c}	92±9 ^{bd}
Hydrolycus scomberoides	12	Carnivore (piscivore)	667±173 ^{cd}	696±256 ^{b}	103±10 ^{c}
Serrasalmus rhombeus	16	Carnivore (piscivore)	736±382 ^{d}	686±353 ^{b}	95±6 ^{d}
Dolphin					
Sotalia fluviatilis	1	Carnivore (piscivore)	7410	6979	94

$^{a, b, c, d}$Different letters in fish values, within each column, indicate statistical differences (Kruskal-Wallis test and post hoc Dunn). w.w. = wet weight

THg concentrations in phytoplankton and zooplankton were positively correlated with each other ($p < 0.05$), demonstrating that mechanisms that modify the concentrations in one reflect in another. Meanwhile, THg concentrations in phytoplankton and in zooplankton were not correlated to THg concentrations of the sediment ($p > 0.05$) unlike observed in other studies in lakes (e.g., Paterson et al. 1998). That demonstrates that the Hg concentrations in plankton are not influenced only by environmental levels. Other variables can influence on mercury levels in plankton as biodilution, for example, already observed in an Amazon lake (Brito et al. 2017).

Hg concentrations were lower in omnivorous and detritivorous fish than in carnivorous fish and dolphin (Table 3.4) that is expected by the biomagnification of mercury. This pattern was observed for THg and MeHg concentrations and for %MeHg. Almost all carnivorous specimens were above the concentration of Hg in fish recommended by the World Health Organization for human consumption (500 μg.kg^{-1} w.w.). This is based on approximately 400 g weekly intake of fish for a person with 60 kg of body weight. This consumption is underestimated to riverine people of the Amazon, whose diet is primarily composed of cassava and fish.

We conclude that MeHg is formed mainly in the flooded area, where dissolved oxygen in depleted and, probably, the microbial activity of anoxic organisms is increased. The Amazonian lakes are hot spots for MeHg absorption by planktonic communities and can have an important contribution to Hg concentrations of the basin by export large MeHg levels associated with these organisms. Along the food chain of the Negro basin, it was possible to observe the biomagnification of mercury, with increase of mercury concentrations around one order of magnitude from one trophic level to other (Fig. 3.2).

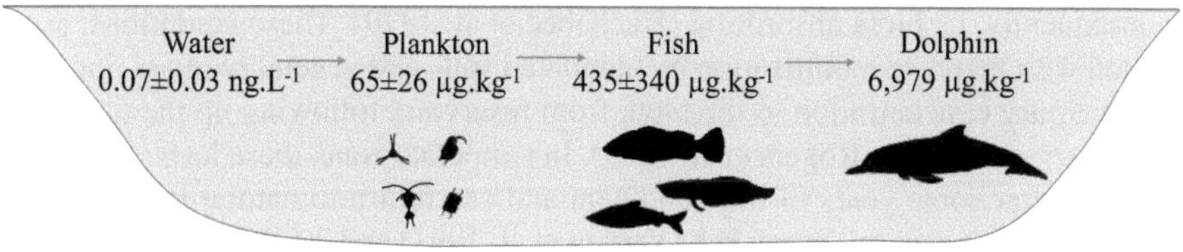

Fig. 3.2 Methylmercury concentrations along food chain of Negro River basin highlighting the increase of about one order of magnitude from one trophic level to the other. Values are mean ± standard deviation (except for dolphin: showed the unique value) observed in Negro Basin (detailed values in Tables 3.2, 3.3, and 3.4)

3.3 New Challenges for Studies in the Brazilian Amazonian Black-Water Rivers in a Changing Landscape

The Amazonian Rivers have been under environmental pressure due to diverse interests of land use. The deforestation arch, which includes Pará, Mato Grosso, and Rondônia states, for example, has many deforested hydrographic basins, and the remaining preserved aquatic systems are generally inside natural reserves. On the other side, the northwest area of the Brazilian Amazon basin has minor disturbances, proving the opportunity to investigate the biogeochemistry of large rivers and their floodplains in their natural condition (e.g., Negro basin). Nowadays, large rivers and their floodplains seldom are preserved. Therefore, this region provides an opportunity to understand natural mercury cycle which may not last for many years because land use is planning (e.g., planning of damming rivers, new roads construction, or mining areas). Beyond goldmining, many other human-driven activities can change the mercury cycle in the Amazon and increase the mercury concentrations in aquatic systems.

The suspended inorganic material generally carries high load of mercury adsorbed; therefore, the intensification of erosion will raise the input of this material to aquatic system, causing an increase in aqueous Hg levels. That is especially concern in aquatic systems with naturally low suspended inorganic material, as blackwaters. Erosion, resulting from deforestation, carried the Hg adsorbed to soil particles to aquatic systems in the Tapajós basin that is naturally poor in suspended solids (clear-water river, Roulet et al. 1998a). Therefore, 40–80% of the THg present in the water of this river was associated with the fine particulate fraction (Roulet et al. 1998a). The deposition of Hg in sediments of a lake in the Meridional Amazon increased with erosion rose by deforestation and land use (for road construction, e.g., Cordeiro et al. 2002). Beyond erosion, burn of land (very common in the region after slash) can generate cationic enrichment in soils, causing a change in the cationic dynamic and, consequently, raising losses of Hg from soils by lixiviation (Farella et al. 2006). Therefore, land use activities that increase erosion and/or lixiviation can raise mercury input in the aquatic systems.

The hydroelectric construction also modifies the Hg dynamics in the Amazonian aquatic systems. The decomposition of organic matter inundated (mainly from soils and vegetal biomass) consumes dissolved oxygen from water (Tundisi et al. 1995). The conditions generated by oxygen deficit favor the Hg methylation and, consequently, its biota absorption (Huchabee et al. 1979). These conditions, associated with mercury mobilization by inundated soils and vegetal biomass, increase the mercury concentration in the biota from reservoirs following up the damming (Forsberg et al. 2013; Rogers et al. 1995). In temperate zone, these levels generally decline after some years of impoundment and can return to natural levels around 5–10 years in noncarnivorous fish (Verdon et al. 1991) and 20–30 years in carnivorous fish (Anderson et al. 1995). In Balbina reservoir, a black-water Amazonian reservoir, Forsberg et al. (2013) observed Hg levels in predatory fish and in hair of fish-feeding people increasing for 10 years following up impoundment and falling after.

In the Madeira River (white-water river in the Brazilian Amazon), two hydroelectric plants were built as of 2011. The high-water flow of the Madeira River (from 5,000 to 50,000 $m^3.s^{-1}$) allowed the adoption of a bulb-type turbine for plants that operate with low falls for hydroelectric generation, so they do not require the formation of large reservoir lakes, such as those found in other hydroelectric dams in the Amazon (Carvalho 2016). Although it is early to reach conclusions about these hydroelectric in Madeira, since their damming is only to be 6 years old, it can be said that some species of fish show a significant increase in Hg concentration after damming (Mussy 2017 pers. comm.). Recent study in the Madeira River basin observed an increase in three fish species (planctivore, *Hypophthalmus edentatus*; omnivore, *Triportheus albus*; and piscivore, *Pinirampus pirinampu*) between 2000 and 2015 (Bastos et al. 2015).

When water passes through the turbines (to electric generation), it transfers the MeHg produced in anoxic hypolimnion of the reservoir lake to river downstream the dam (Canavan et al. 2000; Kasper et al. 2014). This MeHg is mainly associated with suspended matter (Dominique et al. 2007; Schetagne et al. 2000) and can be absorbed by biota and transferred to the downstream food web. Therefore, the biota located downstream the dam can present higher Hg levels than those of reservoir itself. The first studies on this subject were done in temperate reservoirs (e.g., Schetagne et al. 2000), and after, it was also observed in Amazonian reservoirs, such as Tucuruí (Palermo et al. 2004), Samuel (Kasper et al. 2012), and two black-water reservoirs, Petit-Saut (Dominique et al. 2007) and Balbina (Kasper et al. 2014). In these two last reservoirs, consistent changes in MeHg levels in the water were associated with stratification-destratification dynamics (Kasper et al. 2014; Muresan et al. 2008b). Higher MeHg levels in river water were observed just below these dams compared with those in the reservoirs surface only in sampling campaigns where the reservoirs were thermally stratified (Kasper et al. 2014; Muresan et al. 2008b). Consequently, fish living downstream from Petit-Saut (Dominique et al. 2007) and Balbina (Kasper et al. 2014) reservoirs were found to have higher mercury levels than those living in the reservoir, showing a clear link between MeHg export and downstream contamination. Understanding how far from Amazonian dams these high concentrations occur in fish is still a challenge. Studies were done in Balbina (Kasper et al. 2014) and Petit-Saut (Muresan et al. 2008a) reservoirs, but the inflow of tributaries in the river downstream the dam (contributing with reactive mercury and changing the chemical river conditions), and the natural presence of fluvial wetlands (contributing with MeHg to the river since increase Hg methylation) make it difficult to interpret the Hg data.

Among the disturbances that have been discussed in recent years, the climatic changes need to be highlighted. The impacts of climatic changes on environmental mercury levels and human exposure in tropical areas remain uncertain. It is estimated, for example, that the inputs of mercury to the aquatic systems through direct deposition and runoff will increase, and biological mercury levels will change due to modifications in feeding patters (Krabbenhoft and Sunderland 2013). In the Amazon basin, the climatic changes will affect air temperature and precipitation feature that can modify the inundation extent of the aquatic systems (Sorribas et al. 2016). Therefore, the climatic changes can indirectly affect Hg biogeochemical in the Amazonian aquatic systems since mercury cycle in this region is highly influenced by the extension of flooded areas (Belger and Forberg 2006; Kasper et al. 2014). Predicting the impacts of climatic changes in the Hg dynamics in Amazon aquatic systems will be possible only from models that demand data collected in situ and remote sensing information.

In the Amazon basin, the multiple-use scenario (e.g., agriculture, logging, fishing) and its associated impacts (e.g., erosion, biodiversity losses) will be increasingly common. These human-driven impacts (and their impacts on Hg levels of fish) are worrisome on a high fish consumption site. This scenario can be worst if it is considered that black-waters have naturally high mercury levels. The high mercury levels of Amazon basin associated with elevated fish consumption rates result in a chronic exposure of Amazon population to mercury. The intake of mercury by riverine populations from Tapajós basin was estimated, on average, 0.9 $\mu g.kg^{-1}.day^{-1}$ (Passos et al. 2008) that is higher than maximum dose recommended of 0.3 $\mu g.kg^{-1}.day^{-1}$ (EPA 1998). Neurological, immunological, cardiovascular, and cytogenetic effects were observed in Amazonian riverine and associated with the higher fish consumption with elevated mercury levels (e.g., Amorim et al. 2000; Lebel et al. 1998).

It is urgent to know possible synergisms and antagonisms of the mercury availability taking diverse human impacts into account (specially in black-waters due to their naturally high Hg levels). Nowadays, studies have used refined tools as remote sensing, analytical techniques, and hydrologic models that can improve understanding the cycling and health impacts of environmental mercury. These can allow to assess multiple impacts in local and regional scales, with elaboration of models that express the actual or future perspectives of river basin conditions.

Acknowledgments The authors are thankful for the financial support of CAPES, CNPq, FAPEAM, and FAPERJ and for the logistical support of INPA and UFRJ. We also thank the staff of Laboratório de Biogeoquímica Ambiental (UNIR) and Laboratório de Radioisótopos (UFRJ) for their help with mercury analyses.

References

Akagi H, Malm O, Kinjo Y, Harada M, Branches FJP, Pfeiffer WC, Kato H (1995) Methylmercury pollution in the Amazon, Brazil. Sci Total Environ 175:85–95

Amorim MIM, Mergler D, Bahia MO, Dubeau H, Miranda D, Lebel J, Burbano RR, Lucotte M (2000) Cytogenetic damage related to low levels of methylmercury contamination in the Brazilian Amazon. An Acad Bras Cienc 72:497–507

Anderson MR, Scruton DA, Williams UP, Payne JF (1995) Mercury in fish in the Smallwood reservoir, Labrador, twenty one years after impoundment. Water Air Soil Pollut 80:927–930

Arcagni M, Juncos R, Rizzo A, Pavlin M, Fajon V, Arribére MA, Horvat M, Guevara SR (2018) Species- and habitat-specific bioaccumulation of total mercury and methylmercury in the food web of a deep oligotrophic lake. Sci Total Environ 612:1311–1319

Azevedo-Silva CE, Almeida R, Carvalho DP, Ometto JPHB, Camargo PB, Dornele PR, Azeredo A, Bastos WR, Malm O, Torres JP (2016) Mercury biomagnification and the trophic structure of the ichthyofauna from a remote lake in the Brazilian Amazon. Environ Res 151:286–296

Barbosa AC, Souza J, Dórea JG, Jardim WJ, Fadini PS (2003) Mercury biomagnification in a tropical black water, Rio Negro, Brazil. Arch Environ Contam Toxicol 45:235–246

Bastos WR, Malm O, Pfeiffer WC, Cleary D (1998) Establishment and analytical quality control of laboratories for Hg determination in biological and geological samples in the Amazon, Brazil. Cienc Cult 50:255–260

Bastos WR, Gomes JPO, Oliveira RC, Almeida R, Nascimento EL, Bernardi JVE, Lacerda LD, Silveira EG, Pfeiffer WC (2006) Mercury in the environment and riverside population in the Madeira river basin, Amazon, Brazil. Sci Total Environ 368:344–351

Bastos WR, Dórea JG, Bernardi JVE, Lauthartte LC, Mussy MH, Lacerda LD, Malm O (2015) Mercury in fish of the Madeira river (temporal and spatial assessment), Brazilian Amazon. Environ Res 140:191–197

Belger L, Forsberg BR (2006) Factors controlling hg levels in two predatory fish species in the negro river basin, Brazilian Amazon. Sci Total Environ 367:451–459

Boischio AAP, Henshel D (2000) Fish consumption, fish lore, and mercury pollution – risk communication for the Madeira river people. Environ Res 84:108–126

Branfireun BA, Heues A, Roulet NT (1996) The hydrology and methylHg dynamics of Precambrian shield headwater peatland. Water Resour Res 32:1785–1794

Brito BC, Forsberg BR, Kasper D, Amaral JHF, Vasconcelos MRR, Sousa OP, Cunha FAG, Bastos WR (2017) The influence of inundation and lake morphometry on the dynamics of mercury in the water and plankton in an Amazon floodplain lake. Hydrobiologia 790:35–48

Brouard D, Doyon JF (1991) Recherches exploratoires sur le mercure au complexe La Grande. Rapport du Groupe Environnement Shooner Inc. à la vice-présidence Environnement Hydro-Quábec, Montréal

Canavan CM, Caldwell CA, Bloom NS (2000) Discharge of methylmercury-enriched hypolimnetic water from a stratified reservoir. Sci Total Environ 260:159–170

Carvalho DP (2016) Dinâmica e especiação de mercúrio em compartimentos abióticos na formação do reservatório da hidrelétrica de Santo Antônio do Rio Madeira, Rondônia. Ph.D. thesis. Universidade Federal do Rio de Janeiro, UFRJ, Brasil

Compeau G, Bartha R (1985) Sulfate reducer bacteria: principal methylators of mercury in anoxic estuarine sediments. Appl Environ Microbiol 50:498–502

Cordeiro RC, Turcq B, Ribeiro MG, Lacerda LD, Capitâneo J, Silva AO, Sifeddine A, Turcq PM (2002) Forest fire indicators and mercury deposition in an intense land use change region in the Brazilian Amazon (Alta Floresta, MT). Sci Total Environ 293:247–256

Correia RRS, Miranda MR, Guimarães JRD (2012) Mercury methylation and the microbial consortium in periphyton of tropical macrophytes: effect of different inhibitors. Environ Res 112:86–91

Dominique Y, Maury-Brachet R, Muresan B, Vigouroux R, Richard S, Cossa D, Mariotti A, Boudou A (2007) Biofilm and mercury availability as key factors for mercury accumulation in fish (*Curimata cyprinoides*) from a disturbed Amazonian freshwater system. Environ Toxicol Chem 26:45–52

Dorea JG, Barbosa AC, Silva GS (2006) Fish mercury bioaccumulation as a function of feeding behavior and hydrological cycles of the Rio Negro, Amazon. Comp Biochem Physiol 142:275–283

Durrieu G, Maury-Brachet R, Boudou A (2005) Goldmining and mercury contamination of the piscivorous fish *Hoplias aimara* in French Guiana (Amazon basin). Ecotoxicol Environ Saf 60:315–323

EPA-Environmental Protection Agency (1998) Guidelines for neurotoxicity risk assessment. Federal Register 63, Washington

EPA-Environmental Protection Agency (2001) EPA Method 1630. Methyl mercury in water by distillation, aqueous ethylation, purge and trap, and CVAFS. United States Environmental Protection Agency, Washington

EPA-Environmental Protection Agency (2002) EPA Method 1631, revision E. Mercury in water by oxidation, purge and trap, and Cold Vapor Atomic Fluorescence Spectrometry. United States Environmental Protection Agency, Washington

Fadini PS, Jardim WF (2001) Is the negro river basin (Amazon) impacted naturally occurring mercury? Sci Total Environ 275:71–82

Farella N, Lucotte M, Davidson R, Daigle S (2006) Mercury release from deforested soils triggered by base cation enrichment. Sci Total Environ 368:19–29

Forsberg BR, Kasper D, Peleja JRP, Weisser SC, Marshall BG, Torres SS (2013) History of mercury contamination in Balbina Reservoir, Central Amazon, Brazil. 11th international conference on mercury as global pollutant, Edinburgh

Fréry N, Maury-Brachet R, Maillot E, Deheeger M, Mérona B, Boudou A (2001) Gold-mining activities and mercury contamination of native Amerindian communities in French Guiana: key role of fish in dietary uptake. Environ Health Perspect 109:449–456

Gilmour CC, Podar M, Bullock AL, Graham AM, Brown SD, Somenahally AC, Johs A, Hurt RA, Bailey KL, Elias DA (2013) Mercury methylation by novel microorganisms from new environments. Environ Sci Technol 47:11810–11820

Guimarães JRD, Roulet M, Lucotte M, Mergler D (2000) Mercury methylation along a lake-forest transect in the Tapajós river floodplain, Brazilian Amazon: seasonal and vertical variations. Sci Total Environ 261:91–98

Huchabee JW, Elwood JW, Hildebrand SC (1979) Accumulation of mercury in freshwater biota. In: Nriagu JO (ed) The biogeochemistry of mercury in the environment. Elsevier, Amsterdam, pp 277–302

Ikingura JR, Akagi H (2003) Total mercury and methylmercury in fish from hydroelectric reservoirs in Tanzânia. Sci Total Environ 304:355–368

Junk WJ (1993) Wetlands of tropical South America. In: Whigham DF, Dykyjová D, Hejny S (eds) Wetlands of the World I: inventory, ecology and management. Kluwer Academic Publisher, Dordrecht, pp 679–739

Kasper D, Palermo EFA, Branco CWC, Malm O (2012) Evidence of elevated mercury levels in carnivorous and omnivorous fishes downstream from an Amazon reservoir. Hydrobiologia 694:87–98

Kasper D, Forsberg BRF, Amaral JHF, Leitão RP, Py-Daniel SS, Bastos WR, Malm O (2014) Reservoir stratification affects methylmercury levels in river water, plankton, and fish downstream from Balbina hydroelectric dam, Amazonas, Brazil. Environ Sci Technol 48:1032–1040

Kasper D, Forsberg BRF, Almeida R, Bastos WR, Malm O (2015) Metodologias de coleta, preservação e armazenamento de amostras de água para análise de mercúrio – uma revisão. Quim Nova 38:410–418

Kasper D, Forsberg BRF, Amaral JHF, Py-Daniel SS, Bastos WR, Malm O (2017) Methylmercury modulation in Amazon rivers linked to basin characteristics and seasonal flood-pulse. Environ Sci Technol 51:14182–14191

Kehrig HA, Malm O (1999) Methylmercury in fish as a tool for understanding the Amazon Hg contamination. Appl Organomet Chem 13:689–696

Kehrig HA, Malm O, Akagi H, Guimarães JRD, Torres JPM (1998) Methylmercury in fish and hair samples from the Balbina reservoir, Brazilian Amazon. Environ Res 77:84–90

Kehrig HA, Palermo EFA, Seixas TG, Santos HSB, Malm O, Akagi H (2009) Methyl and total mercury found in two man-made Amazonian reservoirs. J Braz Chem Soc 20:1142–1152

Kerin EJ, Gilmour CC, Roden E, Suzuki MT, Coates JD, Mason RP (2006) Mercury methylation by dissimilatory iron-reducing bacteria. Appl Environ Microbiol 72:7919–7921

Krabbenhoft DP, Sunderland EM (2013) Global change and mercury. Science 341:1457–1458

Kwon SY, McIntyre PB, Flecker AS, Campbell LM (2012) Mercury biomagnification in the food web of a neotropical stream. Sci Total Environ 417-418:92–97

Lacerda LD (2003) Updating global hg emissions from small-sale gold mining and assessing its environmental impacts. Environ Geol 43:208–314

Lebel J, Mergler D, Branches F, Lucotte M, Amorim M, Larribe F, Dolbec J (1998) Neurotoxic effects of low-level methylmercury contamination in the Amazonian basin. Environ Res 79:20–32

Lechler PJ, Miller JR, Lacerda LD, Vinson D, Bozongo JC, Lyons WB, Warwick JJ (2000) Elevated mercury concentrations in soils sediments, water, and fish of the Madeira River basin, Brazilian Amazon: a function of natural enrichments? Sci Total Environ 260:87–96

Malm O, Pfeiffer WC, Souza CMM, Reuther R (1990) Mercury pollution due to gold mining in the Madeira river basin, Brazil. Ambio 19:11–15

Malm O, Castro MB, Bastos WR, Branches FJPB, Guimarães JRD, Zuffo CE, Pfeiffer WC (1995) An assessment of hg pollution in different goldmining areas, Amazon Brazil. Sci Total Environ 175:127–140

Mason RP, Benoit JM (1986) Organomercury compounds in the environment. In: Craig PJ (ed) Organometallic compounds in the environment. Longman, New York, pp 57–100

Mauro JBN, Guimarães JRD, Melamed R (1999) Aguapé agrava contaminação por mercúrio. Ciência Hoje 25:68–72

Maury-Brachet R, Durrieu G, Dominique Y, Boudou A (2006) Mercury distribution in fish organs and food regimes: significant relationships from twelve species collected in French Guiana (Amazonian basin). Sci Total Environ 368:262–270

Mol JH, Ramlal JS, Lietar C, Verloo M (2001) Mercury contamination in freshwater, estuarine, and marine fishes in relation to small-scale gold mining in Suriname, South America. Environ Res 86:183–197

Muresan B, Cossa D, Coquery M, Richard S (2008a) Mercury sources and transformations in a man-perturbed tidal estuary: the Sinnamary estuary, French Guiana. Geochim Cosmochim Acta 72:5416–5430

Muresan B, Cossa D, Richard S, Dominique Y (2008b) Monomethylmercury sources in a tropical artificial reservoir. Appl Geochem 23:1101–1126

Mussy MH (2017) Dinâmica das concentrações de Hg em peixes no antes e pós-enchimento da Usina hidrelétrica de Santo Antônio, Rio Madeira, Rondônia. Ph. D. thesis. Universidade Federal do Rio de Janeiro, UFRJ, Brasil (in progress)

Olivero J, Solano B (1998) Mercury in environmental samples from a waterbody contaminated by gold mining in Colombia, South America. Sci Total Environ 217:83–89

Olivero-Verbel J, Johnson-Restrepo B, Mendoza-Marín C, Paz-Martinez R, Olivero-Verbel R (2004) Mercury in the aquatic environment of the village of Caimito at the Mojana region, north of Colombia. Water Air Soil Pollut 159:409–420

Padovani CR, Forsberg BR, Pimentel TP (1995) Contaminação mercurial em peixes do rio Madeira: Resultados e recomendações para consumo humano. Acta Amaz 25:127–136

Palermo EFA (2008) Acúmulo e transporte de mercúrio em reservatórios tropicais. Ph.D. thesis. Universidade Federal do Rio de Janeiro, UFRJ, Brasil

Palermo EFA, Kasper D, Reis TS, Nogueira S, Branco CWC, Malm O (2004) Mercury level increase in fish tissues downstream the Tucuruí reservoir, Brazil. RMZ Mater Geoenviron 51:1292–1294

Passos CJS, Silva DS, Lemire M, Fillion M, Guimarães JRD, Lucotte M, Mergler D (2008) Daily mercury intake in fish- eating populations in the Brazilian Amazon. J Expo Sci Environ Epidemiol 18:76–87

Paterson MJ, Rudd JWM, St. Louis V (1998) Increases in total and methylmercury in zooplankton following flooding of a peatland reservoir. Environ Sci Technol 32:3868–3874

Pfeiffer WC, Lacerda LD (1988) Mercury inputs into the Amazon region, Brazil. Environ Technol Lett 9:325–330
Plourde Y, Lucotte M, Pichet P (1997) Contribution of suspended particulate matter and zooplankton to MeHg contamination of the food chain in midnorthern Quebec (Canada) reservoirs. Can J Fish Aquat Sci 54:821–831
Porvari P (1995) Mercury levels of fish in Tucuruí hydroelectric reservoir and in river Mojú in Amazonia, in the state of Pará, Brazil. Sci Total Environ 175:109–117
Pouilly M, Pérez T, Rejas D, Guzman F, Crespo G, Duprey JL, Guimarães JRD (2012) Mercury bioaccumulation patterns in fish from the Iténez river basin, Bolivian Amazon. Ecotoxicol Environ Saf 83:8–15
Rogers DW, Dickman M, Han X (1995) Stories from old reservoirs: sediment hg and hg methylation in Ontario hydroelectric developments. Water Air Soil Pollut 80:829–839
Roulet M, Lucotte M, Aubin AS, Tran S, Rhéault I, Farella N, Silva EJ, Dezencourt J, Passos CJ, Soares GS, Guimarães JRD, Amorim DM (1998a) The geochemistry of mercury in central Amazonian soils developed on the Alter-do-Chão formation of the lower Tapajós river valley, Pará state, Brazil. Sci Total Environ 223:1–24
Roulet M, Lucotte M, Farella N, Serique G, Coelho H, Passos CJS, Silva EJ, Andrade PS, Mergler D, Guimarães JRD, Amorim M (1998b) Effects of recent human colonization on the presence of mercury in Amazonian ecosystems. Water Air Soil Pollut 112:297–313
Roulet M, Lucotte M, Guimarães JRD, Rheault I (2000) Methylmercury in water, seston, and epiphyton of an Amazonian river and its floodplain, Tapajós river, Brazil. Sci Total Environ 261:43–59
Roulet M, Lucotte M, Canuel R, Farella N, Goch YGF, Peleja JRP, Guimarães JRDG, Mergler D, Amorim M (2001) Spatio-temporal geochemistry of mercury in waters of the Tapajos and Amazon rivers, Brazil. Limnol Oceanogr 46:1141–1157
Sampaio da Silva D, Lucotte M, Paquet S, Davidson R (2009) Influence of ecological factors and of land use on mercury levels in fish in the Tapajós river basin, Amazon. Environ Res 109:432–446
Santos LSN, Muller RCS, Sarkis JES, Alves CN, Brabo ES, Santos EO, Bentes MHS (2000) Evaluation of total mercury concentrations in fish consumed in the municipality of Itaituba, Tapajós River basin, Pará, Brazil. Sci Total Environ 261:1–8
Schetagne R, Doyon JF, Fournier JJ (2000) Export of mercury downstream from reservoirs. Sci Total Environ 260:135–145
Silva-Forsberg MC, Forsberg BR, Zeidemann VK (1999) Mercury contamination in humans linked to river chemistry in the Amazon basin. Ambio 28:519–521
Sorribas MV, Paiva RCD, Melack JM, Bravo JM, Jones C, Carvalho L, Beighley E, Forsberg B, Costa MH (2016) Projections of climate change effects on discharge and inundation in the Amazon basin. Clim Chang 136:555–570
Sousa OP (2015) O papel da matéria orgânica e do hidromorfismo na dinâmica do mercúrio em diferentes solos da Amazônia Central. MSc. thesis. Instituto Nacional de Pesquisas da Amazônia, INPA, Brasil
Stewart AR, Saiki MK, Kuwabara JS, Alpers CN, Marvin-DiPasquale M, Krabbenhoft DP (2008) Influence of plankton mercury dynamics and trophic pathways on mercury concentrations of top predator fish of a mining-impacted reservoir. Can J Fish Aquat Sci 65:2351–2366
Svobodová Z, Dusek L, Hejtmánek M, Vykusová B, Smíd R (1999) Bioaccumulation of mercury in various fish species from Orlík and Kamýr reservoirs in the Czech Republic. Ecotoxicol Environ Saf 43:231–240
Tundisi JG, Bicudo CE, Matsumura-Tundisi T (1995) Limnology in Brazil. Academia Brasileira de Ciências e Sociedade Brasileira de Limnologia, Rio de Janeiro
Uryu Y, Malm O, Thornton I, Payne I, Cleary D (2001) Mercury contamination of fish and its implications for other wildlife of the Tapajós basin, brazilian amazon. Conserv Biol 15:438–446
Verdon R, Brouard D, Demers C, Lalumiere R (1991) Mercury evolution (1978-1988) in fishes of the La Grande hydroelectric complex, Quebec, Canada. Water Air Soil Pollut 56:405–417

Wang Q, Feng X, Yang Y, Yan H (2011) Spatial and temporal variations of total and methylmercury concentrations in plankton from a mercury-contaminated and eutrophic reservoir in Guizhou Province, China. Environ Toxicol Chem 30:2739–2747

Watras CJ, Back RC, Halvorsen S, Hudson RJM, Morrison KA, Wentw SP (1998) Bioaccumulation of mercury in pelagic freshwater food webs. Sci Total Environ 219:183–208

Winfrey MR, Rudd JWM (1990) Environmental factors affecting the formation of methylmercury in low pH lakes. Environ Toxicol Chem 9:853–869

Part III
Soil and the Carbon Cycle

Chapter 4
Soil Carbon and the Carbon Cycle in the Central Amazon Forest

Fabrício Berton Zanchi

4.1 Introduction

The Amazon region contains more than half of Earth's remaining tropical forest and is the largest contiguous tropical forest ecosystem in the world (6.6 million km^2). Eighty percent of the region is still in a pristine state (INPE 2008), and 5.4 million km^2 is located in Brazil with the remaining areas in Bolivia, Colombia, Ecuador, Guyana, Peru, Surinam and Venezuela (Houghton et al. 1995; IBGE 1997; Waterloo et al. 2006).

The region is a mosaic of landscapes, formed in the Tertiary and Quaternary geological period, and comprises at present 13% cattle pasture and agricultural lands, concentrated around the Brazilian states of Acre, Rondônia, Mato Grosso, Tocantins and Pará (IBGE 1997; Soares-Filho et al. 2006). 6% consist of savannah-like vegetation (cerrado, campina and campinarana) and is located in areas around Roraima and the North and South of Amazonas state (McClain et al. 1997; IBGE 1997; Luizão et al. 2007); 70% is evergreen forest (terra firme forest) and can be found throughout the whole Amazon region (Fig. 4.1) including areas of swamp, Igapó and Várzea forests (about 6–8%) located near along the large rivers (Prance and Schubart 1978; Klinge and Medina 1979; Prance 1979; Anderson 1981; Luizão et al. 2007). According to Junk (1993) and Junk et al. (2013), the total wetland areas correspond to 20–25% of the Amazon region. These areas are located along the lakes, streams and low-order rivers, and these valley forests are periodically flooded.

F. B. Zanchi, PhD. (✉)
Centro de Formação em Ciências Ambientais-CFCAm,
Universidade Federal do Sul da Bahia-UFSB,
Porto Seguro, BA, Brazil

Doutor pela Vrije Universiteit van, Amsterdam, Netherlands
e-mail: fabricio.berton@csc.ufsb.edu.br

R. W. Myster (ed.), *Igapó (Black-water flooded forests) of the Amazon Basin*,
https://doi.org/10.1007/978-3-319-90122-0_4

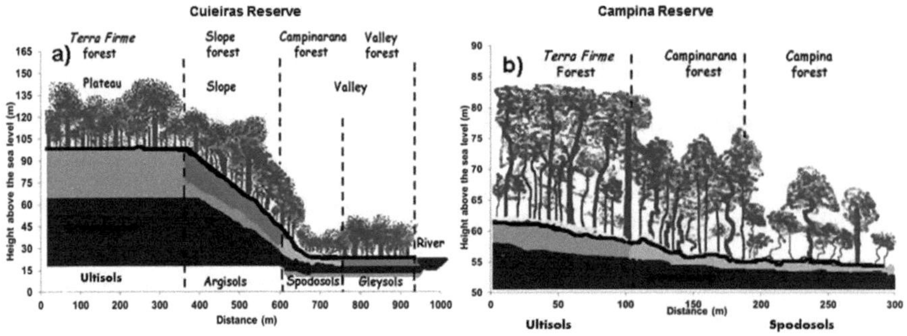

Fig. 4.1 Typical topographic transect for tropical rainforest (central Amazonia). (**a**) Cuieiras Reserve (near Manaus city, Brazil), which has a steep inclination forest and is composed of a mosaic of riparian (valley forest), campinarana, slope and plateau forests (swampy area) with their respective soil types; (**b**) Campina Reserve (near Manaus city, Brazil), which presents the transition from plateau forest via campinarana to campina forest, with different in soil types and properties. (After, Chauvel et al. (1987) and Zanchi (2013))

As in the central Amazonia, the climate does not show much seasonality. Permanently wet soils in the valley area are maintained by a continuous supply of groundwater from plateau and slope areas (Zanchi 2013). Whilst Oxisols and Ultisols on the slopes and plateau have a significant clay fraction, valley soils generally consist of strongly leached quartz sand (Spodosols) with little capacity to retain water or nutrients (Chauvel et al. 1987). Nutrients may be retained in the valley soil by the soil carbon fraction (humus) and the high groundwater table, which may also buffer the pH to some extent, allowing tall forests to grow on these white sands (Guillaumet 1987; Luizão 1996). This might trigger a change from an existing tall valley forest to a heath forest with more sclerophyllous leaves (Zanchi et al. 2012). This forest is locally known as *campinarana*, *campina* or *bana* forest (Anderson 1981; Sobrado and Medina 1980; Reich et al. 1994). The presence of *campinarana* forest between the tall forest in the valley bottom and that on the slope (Waterloo et al. 2006) is an indication that a relatively small change in the valley groundwater level (rainfall regime) may be sufficient to cause an extension of the *campinarana* zone towards the river.

Although these *campina or campinarana* forests are quite common in the tropics, little is known about their hydrologic functioning (e.g. evapotranspiration rate) and their carbon balance (Becker 1996; Tyree et al. 1998; Zanchi et al. 2012; Zanchi et al. 2015).

4.2 Total Amazon Carbon Estimative

The total global soil carbon pool of 2000–3800 Pg contains about 1550 Pg of organic carbon and 950 Pg of inorganic carbon. This soil C pool is about three times the size of the atmospheric pool (760 Pg) and 4.5 times that of the biotic pool (500–650 Pg)

Table 4.1 Soil temperature, soil efflux and release of carbon from plateau, campinarana and valley forest

Site	Soil temp. (°C) at 5 cm	Soil efflux (μmol CO_2 m^{-2} s^{-1})	Release of C (t C ha^{-1} $year^{1}$)
Plateau	25.7 ± 0.6	2.4 ± 0.4	9.0
Campinarana	25.6 ± 0.5	6.0 ± 1.6	22.7
Valley	25.8 ± 0.6	4.1 ± 1.2	15.5

(Gruber et al. 2004, Janzen 2004; Lal 2004). Schlesinger (1997) estimates that 13–17% of the total soil carbon in tropical forests resides in the upper metre of soil and has a fast turnover time. The estimated total stock of carbon in the old growth Amazon forests, represented in Fig. 4.1, is still uncertain; some authors found values which vary from 81.9 to 91.2 Gt C (Saatchi et al. 2011; Baccini et al. 2012).

Floodplain inundated forests store 44 Mg C ha^{1}, woodland savanna 31 Mg C ha^{1} and open shrubs 23 Mg C ha^{1} (Baccini et al. 2012). The net effect of Amazonian deforestation and reforestation results in an annual net C source of 0.15–0.35 Pg C, and to the tall forest fire C emission is about 0.23–0.54 Pg C yr^{-1} (van der Werf et al. 2009; Davidson et al. 2012). Tropical rainforest fires and deforestation activities in the State of Mato Grosso, Brazil, have been releasing about 67 Tg C yr^{-1} (DeFries et al. 2008). The combined carbon emissions from fire and from logging would annually release 0.2–0.8 Pg C. The rivers in the Amazon region also release carbon to the atmosphere; it happens because they are supersaturated with dissolved CO_2, which may eventually release to the atmosphere an estimated 0.5 Pg C yr^{-1} (Richey et al. 2002).

There exists considerable uncertainty about the estimates of the carbon balance across the Amazonian ecosystem (NEP). According to Phillips et al. (2009), mature forests are accumulating on average about 0.4 Pg C yr^{-1}. According to Malhi and Grace (2000), the gross productivity from tropical forest is about 30.4 t C ha^{-1} $year^{-1}$, where 14.8 t C ha^{-1} $year^{-1}$ is the mean allocation carbon in leaves respiration, wood and roots and the remained carbon (15.6 t C ha^{-1} $year^{-1}$) is addressed to the net primary productivity carbon. Zanchi et al. (2012) determine for soil emission in plateaus (terra firme forest), campinarana and valley (riparian) forests a total annual emission of 9.0, 22.7 and 15.5 t C ha^{-1} $year^{-1}$, respectively (Table 4.1). Trumbore et al. (1996) suggested that the greatest losses of soil carbon due to climate change would be in tropical regions, where their measurements of radiocarbon content indicated the presence of a large pool of soil organic matter with a relatively rapid turnover time. The uncertainties of the total carbon stock estimates will eventually be reduced when field inventories and the geographical distance in the Amazon region match and provide more reliable data sets to improve the remote sensing techniques for determining local carbon stock (Zanchi et al. 2012).

The difference in the emission is about the root content in each soil. Campinarana forest is not a flooded area, but has a shallow water table depth. Devol et al. (1988) report a carbon emission during the high water period, where the flooded forest emitted 3.4 t C ha^{-1} $year^{-1}$ emission. Zanchi et al. (2012) also found a low soil C

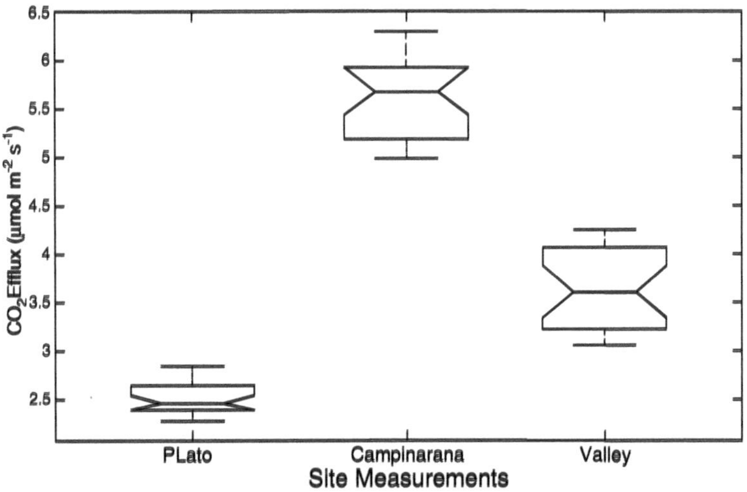

Fig. 4.2 Topographical gradients from the rainforest in the central Amazonia along which CO_2 were measured in different locations of the *Cuieiras* reserves. The boxplot describes the minimum sample, lower quartile, median, upper quartile and maximum emission sampled from each location

emission in shrub forests, also known as Campina forests, where the emission ranged from 3.4 to 4.5 t C ha^{-1} year^{-1} (Fig. 4.2). According to Zanchi et al. (2011), campinarana forest, located between the plateau/slope and riparian area, experienced lower phreatic levels ranging between depths of 0.39 m and 1.30 m below the surface, with a mean depth of 0.81 ± 0.21 m (n = 32). However, valley (riparian) forest has a shallow groundwater table; the mean water level depth was 0.12 ± 0.06 m (n = 36).

4.3 Litter Decomposition Rates

Luizão and Schubart (1987) studied the litter decomposition rates in the Amazon region. This study observed an estimate of 218 days (50% of weight loss) and 1006 days (95% of weight loss), for dry season conditions based on a 150-day measurement period. Zanchi et al. (2011) observed in the riparian forest region a slower litter decomposition rate compared to plateau forest in Luizão and Schubart (1987). Zanchi et al. (2011) found a rapid initial loss of leaf mass in the first month (up to 25% of the initial mass loss for individual samples), followed by a more gradual decline to about 73% mass loss a year after installation (Fig. 4.3). The decomposition rate was much lower in the campina forest, where only 35% of litter mass was lost in the first year. About 50% of weight loss occurred after 210 days in riparian forest, whereas in the campina forest, the 50% mass loss occurred only after

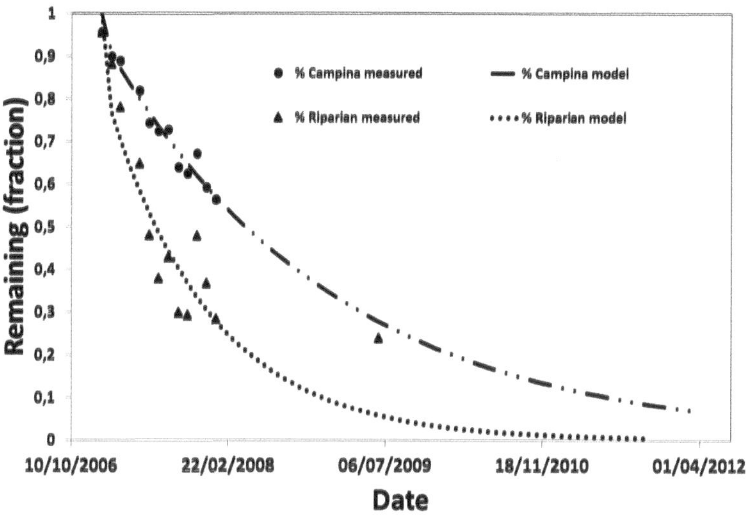

Fig. 4.3 Remaining mass fraction of organic material in litterbags installed in riparian forest and in the campina forest plot

480 days. The data indicated that 95% mass loss would occur after 925 days in the riparian forest and only after 1825 days in the campina forest.

In the plateau forest, Luizão and Schubart (1987) found that litter decomposition was faster on the plateau than in the valley of this *terra firme* landscape and that decomposition rates were higher during the wet season (50% mass loss in 32 days) than during the dry season. According to Luizão et al. (2004), litter turnover rates on plateau and slope were not significantly different from those in the valley in spite of lower N concentrations in the upper soil layer in the valley, which they attributed to other processes, such as leaching, removing litter from the soil surface.

There may be several reasons for the slow decomposition observed in campina compared to the plateau forest. Tree species in campina forest do have sclerophyllous leaves that may be more difficult to decompose, and studies have shown that the decomposition activity of fungi is also suppressed in this environment, where ectomycorrhizae are instrumental in the cycling of nutrients (Singer and Aguiar 1986). Walker (1987), Luizão and Schubart (1987), Luizão et al. (2004) and Zanchi et al. (2011) have reported that even the high annual rainfall, which percolates rapidly through the nutrient-poor sandy soils, may not be sufficient to transport sufficient nutrients for the synthesis of adequate quantities of structural proteins (endoplasmic reticulum, membranes, histones, etc.) in order to balance intense assimilation. This could be a cause for a relative excess of cellulose and lignin in the leaves. The first step in litter decomposition is leaching, when nutrients and organic compounds dissolve in water and move into the soil (Zanchi et al. 2011). Compounds that are easily leached from litter include potassium, sugars and amino acids. Subsequent leaf litter weight loss is related to termite activity in terra firme rainforest (Luizão and Schubart 1987), which can also degrade such resistant substances as lignin (Butler and Buckerfield 1979). When litter becomes increasingly fragmented over time, it reveals new surfaces for bacteria and fungi to attack (Chapin et al. 2002). The presence and activity of termites may be low in nutrient-poor campina forest as compared to those in other rainforest types, which may slow down litter fragmentation and decomposition in campina forest, causing low carbon levels in the soil and high DOC concentrations in soil moisture and groundwater, as was observed in this and other studies in central Amazon (McClain and Richey 1996; McClain et al. 1997). Litter fragment decomposition contributes significantly to CO_2 production in the soil, accounting for about 27% of soil emission (Wang et al. 1999), which is low at Campina Reserve. Su (2005) reported that litter decomposition rates decrease over time due to labile compounds leaving the litter being attacked quickly. This process leaves a residue of less readily decomposable substances (Su 2005), with potentially high C:N ratios (McClain et al. 1997). High nitrogen content in litter is considered a factor affecting the speed of decomposition by microorganisms. Pate and Lyzell (1990) show that cellulose degradation is also a nitrogen-limited process and will increase with the nitrogen content of litter. Vitousek and Sanford Jr. (1986) compared foliar and fine litterfall nutrients of various rainforests and showed that nitrogen and phosphorus appear to cycle less in campina forest than in other lowland forests.

References

Anderson AB (1981) White-sand vegetation of Brazilian Amazonia. Biotropica 13(3):199–210

Baccini A, Goetz SJ, Walker WS, Laporte NT, Sun M, Sulla-Menashe D, Hackler J, Beck PSA, Dubayah R, Friedl MA, Samanta S, Houghton RA (2012) Estimated carbon dioxide emissions from tropical deforestation improved by carbon–density maps. Nat Clim Chang 2(3):182–185

Becker P (1996) Sap flow in Bornean heath and dipterocarp forest trees during wet and dry periods. Tree Physiol 16:295–299

Butler JHA, Buckerfield JC (1979) Digestion of lignin by termites. Soil Biol Biochem 11(5):507–513

Chapin FSIII, Matson PA, Mooney HA (2002) Principles of terrestrial ecosystem ecology, 1st edn. Springer-Verlag, Inc., New York

Chauvel A, Lucas Y, Boulet R (1987) On the genesis of the soil mantle of the region of Manaus, Central Amazonia, Brazil. Experientia 43:234–241

Davidson EA, de Araujo AC, Artaxo P, Balch JK, Brown IF, Bustamante MMC, Coe MT, DeFries RS, Keller M, Longo M, Munger JW, Schroeder W, Soares-Filho BS, Souza CM, Wofsy SC (2012) The amazon basin in transition. Nature 481:321–328

DeFries RS, Morton DC, van der Werf GR, Giglio L, Collatz GJ, Randerson JT, Houghton RA, Kasibhatla PK, Shimabukuro Y (2008) Fire-related carbon emissions from land use transitions in southern Amazonia. Geophys Res Lett 35:L22705

Devol AH, Richey JE, Clark WA, King SL, Martinelli LA (1988) Methane emissions to the troposphere from the Amazon floodplain. J Geophys Res 93(D2):1583–1592

Gruber N, Friedlingstein P, Field CB, Valentini R, Heimann M, Richey JE, Lankao PR, Schulze ED, Chen CTA (2004) The vulnerability of the carbon cycle in the 21st century: an assessment of carbon-climate- human interactions. In: Field CB, Raupach MR (eds) SCOPE 62: the global carbon cycle. Washington, Covelo. Island Press, London

Guillaumet JL (1987) Some structural and floristic aspects of the forest. Experientia 43(3):241–251

Houghton JT, Meira Filho LG, Callander BA, Harris N, Kattenberg A, Maskell K (1995) Climate change 1995, the science of climate change. Cambridge University Press, Cambridge, UK
IBGE. Instituto Brasileiro de Geografia e Estatística (1997) Diagnostico Ambiental da Amazonia Legal. CD-ROM
INPE. Instituto Nacional de Pesquisas espaciais (2008) Monitoramento da cobertura florestal da Amazônia por satélites sistemas prodes, deter, degrad e queimadas 2007–2008 (Monitoring the Brazilian Amazonia Forest by Satellite). Instituto Nacional de Pesquisas Espaciais. Accessed on November of 2017
Janzen HH (2004) Carbon cycling in earth systems–a soil science perspective. Agric Ecosyst Environ 104(3):399–417
Junk WJ (1993) Wetlands of tropical South America. In: Whigham D, Hejny S, Dykyjova D (eds) Wetlands of the world. Junk Publications, Dordrecht, pp 679–739
Junk WJ, Piedade MTF, Lourival R, Wittmann F, Kandus P, Lacerda LD, Bozelli RL, Esteves FA, Nunes da Cunha C, Maltchik L et al (2013) Brazilian wetlands: their definition, delineation, and classification for research, sustainable management, and protection. Aquat Conserv Mar Freshwat Ecosyst 24:5–22
Klinge H, Medina E (1979) Rio Negro caatingas and Campinas, Amazonas states of Venezuela and Brazil. In: Specht RL (ed) Ecosystems of the world: heathlands and related shrublands. Ecosystems of the world, vol 9A. Elsevier, Amsterdam, pp 483–488
Lal R (2004) Soil carbon sequestration impacts on global climate change and food security. Science 304:1623–1627. https://doi.org/10.1126/science.1097396
Luizão FJ (1996) Ecological studies in three contrasting forest types in central Amazonia. PhD thesis, University of Stirling, Scottland, UK
Luizão FJ, Schubart HOR (1987) Litter production and decomposition in a terra-firme forest of Central Amazonia. Experientia 43:259–265
Luizão RCC, Luizão FJ, Paiva RQ, Monteiro TF, Sousa LS, Kruijt B (2004) Variation of carbon and nitrogen cycling processes along a topo- graphic gradient in a central Amazonian forest. Glob Chang Biol 10(5):592–600
Luizão FJ, Luizão RCC, Proctor J (2007) Soil acidity and nutrient deficiency in central Amazonian heath forest soils. Plant Ecol 192(2):209–224, 10
Malhi Y, Grace J (2000) Tropical forests and atmospheric carbon dioxide. Trends Ecol Evol 15:332–337
McClain ME, Richey JE (1996) Regional-scale linkages of terrestrial and lotic ecosystems in the Amazon basin: a conceptual model for organic matter. Arch Hydrobiol., Suppl. 113, Large Rivers 10(1–4):111–125
McClain ME, Richey JE, Brandes JA, Pimentel TP (1997) Dissolved organic matter and terrestrial-lotic linkages in the Central Amazon Basin of Brazil. Glob Biogeochem Cycles 11(3):295–311
Pate JS, Lyzell DB (1990) Energetic and biological cost of nitrogen assimilation. In: Biochemistry of plants, Intermediary nitrogen metabolism, vol 16. Academic Publishing, San Diego, pp 1–42
Phillips OL, Aragão LEOC, Lewis SL et al (2009) Drought sensitivity of the Amazon rainforest. Science 323(5919):1344
Prance GT (1979) Notes on the vegetation of Amazonia iii. The terminology of amazonian forest types subject to inundation. Brittonia 31(1):26–38
Prance GT, Schubart HOR (1978) Notes on the vegetation of Amazonia I. A preliminary note on the origin of the open white sand Campinas of the lower Rio Negro. Brittonia 30(1):60–63
Reich PB, Walters MB, Ellsworth DS, Uhl C (1994) Photosynthesis-nitrogen relations in Amazonian tree species. I. Patterns among species and communities. Oecologia 97:62–72
Richey JE, Melack JM, Aufdenkampe AK, Ballester VM, Hess L (2002) Outgassing from Amazonian rivers and wetlands as a large tropical source of atmospheric CO_2. Nature 416:617–620
Saatchi SS, Harris NL, Brown S, Lefsky M, Mitchard ETA, Salas W, Zutta BR, Buermann W, Lewis SL, Hagen S, Petrova S, White L, Silman M, Morel A (2011) Benchmark map of forest carbon stocks in tropical regions across three continents. Proc Natl Acad Sci 108(24):9899–9904

Schlesinger WH (1997) Biogeochemistry: an analysis of global change, 2nd edn. Academic Press, San Diego

Singer R, Aguiar IA (1986) Litter decomposing and ectomycorrhizal Basid- iomycetes in an igapó forest. Plant Syst Evol 153(1–2):107–117

Soares-Filho BS, Nepstad DC, Curran LM, Cerqueira GC, Garcia RA, Ramos CA, Voll E, McDonald A, Lefebvre P, Schlesinger P (2006) Modelling conservation in the Amazon basin. Nature 440(7083):520–523

Sobrado MA, Medina E (1980) General morphology, anatomical structure, and nutrient content of sclerophyllous leaves of the 'bana' vegetation of Amazonas. Oecologia 45:341–345

Su B (2005) Interactions between ecosystem carbon, nitrogen and water cycles under global change: results from field and mesocosm experiments, PhD thesis, The University of Oklahoma

Trumbore SE, Chadwick OA, Amundson R (1996) Rapid exchange between soil carbon and atmospheric carbon dioxide driven by temperature change. Science 272:393–396

Tyree MT, Patiño S, Becker P (1998) Vulnerability to drought-induced embolism of Bornean heath and dipterocarp forest trees. Tree Physiol 18:583–588

Van Der Werf GR, Morton DC, DeFries RS, Olivier JGJ, Kasibhatla PS, Jackson RB, Collatz GJ, Randerson JT (2009) CO_2 emissions from forest loss. Nat Geosci 2:737–738

Vitousek PM, Sanford RL (1986) Nutrient cycling in moist tropical forest. Annu Rev Ecol Syst 17:137–167

Walker I (1987) Conclusion. The forest as a functional entity. Experientia 43(3):287–290

Wang Y, Amundson R, Trumbore S (1999) The impact of land use change on C turnover in soils. Glob Biogeochem Cycles 13(1):47–57

Waterloo MJ, Oliveira SM, Drucker DP, Nobre AD, Cuartas LA, Hodnett MG, Langedijk I, Jans WWP, Tomasella J, de Araújo AC, Pimentel TP, Munera Estrada JC (2006) Export of organic carbon in run- off from an Amazonian rainforest Blackwater catchment. Hydrol Process 20:2581–2597

Zanchi FB (2013) Vulnerability to drought and soil carbon exchange of valley forest in Central Amazonia (Brazil). PhD. Thesis, Vrije Universiteit Amsterdam

Zanchi FB, Waterloo MJ, Dolman AJ, Groenendijk M, Kesselmeier J, Kruijt B, Bolson MA, Luizão FJ, Manzi AO (2011) Influence of drainage status on soil and water chemistry, litter decomposition and soil respiration in central Amazonian forests on sandy soils. Ambi-Agua, Taubat 6(1):6–29

Zanchi FB, Waterloo MJ, Kruijt B, Kesselmeier J, Luizão FJ, Dolman AJ (2012) Soil CO_2 efflux in Central Amazonia: environmental and methodological effects. Acta Amaz 42(2):173–184

Zanchi FB, Waterloo MJ, Peralta Tapia A, Alvarado Barrientos MS, Bolson MA, Luizão FJ, Manzi AO, Dolman AJ (2015) Water balance, nutrient and carbon export from a heath forest catchment in central Amazonia, Brazil, Hydrol. Processes, https://doi.org/10.1002/hyp.10458

Chapter 5
Igapó Ecosystem Soils: Features and Environmental Importance

Maria de Lourdes Pinheiro Ruivo, Denise de Andrade Cunha, Rosecelia Moreira da Silva Castro, Elessandra Laura Nogueira Lopes, Darley C. Leal Matos, and Rita Denize de Oliveira

5.1 Amazonia Wetlands

Wetlands are defined as those areas episodically or periodically flooded by the lateral overflow of rivers or lakes and/or by direct precipitation or outcropping of the water table, so that the biota responds to the physical-chemical environment with morphological, anatomical, physiological, and etiological specific structures and characteristics of those communities (Junk et al. 1989; Cunha et al. 2015). A total of 8.3 and 10.2 million km^2 of the Earth's surface is expected to be made up of these environments (Mitsch et al. 2009; Oliveira 2017), about 30% concentrated in the tropical regions of the planet.

The vegetation in the Amazon is heterogeneous since it is formed by a diversity of highly distinct habitats; it is a unique territory due to the indescribable variety of its flora and fauna. It includes nine countries in South America where Brazil contains the largest share, 63.4% of the total (Ayres 2006). It is bordered by the massifs of the Guianas and Central Brazil to the north and south, respectively, and to the west by the Andes. It shelters the longest river system and with the greatest liquid mass in the world, covered by the largest tropical rainforest. The Amazonas drains more than seven million square kilometers of land and is, by a large margin, the river with the highest liquid mass, with an annual average flow of 200,000 cubic

M. de Lourdes Pinheiro Ruivo (✉) · R. M. da Silva Castro · D. C. Leal Matos
Museu Paraense Emilio Goeldi, Belém, PA, Brazil
e-mail: ruivo@museu-goeldi.br

D. de Andrade Cunha
Instituto Federal do Pará, Castanhal, PA, Brazil

E. L. N. Lopes · R. D. de Oliveira
Universidade Federal do Pará, Belém, PA, Brazil

R. W. Myster (ed.), *Igapó (Black-water flooded forests) of the Amazon Basin*,
https://doi.org/10.1007/978-3-319-90122-0_5

meters per second (Molinier et al. 1995). This region corresponds to 1/20 of the Earth's surface, 2/5 of South America's surface, 1/5 of the world's availability of fresh water, 1/3 of the world reserves of broad-leaved forests, and only 3.5 thousandths of the world population, with a density of 2 inhabitants/km^2 (MPEG 2017).

About 20% of the Amazon is occupied by the rivers or flooded in a permanent or seasonal character, whose vegetation is denominated floodplain forest: *várzea* and *igapó*. This system is ruled by the "flood pulse" and is responsible for the seasonal variation, resulting from the rainy season and the Andean ice melt (Junk 2010). This arrival of waters bringing nutrients and changing the environment periodically is also responsible for the formation of the soils, either gleysol or luvisols, the main classes of soils related to Amazon dry land flooded ecosystems.

According to Daniels and Nelson (1987), we must abandon the idea that soils are independent entities occurring at specific points and consider that all parts of the landscape are interrelated. Each of these parts is affected and affects adjacent parts, especially those of a slope toward a water gradient. It is required a better understanding of soil-plant environmental relationships, including the physical basis of soil variability as well as temporal changes in the conditions existing in a particular landscape.

The findings that the fertility level of the rivers reflected in the vegetation led most botanists to adopt the criterion "color of the rivers" to separate the flooded forests into *várzea* and *igapó*. Hence, *várzea* forests are those vulnerable to floods by white water rivers, and the *igapó* forest is that flooded by rivers of clear or black water (Prance 1980). The *várzea* forest is characterized by greater richness in nutrients, while the *igapó* is characterized by the acidity and nutrient poverty, which may be related to the quality of its substrate or soil, since the movement of waters, often daily, carries sediment to these areas or removes sediment from these areas.

The main factors for the maintenance of diversity in these environments are the physical and biological processes, among which the hydrological cycle is one of the basic factors (Parolin 2001) since the vegetation of these environments is adapted to survive long periods of complete or partial submersion (Ferreira 2000).

Plant species occurring in flooded forests present different types of phenological, anatomical, physiological, and morphological adaptations in leaves, stems, and roots to survive the periodic flood gradient caused by the fluctuation in the level of the rivers, such as the development of spaces of air in the roots and stems that allow the diffusion of oxygen from the aerial parts of the plant to the roots, formation of adventitious roots, aerenchyma development, morphological changes in the leaves probably to facilitate the underwater exchange of gases, etc. (Parolin et al. 2004; Mommer and Visser 2005).

The combination of adaptations considering seed germination, seedling development, and structure of the roots, stem, and leaves results in a variety of growth establishment strategies among species. This related to the duration of the flood period, types of soil, flooded area types, and tolerance of the plants to flooding leads to specific distributions of the plant species (Parolin 2012).

To understand the dynamics of the Amazon depends on knowledge of this ecosystem and the way plants and animals adapt to floodplains, such as the *igapó* forests, which are areas of the Amazonian biome, regularly flooded. According to Ducke and Black (1954) cited by Prance (1980), *igapó* area is characterized as forest on a soil that never gets dry.

5.2 Igapó Ecosystem: Soil Characteristics and Study Case

Igapó forests cover an area of approximately 180,000 km^2 of the Amazon basin and are in areas with low geomorphological dynamics originated from the pre-Cambrian of the Brazilian Shields and Guianas (Melack and Hess 2010; Iion et al. 2010; Furch and Junk 1997). Historically, those forests are associated with rivers that contain low amounts of dissolved sediment, and therefore low fertility, but with high amounts of humic acids and acidic pH (Prance 1980; Furch and Junk 1997; Junk et al. 2015).

The geology of the substrates over which the rivers flow results in differences in the physical-chemical properties of water, which have a direct influence on the vegetation of the floodplain in the Amazon basin (Sioli 1956). In addition to organizing the confusion in the terminologies used in the past, Prance (1979) defined seven major types of forests subject to flooding and restricted the use of the term "igapó" to forests flooded by rivers of black or clear water.

The most important black water river in Brazil is the Negro River, a large tributary on the left side of the Amazon River, which drains tertiary sediments composed of kaolinitic soils. Its waters have a dark color that derives from the high content of humic substances leached from podzolic soils (Sioli 1968, 1975; Furch 1984).

Clear water rivers, such as the Xingu and the Tapajós, drain severely degraded tertiary sediments, composed of kaolinitic soils, presenting large physicochemical variability and considered intermediates, in relation to the nutritive content, between those of black and white water (Sioli 1975; Prance 1979; Ayres 1993).

Overall, all ecosystems characterized as igapó display a low nutritional status due to the small load of nutrient dissolved and suspended by annual floods (Junk et al. 2015). The so-called flooded soils are characterized by presenting varied natural fertility and the occurrence near the riverbanks. These soils are still poorly explored; some still maintain vegetation cover in the natural forest of the várzeas and igapó with their variations related to the altitude of the terrain and duration of the annual flooding period (Table 5.1). These environments are susceptible to periodic inundations caused by flooding of the rivers (flood pulse) that, while helping to maintain its fertility, makes difficult its use since it requires different management and specific crops.

Table 5.1 Main characteristics of Amazon flooded soil

Soil characterization	Land use characterization	Author
Varzea soils	Areas subject to periodic floods caused by river floods, always offering a new and fertile layer. Those rivers carry fertile sediments, rich in nutrients	Cravo et al. (2002); Carim (2016)
Igapó	Those soils are permanently flooded, narrow, and located in low relief near the rivers Relatively low fertility and low productivity environments. Its soil and water contain a very acidic composition due to the large amount of organic matter in transformation present in their area	Santiago (2017); Carim (2016)

5.2.1 Case Study: National Forest of Caxiuanã

The Caxiuanã National Forest (FLONA) is located in the municipalities of Melgaço and Portel in the state of Pará, about 400 km from Belém, near Caxiuanã Bay, located in the morphotectonic compartment of Gurupá, in the channel of the Amazon River, between the Xingu and Tapajós rivers (Bemerguy 1997) (Fig. 5.1).

The relief of the area is flat and undulating without large elevations (Radam Brasil 1974). The predominant vegetation cover is a lowland dense ombrophilous forest with a canopy ranging between 30 and 40 m in height, and to a lesser extent, other vegetation types, such as periodically flooded forests (igapó and várzeas), open vegetation enclaves of savannahs, campinaranas, and anthropic areas, denominated secondary forests (capoeiras) of different ages and succession stages (Ferreira 2011). The flooded forests of igapó and várzea occupy 12% of the Caxiuanã FLONA area (Pereira et al. 2012) bordering water courses in relatively flat areas (Pereira et al. 2012; Behling 2011).

Pollen analyses of lacustrine deposits indicate that the flooded forests in the Caxiuanã FLONA appeared in the Holocene due to paleoenvironmental changes caused by the neotectonic activities associated with the increase of the level of the Atlantic sea and consequent flooding of terra firme areas (Costa 2002; Behling 2011). Additionally, at the end of the Holocene, the increase in annual precipitation also contributed to the formation of a passive fluvial system and a marked expansion of those flooded forests. These events profoundly modified the geomorphology and hydrodynamics of the Anapú River and its effluents, promoting the blockage of river drainage and the consequent emergence of the Caxiuanã Bay and the chemical change of the rivers in the region (Behling 2011; Costa 2002; Behling and da Costa 2000).

Currently, the Caxiuanã River Basin is classified as a fluviolacustrine environment (Melo et al. 2002), characterized by the absence or low presence of suspended matter (Costa 2002), with a hydrological system classified as monomodal in which fluctuation of the water level is seasonal, influenced by annual and polymodal precipitation regimes influenced by daily tidal regimes with annual fluctuations of 17–21 cm between low and high tide and 33 cm in intertidal regimes (IBGE 1977; Hida et al. 1998).

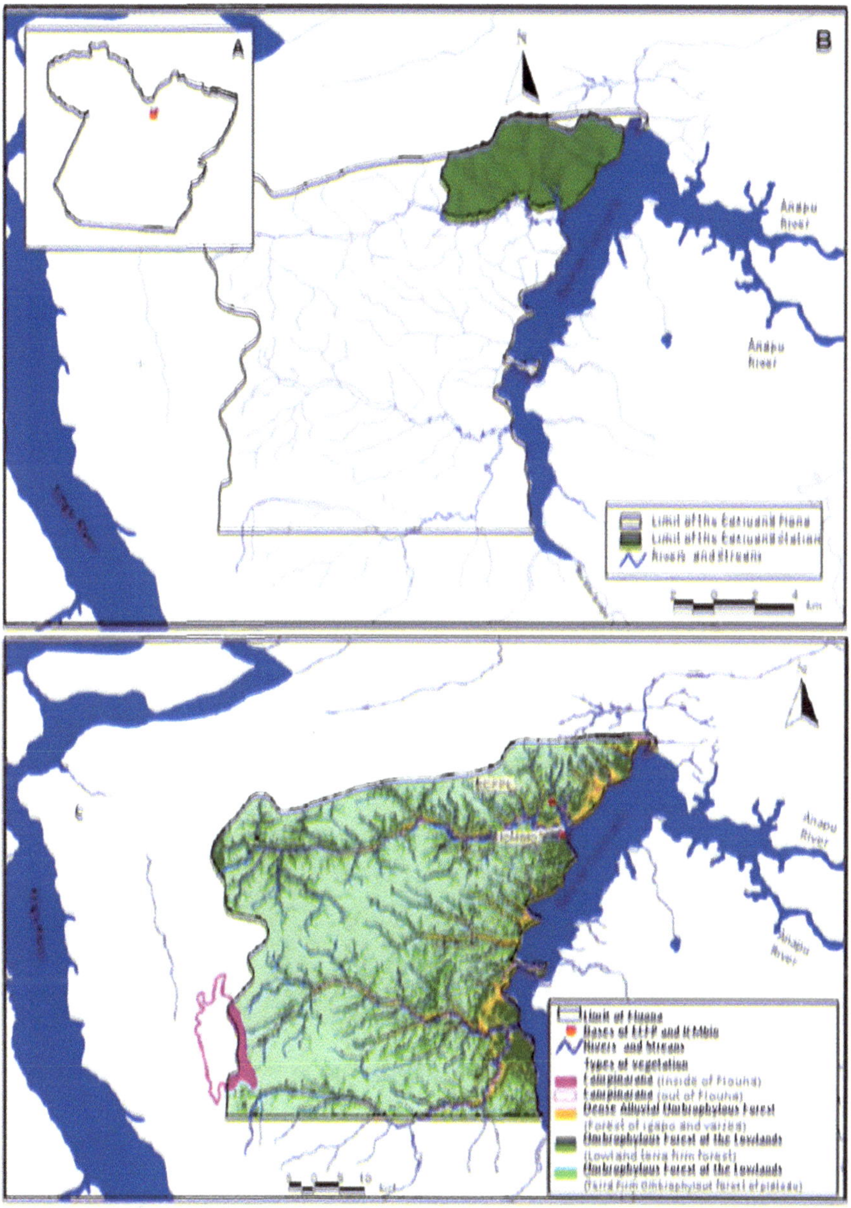

Fig. 5.1 Caxiuanã National Forest in the state of Pará (**a**), the location of Estação Científica Ferreira Penna in relation to Caxiuanã National Forest (**b**) and the vegetation types at Estação Científica Ferreira Penna (**c**)

Because of the geological and hydrological characteristics, the floodplain forests of Caxiuanã present features, like in other regions, related with hydro-geochemistry, soil, and associated vegetation.

Fig. 5.2 General (**a**) and inner (**b**) aspect of igapó forest in Caxiuanã National Forest

Igapó forests in Caxiuanã National Forest are mainly located on the banks of the Curuá River and Caxiuanã River, both effluents of Caxiuanã Bay (Fig. 5.2). These rivers contain a high concentration of humic compounds, low concentration of dissolved oxygen due to the low abundance of photosynthetic organisms in the water, low conductivity indicating poor salt water, high amount of dissolved organic matter, and acidic pH due to the high content of organic acids from the decomposition of organic matter of terrestrial origin (ICMBIO 2012; Melo 2013).

Due to these characteristics, the Curuá and Caxiuanã rivers were described as an environment composed of a heterotrophic community with a likely predominance of an input of organic matter of terrestrial origin (Melo 2013). In addition, the hydro-geochemistry of these rivers is strongly influenced by the intense soil weathering of the drainage basin, mainly consisting of kaolinite, resulting in the prevalence of Na2 + ions in these rivers (Melo et al. 2002; Costa 2005).

The soil of the igapó forest in Caxiuanã is classified as silt-textured gleysol, with low drainage, poor in nutrients, and highly fragile (Piccinin and Ruivo 2012). Gleysol is formed of recent sediments from the Holocene period, being partially or completely flooded during a great part of the year, which facilitates the constant and progressive deposition of suspended particles in the water (Vieira 1975). Morphologically, they are deep soils and imperfectly drained, gray colored, and with silt texture (Costa 2005).

The granulometry of this soil is predominantly formed with clay and mainly silt fractions and is chemically poor since it presents low contents of sum of bases and base saturation, carbon, and cation exchange capacity. Tables 5.2 and 5.3 show average values of the chemistry and grain size of studies of soils carried out in igapó forest in the Brazilian Amazonas. Studies indicate this contribution of nutrients as a determinant factor of species composition and structure of communities in flooded forests (Ferreira et al. 2013, 2005).

Caxiuanã igapó forests are characterized by the great abundance of *Ruizterania albiflora* (Warm.) Marc.-Berti and *Carapa guianensis* Aubl. individuals (Ferreira et al. 2005) and for presenting a higher density of individuals, richness, and diversity of species when compared to Caxiuanã várzea forests (Ferreira et al. 2013) which is uncommon in the Central Amazonia where várzea forests present greater richness and diversity (Prance 1980; Kubitzki 1989; Wittmann et al. 2010; Assis et al. 2014). Results of the monitoring in permanent plots indicate that the Caxiuanã igapó forests presented a higher annual rate of recruitment than mortality, resulting in an increase in the number of individuals per plot over the years inventoried.

5.3 Social Importance of Igapó and the Ecosystem Protection Policies

The Amazon is formed by a diversity of habitats thoroughly distinct and is a unique territory by the indescribable variety of its flora and fauna. About 20% of the Amazon is occupied by the rivers or flooded in a permanent or seasonal character, whose vegetation is denominated of forests of flooded areas: the várzea and the igapó. The findings on the impact of the fertility level of the rivers on the vegetation led most botanists to adopt the criterion "color of the rivers" to separate the flooded forests into várzea and igapó. Hence, flooded forests are those forests likely to be flooded by white water rivers, and forests of igapó are those flooded by rivers of clear or black water (Sioli 1985). The várzea forest is characterized by greater richness in nutrients, while the igapó is characterized by the acidity and poor level of nutrients.

Table 5.2 Mean of the chemical components of soils collected in igapó forests of some studies carried out in the Brazilian Amazon

Chemical composition				mg/dm³		cmol$_c$/dm³				mg/kg			
Local	Reference	Depth	pH	P	K	Ca	Ca + mg	Al	H + Al	Cu	Mn	Fe	Zn
Uauaçú Lake, Purus river region, central-western	Haugaassen and Peres (2006)	0–20	4.6	2.2	59.7	0.14		3.28	8.08		40.5	418.8	10.4
Jari River, Amapá	Carim (2016)	0–20	4.4	9.75	0.11	0.052	1.71	1.24	5.9	3.7	370.96	551.08	6.11
Curuá River, Caxiuanã National Forest, Pará	This study	0–20	4.5	7.25	10.19	0.13	0.25	3.48	26.30	15.94	0.89	1095.98	4.79

Table 5.3 Average values of the major physical characteristics of soils collected in igapó forests of some studies carried out in the Brazilian Amazon

Particle size			g/Kg				
Local	Reference	Depth	Thick sand	Fine sand	Sand	Silt	Clay total
Caxiuanã National Forest, Pará	Costa (2002)	0–12	5	90		785	120
		12–20	5	80		755	160
Uauaçú Lake, Purus river region, western Amazônia	Haugaassen and Peres (2006)	0–20	7.5	28		47	16
Jari River, Amapá	Carim (2016)	0–20			80.14	16.78	2.96
Curuá River, Caxiuanã National Forest, Pará	This study	0–20	13	10.375		873.4	103.75

To understand the dynamics of the Amazon depends on knowledge of this ecosystem and how plants and animals adapt to floodplains, such as the igapó forests, which are regularly flooded areas in the Amazonian biome. The igapó areas represent the banks of clear water rivers such as the Xingu River, at the height of Altamira, southwestern Pará, a source of survival and socioeconomic utilization by the river people, fishermen, natives, extractive communities, and potters who give territoriality to it.

Prior to the installation of the Belo Monte Hydroelectric Plant, these activities were well marked by seasonality. During the Amazonian winter, the flood pulse of the river made the igapó an environment that promotes artisanal fishing of species such as tucumaré (*Cichla melaniae*) and surubim (*Pseudoplatystoma fasciatum*). It should be reminded that many fish species depend on floods for feeding and reproduction, invading the flooded forest (igapó) to eat fruits and spawn in protected environments. In the Amazonian summer, management of palm trees such as açaí (*Euterpe oleracea*, Mart) and buriti (*Mauritia flexuosa*) and clay exploitation in the Gleissolos of the fluvial plains of Xingu effluents are carried out for the manufacture and commercialization of bricks in the potteries (Oliveira 2017).

5.4 Final Considerations

The wetlands in the Amazon are associated with the dynamics of the seasonal rise and fall of river waters and mark a morphological adaptation, reproduction, and dispersal by the biota. In the Caxiuanã basin, the study of the igapó areas presents greater complexity since the flood pulse is monomodal and polymodal, responding to the seasonal variation of the precipitation and daily variation of the influence of the tides.

This dynamics originates soils with high humidity, predominantly gleysols, in the Curuá River National Forest, area where the study was developed. Morphologically, the soils present grayish colors and a massive structure. Physically, fine fractions such as silt and clay predominate. Chemically, it presents low levels of K, Na, Ca, and Mg and consequently low sum of bases and base saturation, giving them a dystrophic character. Similar results were obtained by Costa (2002) in Caxiuanã National Forest, Pará, and by Haugaassen and Peres (2006) in Uauçú Lake, in the Purus river region with predominantly silt-textured and chemically poor soils, differing from soils along the Jari River in Amapá with more intense geomorphological dynamics, with the predominance of soils with sandy fractions.

The importance of this type of study is indispensable for the use of these environments by the riverside population, who depends economically on these and areas by means of vegetal extractivism and artisanal fishing, and for the conservation of the igapó that comprise important biological reserves and important centers of interdisciplinary research in the Amazon.

References

Assis RL, Haugaasen T, Schöngart J, Montero JC, Piedade MTF, Wittmann F (2014) Patterns of tree diversity and composition in Amazonian floodplain paleo-varzea forest. J Veg Sci 26:1–11

Ayres JMC (1993) As matas de várzea do Mamirauá (MCT-CNPq- Programa do trópico úmido). Sociedade civil de Mamirauá, Brasil

Ayres JM (2006) As Matas de Várzea do Mamirauá: Médio Rio Solimões, 3rd edn. Sociedade Civil Mamirauá, Belém, 123 p. il. (Estudos do Mamirauá, 1)

Behling H (2011) Holocene environmental dynamics in coastal, eastern and Central Amazonia and the role of the Atlantic Sea-level change. Geographica Helvetica 66:208–216

Behling H, da Costa ML (2000) Holocene environmental changes from the Rio Curuá record in the Caxiuanã region, eastern Amazon Basin. Quat Res 53:369–377

Bemerguy RL (1997) Morfotectônica e evolução paleogeográfica da região da calha do Rio Amazonas. Belém: Universidade Federal do Pará. Tese de Doutorado, 200 p.

Carim MJV (2016) Estrutura, composição e diversidade em florestas alagáveis de várzea de maré e de igapó e suas relações com variáveis edáficas e o período de inundação no Amapá, Amazônia oriental, Brasil. Tese (doutorado). INPA, Manaus, 95f

Costa JA (2002) Caracterização e classificação dos solos e dos ambientes da Estação Científica Ferreira Penna, Caxiuanã, Pará. Faculdade de Ciências Agrárias do Pará. Departamento de Ciência do solo. Dissertação de Mestrado. Belém, FCAP, 63f

Costa JA (2005) Classificação e distribuição dos padrões pedomorfológicos da Estação Científica Ferreira Penna, na Região de Caxiuanã, no Estado do Pará. Boletim do Museu Paraense Emílio Goeldi, Ciências Naturais 1:117–128

Cravo MS, Xavier JJBN, Dias MC, Barreto JF (2002) Características, Uso Agrícola Atual e Potencial. Acta Amazônica 32:351–365

Cunha CN, Piedade MTF, Junk WJ (2015) Classificação e delineamento das áreas úmidas brasileiras e de seus macrohabitats. Ed. UFMT, Cuiabá, 165p

Daniels RB, Nelson LA (1987) Soil variability and productivity: future developments. In: Future developments in soil science researcher. Soil Sci Soc Am 30:279–291

Ducke A, Black GA (1954) Nota sobre a fitogeografia da Amazônia brasileira. Boletim Técnico Instituto Agronômico do Norte 29:3–48

Ferreira LV (2000) Effect of flooding duration on species richness, floristic composition and forest structure in river margin habitats in Amazonian blackwater floodplain forests: Implications for future Design of protectedareas. Biodivers. Conserv 9:1–14

Ferreira LV (2011) Os tipos de vegetação da Floresta Nacional de Caxiuanã. In: Instituto Chico Mendes de Conservação da Biodiversidade (Org.). Plano de Manejo da Floresta Nacional de Caxiuanã 1: 25–42

Ferreira FV, Almeida SS, Amaral DD, Parolin P (2005) Riqueza e composição de espécies da floresta de igapó e várzea da Estação Científica Ferreira Penna: subsídios para o plano de manejo da Floresta Nacional de Caxiuanã. Pesquisas Botânica 56:103–116

Ferreira LV, Chaves PP, Cunha DA, Matos DCL, Parolin P (2013) Variação da riqueza e composição de espécies da comunidade de plantas entre as florestas de igapós e várzeas na Estação Científica Ferreira Penna-Caxiuanã na Amazônia Oriental. Pesquisa Botânica 64:175–195

Furch K (1984) Water chemistry of the Amazon basin: the distribution of chemical elements among freshwaters. In: Sioli H, Junk W (eds) The Amazon: limnology and landscap ecology of a mighty tropical river and its basin. Springer, Dordrecht, pp 99–167

Furch K, Junk WJ (1997) Physicochemical conditions in floodplains. In: Junk WJ (Org.). The central Amazon floodplain: ecology of a pulsing system. Springer, Berlin, pp 69–108

Haugaasen T, Peres CA (2006) Floristic, edaphic and structural characteristics of flooded and unflooded forests in the lower Rio Purús region of Central Amazonia, Brazil. Acta Amazônica 36:25–36

Hida N, Maia JG, Shimmi O, Hiraoka M, Mizutani N (1998) Annual and daily changes of river water level at Breves and Caxiuanã, Amazon Estuary. Geogr Rev Jpn 71:100–105

Instituto Brasileiro de Geografia e Estatística (IBGE) Geografia do Brasil (1977) Região Norte. Rio de Janeiro, 466 p

Instituto Chico Mendes de Conservação da Biodiversidade (ICMBIO) (2012) Plano de Manejo da Floresta Nacional de Caxiuanã, ICMBIO

Irion G, Mello J, Morais J, Piedade MTF, Junk WJ, Garming L (2010) Development of the Amazon valley during the middle to late Quaternary: sedimentological and climatological observations. In: Junk WJ, Piedade MTF, Wittmann F, Schocngart J, Parolin P (Orgs.) Ecology and management of Amazonian floodplain forests. Springer, Berlin, pp 27–42

Junk WJ (2010) Amazonian floodplain forests: ecophysiology, biodiversity and sustainable management. Springer, Dordrecht/Heidelberg/London/New York

Junk WJ, Bayley PB, Sparks RE (1989) The flood pulse concept in river-floodplain systems. Can J Fishers Aquatic 106:110–127

Junk WJ, Wittmann F, Schöngart J, Piedade MTF (2015) A classification of the major habitats of Amazonian black-water river floodplains and a comparison with their white-water counterparts. Wetlands Ecology Manage 23:677–693

Kubitzki K (1989) The ecogeographical differentiation of Amazonian inundation forests. Plant Syst Evol 162:285–304

Melack JM, Hess LL (2010) Remote sensing of the distribution and extent of wetlands in the Amazon basin. In: Junk WJ, Piedade MTF, Wittmann F, Schongart J, Parolin P (Orgs.) Amazon floodplain forests: ecophysiology, biodiversity and sustainable management. Springer, Berlin, pp 43–59

Melo DMB (2013) Aspectos físico-químicos dos ambientes fluviolacustres de Caxiuanã. In: Lisboa, Pedro Luiz Braga (Org) Caxiuanã: paraíso ainda preservado, 1st edn. Museu Paraense Emílio Goeldi, Belém, pp 91–103

Mitsch WJ, Gosselink JG, Anderson CJ, Zhang L (2009) Wetland ecosystems. Wiley, Hoboken

Molinier M, Guyot JL, Oliveira E, Guimaraes V, Chaves A (1995) Hidrologia da bacia do Rio Amazonas. Ciência e Tecnologia Orstom Fonds Documentarie A Água em Revista CPRM, pp 31–35

Mommer L, Visser EJW (2005) Underwater photosynthesis in flooded terrestrial plants: a matter of leaf plasticity. Ann Bot 96:581–589

MPEG Fundamentos da ecologia da maior região de florestas tropicais: 2017 Acesso em 16/05/2017 http://marte.museu-goeldi.br/marcioayres/index.php?option=com_content&view=article&id=7&Itemid=8

Oliveira RD (2017) Regime hidrológico do rio Xingu e dinâmica de inundação nas planícies fluviais no entorno de Altamira, Sudoeste do Estado do Pará. (Tese de Doutorado Faculdade de Geografia, Universidade Estadual Paulista Julio de Mesquista Filho), São Paulo, Brasil, p 96

Parolin P (2001) Morphological and physiological adjustments to waterlogging and drought in seedlings of Amazonian floodplain trees. Oecologia 128:326–335

Parolin P (2012) Diversity of adaptations to flooding in trees of Amazonian floodplains. Pesquisas, Botânica 63:7–28
Parolin P, Simone O, Haase K, Waldhoff D, Rottenberger S, Kuhn U, Kesselmeier J, Schmidt W, Piedade MTF, Junk WJ (2004) Central Amazon floodplain forests: tree survival in a pulsing system. Bot Rev 70:357–380
Penna, na Região de Caxiuanã, no Estado do Pará. Boletim do Museu Paraense Emílio Goeldi, Ciências Naturais 1:117–128
Perreira JLG, Rennó CD, Silveira OT, Ferreira LV (2012) Classificação da cobertura da terra na Amazônia com base em imagens de satélite e caracterização das classes com relação à superfície do terreno. Geografia 21:115–131
Piccinin J, Ruivo ML (2012) Os solos da Floresta Nacional de Caxiuanã. In: Instituto Chico Mendes de Conservação da Biodiversidade (Org.). Plano de Manejo da Floresta Nacional de Caxiuanã, ICMBIO, pp 120–127
Prance GT (1979) Notes on vegetation of Amazonia III. The terminology of Amazonian forest types subject to inundation. New York, Brittonia 31:26–38
Prance GA (1980) Terminologia dos tipos de florestas Amazônicas sujeitos à inundação. Acta Amazônica 10:495–504
Radam Brasil (1974) Geologia, Geomorfologia, Pedologia, Vegetação e Uso Potencial da Terra. Rio de Janeiro: Departamento Nacional de Produção Mineral, AS.22-Belém, p 510
Santiago E Igapó. http://www.infoescola.com/biomas/igapo/ acessado em 21 de junho de 2017
Sioli H (1956) Über Natur und Mensch im brasilianischen Amazonasgebiet. Erdkunde 10:89–109
Sioli H (1968) Hydrochemistry and geology in the Brazilian Amazon region. Amazoniana 3:267–277
Sioli H (1975) Tropical River: the Amazon. In: Whitton BA (ed) River ecology. Blackwell Sci. Publ, Cambridge, pp 461–488
Sioli H (1985) Amazônia: fundamentos da ecologia da maior região de florestas tropicais. Editora Vozes, Petropolis, p 72 (Tradução de John Becker)
Vieira LS (1975) Manual de ciência do solo. Agroceres, Belém, 375p
Wittmann F, Schöngart J, Junk WJ (2010) Phytogeography, species diversity, community structure and dynamics of central Amazonian Floodplain forests. In: Junk WJ, Piedade MTF, Wittmann F, Schöngart J, Parolin P (Orgs.). Amazonian Floodplain forests: Ecophysiology, biodiversity and sustainable management. London/New York, pp 61–101

Part IV
Litter, Fungi and Invertebrates

Chapter 6
Diversity and Phenology of Arachnids in Igapó Forests

Lidianne Salvatierra

6.1 Arachnids as Potential Bioindicators

Changes in environmental conditions, whether intentional or unintentional, affect all abiotic and biotic components. Studying ecological guilds (or functional groups) can be useful to investigate assemblage response to climate change, habitat disturbance (transformation, use and management of lands), and its consequence (anthropogenic pollution) among many other areas (Diaz 1995; Chapin et al. 1996; Voigt et al. 2007). Functional diversity is one of the most important parameters used to explain how ecosystems work and adapt to change (Tilman et al. 2001; Petchey and Gaston 2006). Thus, the use of indicator taxa is becoming more and more important in the context of the growing anthropogenic pressure on highly diverse and threatened tropical ecosystems. For the evaluation of the conservation potential and state of secondary and old-growth tropical forests, precise but quick and cheap tools such as indicators are needed (Uehara-Prado et al. 2009).

The Amazonian floodplain forests are a unique and endangered ecosystem due to intense deforestation and overexploitation. In fact, recent evidences imply that Amazonian forests on floodable terrains have a lower resilience when it comes to the risk of transition into a fire-prone vegetation state and the spread of fires from floodplains to adjacent uplands has been shown in central Amazon, with negative impacts on both vegetation structure (Resende et al. 2014; Flores et al. 2017) and biodiversity (Benchimol and Peres 2015). Floodplains are characterized by a remarkable environmental heterogeneity produced by complex interactions and transitions between surface water, subsurface waters (groundwater), and riparian systems (Ward and Tockner 2001). This high spatiotemporal physical complexity provides habitat for a very high biodiversity of fauna and flora (Bonecker et al. 2013).

L. Salvatierra (✉)
State University of Roraima, Postgraduate Program in Education, Boa Vista, Roraima, Brazil

R. W. Myster (ed.), *Igapó (Black-water flooded forests) of the Amazon Basin*,
https://doi.org/10.1007/978-3-319-90122-0_6

The biodiversity is largely controlled by the flood pulse, which encompasses an annual alternation of an aquatic (submerged) phase and a terrestrial, temporarily dry (emerse) phase (Junk 1989). Therefore, individual species exposed to ever-changing environmental conditions may respond either by locally dying or by adapting (physiological and/or morphological) to the new conditions (Arrieira et al. 2016). Understanding how to measure and monitor native diversity in this ecosystem is fundamental to preserve its biodiversity.

Conservation strategies and management in the tropics are often based on large, exotic, and beautiful or rare, endangered vertebrate species. However, the overwhelming part of fauna diversity consists of invertebrates. And, although invertebrates are involved in numerous important ecosystem functions (e.g., nutrient cycling or pollination), the data for its biodiversity and conservation are usually taxonomic (Valdecasas and Camacho 2003). In fact, the analyses of invertebrate diversity for conservation are mainly restricted to species numbers or lists of species of selected taxa (Raub et al. 2014).

Arachnids (class Arachnida) are a diverse group of arthropod orders with an overwhelmed abundance and diversity in the Amazon rainforest. The class Arachnida is divided into eleven orders: Acari, Amblypygi, Araneae, Opiliones, Palpigradi, Pseudoscorpiones, Ricinulei, Schizomida, Scorpiones, Solifugae, and Thelyphonida. Arachnids are instantly recognizable by their eight walking legs (Acari and Ricinulei hatch as six-legged larvae) and presence of the chelicerae (or mouthparts) and pedipalps.

Ecologically important members of the Arachnida are spiders (Araneae), mites and ticks (Acari), scorpions (Scorpiones), and harvestmen (Opiliones), and they are often considered having bioindicative value (Paoletti 1999). For instance, as a taxonomic group, spiders are very well suited for use as bioindicators in studies of habitat quality in terrestrial ecosystems (Maelfait and Baert 1988, Maelfait 1996). It is suggested that any group of arachnid can be used as a bioindicator if they present the following characteristics: sufficient number of species, good knowledge of bionomics, heterogeneous life-history strategies, quick response to abiotic and biotic factors, cost-effectiveness, and easy collecting choice of the best type of bioindicator for the various monitoring objectives (Chobot et al. 2005).

The literature on quantitative arachnofauna inventories of Amazonia is considerable, with several studies available for dryland forest (*terra firme*) but fewer for blackwater floodplain forests (*igapó*) (Adis 1977, 1981; Adis et al. 1988; Adis and Mahnert 1985; Beck 1971; Condé 1993; Cokendolpher and Reddell 2000; Tourinho et al. 2010; Erwin 1983; Franklin 1994; Franklin et al. 1997, 1998, 2006; Friebe and Adis 1983; Höfer 1990; Lourenço 1988, 2002; Lourenço and Cuellar 1999; Lourenço and Francke 1986; Lourenço and Pinto-da-Rocha 2000; Lourenço et al. 2005; Muchmore 1975; Mahnert 1979, 1985). Currently, only 298 arachnid species have been reported to *igapó* forests (Table 6.1).

The Tarumã-Mirim River (03°02'S, 60° 17'W) is probably the most studied site regarding the biodiversity of arachnids in *igapó* areas. It is a small tributary river basin of the Rio Negro near the city of Manaus, State of Amazonas, Brazil, seasonally inundated. Total annual precipitation in the region at an average is about

Table 6.1 Number of arachnid species on *igapó* forests in the Amazon biome

Order	Species	Area	References
Araneae	210	TMR	Höfer (1990)
Acari	45	TMR	Beck (1971), Franklin (1994), Franklin et al. (1997, 1998, 2006)
Pseudoscorpiones	19		Adis (1981), Adis et al. (1988), Adis and Mahnert (1985), Muchmore (1975), Mahnert (1979, 1985)
Scorpiones	10	TMR; IF; CV; IFI	Erwin (1983), Lourenço (1988, 2002), Lourenço and Francke (1986), Lourenço et al. (2005)
Opiliones	9	TMR; CV	Friebe and Adis (1983)
Schizomida	3	TMR	Cokendolpher and Reddell (2000)
Ricinulei	1	JR	Tourinho et al. (2010)
Palpigradi	1	TMR	Condé (1993)
Amblypygi	0	=	=
Solifugae	0	=	=
Thelyphonida	0	=	=
Total	298		

Abbreviations: *TMR* Tarumã-Mirim River, Brazil, *CV* Careiro da Várzea, Brazil, *JR* Jufari River, Brazil, *IFI Igapó* Forest in Iquitos, Peru, *IF* Non-specified *Igapó* Forest, Brazil

2300 mm. The amount of rain is typically irregularly distributed along the year, showing a marked dry season from June through September and a rainy season from December to May. There is a time lag of several months between the peak of rainfall and maximum flooding. The rising phase occurs between late December and early July, while the draining period occurs from late July to late November (Adis 1984; Platnick and Höfer 1990; Ferreira and Parolin 2011).

It should be stressed that *várzea* (white-water floodplain forest) and *igapó* (black-water floodplain forest) are distinct faunistic and floral environments, since they are flooded by rivers with different physical and chemical characteristics (mineral and organic concentrations, pH, transparency) (Junk 1983; Junk and Furch, 1983 1985; Sioli 1984; Ayres 1995; Mertes et al. 1996; Furch and Junk 1997; Irion et al. 1997).

The following descriptive phase will provide a state-of-the-art knowledge of what has been published so far regarding arachnid's diversity in *igapó* forest.

6.2 Diversity of Arachnids in *Igapó* Forests

Spiders are a very significant component of Amazon ecosystems. The order is megadiverse with 46,410 currently valid species, assigned to 4027 genera and 113 families (World Spider Catalog 2017). Spiders are among the most abundant and diverse predators in the Amazon. Most of the very few studies on spiders in the Amazon deal with taxonomic aspects, and only recently comprehensive studies on spider communities (its guild distribution and functional diversity) have begun (Platnick and Höfer 1990; Höfer 1990, 1997; Höfer et al. 1994a, b; Borges and Brescovit 1996; Silva 1996; Vieira and Höfer 1994).

Spiders have the body segments fused into two tagmata, the cephalothorax and abdomen; the chelicerae have two sections and terminate in fangs that inject venom that kills or paralyzes the victim and also present the spinnerets which emit silk used to build webs and wraps the prey (Fig. 6.1c). They respond physiologically to environmental changes and are the main arthropod predators in many biomes and habitat types and therefore ought to be a good indicator taxon to reflect ecological change (Voigt et al. 2007). Additionally, they have been suggested to be an ideal group for predicting extinction debt in other taxa due to habitat destruction (Cardoso et al. 2010).

In flooded forest, spiders are superficially studied as, so far, there are only six publications that treat the species inhabiting flooded forests in the Brazilian Amazon (Adis et al. 1984; Höfer 1990, 1997; Martius et al. 1994; Borges and Brescovit 1996; Brescovit et al. 2003; Rego et al. 2009), but only one survey was conducted in *igapó* forests and exclusively at Tarumã-Mirim River (Adis et al. 1984; Höfer 1990, 1997).

The only data available, for the *igapó* of Tarumã-Mirim River, reported 210 species, belonging to 39 families, collected using pitfall traps, ground photoeclectors, and arboreal funnel traps (Höfer 1990, for the full list of species). Four families made up 42% of all species: Salticidae (38 spp.), Araneidae (32), Corinnidae (12), and Theridiidae (12). Dominant species comprising more than 10% of all individuals were *Ancylometes* sp. (Pisauridae), *Meioneta* sp. (Linyphiidae), and *Theotima* cf. *minutissima* (Ochyroceratidae) using pitfall traps; *Theotima* cf. *minutissima*, *Anapistula* sp. (Symphytognathidae), *Tricongius amazonicus* (Platnick and Höfer 1990) (Gnaphosidae), *Pseudanapis hoeferi* (Kropf 1995) (Anapidae), and *Meioneta* sp. using ground photoeclectors; and *Xyccarph migrans* (Höfer and Brescovit 1996 (Oonopidae) using arboreal funnel traps.

It was suggested that floodplain forests (*igapó* and *várzea*) are poorer in species than dryland forests (*terra firme*) because of the impact of the annual flood, but further collections are necessary using different sampling methods, for the study and deeper knowledge of the spider fauna in Amazon River floodplains and in the Amazon region (Coddington and Levi 1991; Höfer 1997).

Scorpions possess large paired claws as pedipalps and a "tail" that terminates in the telson at the caudal portion of the abdomen (Fig. 6.1d). This appendage houses two venom glands that exit via a stinger at the end of the telson. They are exceptional nocturnal predators with 2342 valid species and 17 families with 8 families recognized for the neotropics: Bothriuridae, Buthidae, Chactidae, Diplocentridae, Hemiscorpiidae, Iuridae, Troglotayosicidae, and Vaejovidae (Lourenço 2002). Considering the vast territorial dimension and its vegetation diversity, the known Amazonian scorpion fauna cannot be considered megadiverse (Lourenço and Eickstedt 2003).

The inventory of the scorpions of Brazilian Amazonia has greatly increased, but many gaps still exist in the knowledge of the fauna. Only 31 scorpions were taken collected from *igapó* at Tarumã-Mirim River during intense collecting efforts during over 12 years (Lourenço 2005) and only 10 species were recorded: *Tityus adisi* (Lourenço 2002), *Tityus canopensis* (Lourenço and Pézier 2002), *Tityus obscurus*

Fig. 6.1 Arachnids on blackwater forest (*igapó*) in the Amazon. (**a**) Whip scorpions (Amblypygi). (**b**) Harvestman (Opiliones). (**c**) Spiders (Araneae). (**d**) Scorpion (Scorpiones). (**e**) Pseudoscorpions (Pseudoscorpiones). (Photo credit: Sidclay Calaça)

(Gervais 1843), *Tityus dinizi* (Lourenço 1997), *Tityus lokiae* (Lourenço 2005), *Tityus silvestris* (Pocock 1897), *Tityus strandi* (Werner 1939), *Tityus matthieseni* (Lourenço and Pinto-da-Rocha 2000), *Tityus metuendus* (Pocock 1897), and *Chactopsis amazonica* (Lourenço and Francke 1986).

From ecological aspects, several *Tityus* species developed an adaptation to the annual flood pulse by migrating into the trunk/canopy region, presenting an arboricolous guild, till pass the submersion period: *Tityus adisi* (Lourenço 2002), *Tityus canopensis* (Lourenço 2002), and *Tityus lokiae* (Lourenço 2005) were solely obtained in the trunk or canopy region of the *igapó*; and *Tityus silvestris* (Pocock 1897) and *Tityus metuendus* (Pocock 1897) were collected from both regions. Also *Chactopsis amazonica* (Lourenço and Francke 1986), considered terricolous in *terra firme* forest, was found in litter of the forest floor during a non-flooded phase.

The Acari include ticks, mites, and their kin. Many members of the Acari are microscopic, but relatively large members of the Acari reach about 5 mm long. With small size has come morphological change: the segmentation of the opisthosoma and the distinction between prosoma and opisthosoma have been almost completely lost. Some mites and ticks feed on vertebrate hair or blood and often carry disease organisms. Yet most mites are free-living, found in great abundance in soils, plant litter, and even in water.

Oribatid mites (Acari: Oribatida) comprise a major component of soil fauna, where they are predominantly fungivores and saprovores, and occur as well in numerous other microhabitats (Norton and Behan-Pelletier 2009). They have been studied at species level in Amazonia since 1967 (Beck 1967, 1968, 1971; Franklin 1994; Franklin et al. 1996, 2004; Hayek 2000; Woas 2002). A recent monograph listed 576 described species of oribatid mites in the Brazilian fauna, with more than 180 species occurring in the Amazon region (Oliveira et al. 2017). For *igapó* forests, 45 species were reported (Beck 1969, 1971, 1972; Franklin 1994; Franklin et al. 1980, 1997, 1998, 2001, full species list available in Franklin et al. 2006).

Opiliones, or harvestmen, constitute the third largest arachnid order with 6476 species described (Kury 2010). They are characterized by having the five first opisthosomatic somites fused with the prosoma forming a dorsal scute, a pair of prosomatic scent glands which open at lateral margin of carapace, presence of penis and ovipositor, and one pair of median eyes (lateral eyes absent) (Kury and Pinto-da-Rocha 2002, Fig. 6.1b).

Little is known about the Amazonian harvestmen, but its high abundance and diversity with 180 species are recognized (Kury 1995, 2003; Pinto-da-Rocha and Kury 2003; Bonaldo et al. 2009). Only one study was conducted in a blackwater-flooded forest in the Amazon biome, and since then the opilionid fauna of the floodplains has not yet been surveyed with the same intensity as in other locations. Friebe and Adis (1983) collected a total number of nine morphospecies in *igapó* at Tarumã-Mirim River and ten morphospecies in a mixedwater inundation forest (*igapó* and *várzea*) at Lago de Janauarí (Manaus, Amazonas).

Pseudoscorpions (false scorpions) are small arachnids that resemble small scorpions lacking the telson, which is the metasoma commonly known as the scorpion's "tail" (Fig. 6.1e). Nowadays 3200 species of pseudoscorpions are known in the world, grouped in 425 genera and 24 families. Knowledge of pseudoscorpions is still very fragmentary in terms of biology and behavior, and the taxonomic situation, particularly in South America, is unsatisfactory with an estimation of more than 100 of new species, completely unknown by the science, in the Amazon forest (Mahnert and Adis 2002, Tizo-Pedroso and Del-Claro 2007).

Most species inhabit the soil and litter, feed on small arthropods, such as mites, beetle larvae or springtails, but some larger species may also attack ants. The distribution of pseudoscorpions in the Amazonia region differs from one habitat to another, caused by ecological factors such as soil, vegetal cover, temperature, and humidity (Adis 1981; Adis and Mahnert 1990, 1993; Adis et al. 1988; Mahnert and Adis 1985; Morais 1985; Aguiar 2000; Aguiar and Bührnheim 1998, 2003). In fact, it is suggested that Pseudoscorpiones underwent extensive evolution within inundation forests, due to forest isolation caused by climatic shifts and microgeographic changes, and are excellent model for studying drivers of life-history strategy variation between species (Adis and Schubart 1984).

In the soil and litter of Amazonia, a total diversity of 24 species is known (Mahnert and Adis 1985). Pseudoscorpions are the most abundant arachnid in flooded forest, and during 5 months of sampling, a total of 7942 individuals were collected in Tarumã-Mirim River using photoeclectors (Adis 1977).

Using arboreal photoeclectors, Adis (1981) registered 15 species in the flooded forest of "*igapó*," with five of them as litter inhabitants. *Tyrannochthonius amazonicus* (Mahnert 1979) presented high abundance in floodplain forests (Morais et al. 1997) and very low abundance and frequency in the litter and suspended soil of undergrowth vegetation (palm trees) in dryland forests (Aguiar 2000). *Tyrannochthonius amazonicus* and *Tyrannochthonius migrans* are considered endemic of *igapó* forest.

Palpigradi, also known as micro-whip scorpion, are small, fragile, and poorly sclerotized inhabitants of tropic soil (Hammen van der 1982). Micro-whip scorpions have a very distinct flagellum bearing bristles and are represented by 99 species, 6 genera, and 2 families, with a little more than 18 Neotropical species (Harvey 2002; Prendini 2011a, b; Souza and Ferreira 2010, 2013; Mayoral 2015). To date, only the species *Eukoenenia janetscheki* (Condé 1993) have been reported at Tarumã-Mirim River.

Ricinulei is an order of cryptic arachnids with little-known natural history and are also called hooded tickspiders, by their resemblance with ticks. Their most notable feature is the cucullus, a hood covering their chelicerae, which can be raised and lowered over the prossoma. Ricinuleids have no eyes and possess a pair of light-sensitive areas on the prosoma. They are predators of small invertebrates (Cooke 1967) and are currently divided into 3 genera and 84 species (Harvey 2003; Botero-Trujillo and Flórez 2017). The species *Cryptocellus iaci* (Tourinho et al. 2010) is the only known species of Ricinulei collected in an area of both *terra firme* forest (dryland) and *igapó* at the Jufari River, Roraima State, Brazil.

Schizomids are small arachnids with well-developed, raptorial pedipalps, elongated first legs, and abdomen ending in a short flagellum and are commonly named short-tailed whip scorpions (Reddell and Cokendolpher 2002). They are mainly tropical and subtropical environment inhabitants and can be found in moist areas in leaf litter or in cavities underside logs and rock. Two families of schizomids are recognized: Protoschizomidae (Rowland 1975) and Hubbardiidae (Cook 1899).

The genus *Surazomus* (Reddell and Cokendolpher 1995) (Hubbardiidae) contains 21 named species, but only 3 species have been collected in *igapó* forests at Tarumã-Mirim River. *Surazomus arboreus* (Cokendolpher and Reddell 2000) has only been found on trees, and *Surazomus rodriguesi* (Cokendolpher and Reddell 2000) and *Surazomus mirim* (Cokendolpher and Reddell 2000) were collected in a region at Tarumã-Mirim River adjacent to a non-flooded upland forest.

Amblypygi comprises over 164 species, 17 genera, and 5 families, which are widely distributed in the warmer tropical regions of the world (Weygoldt 2000). Commonly known as whip spiders, their greatest diversity occurs in the Neotropical region, which has about 100 described species with many endemic species for the Brazilian Amazon (Harvey 2013, Fig. 6.1d). Their bodies are broad and highly flattened; the front pair of legs are very thin, elongate, and antennae-like feelers acting as sensory organs; and they also have robust raptorial pedipalps armed with spines. Most species are nocturnal and live in moist forests where they are found under rotting logs, between rock breaches, and inside caves (Weygoldt 2000).

Solifugae, vulgarly known as sun spiders, wind spiders, or camel spiders, are considered endemic of warmer zones and drier climate, as desert biomes, but they also occur in the neotropics (Besch 1969). Sun spiders can be easily identified by their large chelicerae with two articles, forming a powerful pincer. The order is composed by 12 families, 139 genera, and 1105 species (Harvey 2003; Prendini 2011a, b) with 70 nominal species described from the neotropics (Rocha and Cancello 2002).

Thelyphonida, or whip scorpions, is current composed by 106 living species described in 18 genera placed in a single family (Harvey 2003). The Neotropical region comprises only 22 species and 6 genera of whip scorpions. They are also called vinegaroons by presenting defensive glands near the end of opisthosoma, which produce a combination of acetic acid and caprylic acid or formic acid or chlorine.

So far, no specimens of Amblypygi, Thelyphonida, and Solifugae have been found or studied in *igapó* forests. The surprisingly low diversity and knowledge of these orders are due to specimen's rarity and difficult to sampling (Rocha 2002; Smrž et al. 2013).

6.3 Life Cycles and Adaptations of Arachnids in *Igapó* Forest

Since terrestrial invertebrates living in Amazon have to cope with annual inundation period of several months, survival strategies are a crucial part of the sociobiology of these organisms (Paarmann et al. 1982; Irmler 1985; Adis 1992; Adis et al. 1993; Ribeiro et al. 1996; Adis and Messner 1997; Amorim et al. 1997a, b).

The monomodal and predictable flood pulse and the stable climatic conditions over long evolutionary periods have enabled the development of morphological, phenological, physiological, and behavioral adaptations to inundation and hypoxic conditions, especially in many invertebrate taxa (Erwin and Adis 1982; Junk 2000; Adis and Junk 2002).

Species combine vertical migrations during flood events, remigration after the event, and reproduction rates as adaptations to live in *igapó* forests. Horizontal migration is less frequent in *igapó* forest for most species. The life cycles of spider in *igapó* forest were never studied due to the lack of juvenile description, but juveniles of some less-agile species (such as Caponiidae, Dipluridae, Gnaphosidae, Palpimanidae, and Pholcidae) tend to move to higher areas long before the water level reaches the site (Platnick and Höfer 1990). The activity density of spiders on the ground increased from the beginning of the terrestrial phase and reached a maximum 2–3 weeks prior to inundation of the ground at Tarumã-Mirim River (Hofer 1990). Density activity on tree trunks was low during the completely noninundated period and increased abruptly a few days before inundation of the trunk base. Large wandering spiders (agile spiders, such as Corinnidae, Ctenidae, Lycosidae, Pisauridae) avoid flood by a steady horizontal movement away from the waterline. The ballooning phenomenon, where spiders throw out silk filaments to catch the wind to disperse, was also observed in some small spider species (Orbiculariae).

Scorpion of the genus *Tityus* presents a high level of phenotypic plasticity and climb on trees during inundation periods. The only known schizomid species in *igapó* forests, *Surazomus arboreus* (Cokendolpher and Reddell 2000), has only been found on trees, and it is the only species known by adults from flooded forest, avoiding drowning by ascending into the trees. Harvestmen species (*Auranus parvus* and *Eucynortula lata*) also move to the trunk/canopy region prior to flooding an escape reaction by changing phototaxis where they pass the 5–7 months lasting aquatic phase (Irmler 1981, Friebe and Adis 1983). Both species are univoltine with main reproduction occurring in the upper organic layer of the forest floor during the terrestrial phase (Adis 1992, 1997).

For Pseudoscorpiones, while in seasonally flooded forests, there is a strong synchronization between the phenology of species and flood periods, with most of the species in these areas being considered arboreal (Adis and Mahnert Adis and Mahnert 1990). Some species, such as *Geogarypus* sp., use the soil and leaf litter only during the non-flooded period and migrate later to high parts of the forest (tree trunks and canopies) before the flood periods, using different habitats on the life cycle (Adis 1997; Adis and Mahnert 1990; Mahnert and Adis 2002). *Brazilatemnus browni* (Muchmore 1975) is multivoltine in dryland forest but univoltine or bivoltine in *igapó* forests (Adis 1997).

Small body-sized species and nymphs of pseudoscorpion start the migration to tree 4–6 weeks before flooded season, and medium to big body-sized species started the migration 2 or 3 weeks before flooded season (Adis 1977; Adis et al. 1988; Castilho et al. 2005). Higher reproductive rate observed in the high water season may reflect the greater availability of resources to offspring during this period in canopies due to the accumulation of organic matter in leaf sheaths that provide places for foraging, shelter, and reproduction for many groups of insects, such as Collembola, Psocoptera, Diptera, and Coleoptera (Battirola et al. 2006; Marques et al. 2009), used as food source for Pseudoscorpiones (Weygoldt 1969).

Compared to the phenological studies on insect communities, those on arachnid communities are scarce. The phenology of most species of arachnids is poorly known, but from what are known so far indicates different patterns of seasonal fluctuations. The results of studies on the continuing biological interactions of arachnids of *igapó* forests may lead to a more satisfactory understanding of seasonal influences and reveal the relative importance of various faunal groups in biological systems such as litter decomposition and provide a framework of ecological information on which to base conservation decisions.

6.4 Conclusion and Future Studies

The number of unidentified species observed in most works is an indication of the lack of knowledge with regard to the arachnofauna of the Amazon floodplains and that of tropical arachnids in general. And since guild classification should ideally be made at the species level, because each species usually has a uniform behavior, which may be different from any other species (even closely related), taxonomic studies are crucial to increase the knowledge on the arachnofauna of the Amazonian floodplain forests. Our knowledge of Neotropical arachnid is generally poor; relatively few groups have been revised, most described species cannot be recognized accurately from their initial descriptions, and the available collections (which are sparse compared to existing Nearctic samples) are often poorly sorted and difficult to access.

Future studies should aim to:

1. Explore the abundance and distribution of arachnids on the Amazon floodplains.
2. Develop an arachnofauna checklist and establish the species composition in the studies areas.
3. Establish the effect of habitat fragmentation (habitat health) on species richness, species evenness, and species diversity.
4. Establish the variation in guild richness, evenness, and diversity in relation to habitat fragments.
5. Establish the most threatened spider species (red-listing of critically threatened species).
6. Establish the value of groups of arachnids as indicators of disturbance and link the study results to a feasible conservation and management strategy for the Amazon.
7. Provide baseline information for long-term monitoring studies on arachnofauna of these areas.

8. Produce technical reports and publications (annually) which will form baseline information for current and future studies.

General effects of habitat fragmentation and destruction on arachnids (as well as other fauna) should be discussed in the main technical report.

More complex and precise measures should be used to explore and quantify guild and functional group diversity, such as (1) dendrogram branch length (to measure functional diversity); (2) phylogeny studies (probably the best predictor of ecology); (3) molecular studies by directing more effectively the application of techniques; (4) investigation of bioindicator role of these groups; (5) morphological variation studies of arachnid species to elucidate many questions involving the ecology of these organisms; and (6) phenological studies to contribute to the understanding of reproduction, dispersal patches, temporal organization of communities resources, plant-animal interactions, and evolution of life history of the animals that depend on plants for feeding.

References

Adis J (1977) Programa mínimo para análises de ecossistemas: Artrópodos terrestres em florestas inundáveis da Amazônia Central. Acta Amaz 7(2):223–229

Adis J (1981) Comparative ecological studies of the terrestrial arthropod fauna in central Amazonian inundation-forests. Amazoniana 7:87–173

Adis J (1984) 'Seasonal igapó'-forests of central Amazonian blackwater rivers and their terrestrial arthropod fauna. In: Sioli H (ed) The Amazon: Limnology and landscape ecology of a mighty tropical river and its basin. Junk, Dordrecht, pp 245–268

Adis J (1992) Überlebensstrategien terrestrischer Invertebraten in Überschwemmungswäldern Zentralamazoniens. Verh naturwiss, Ver Hamburg 33:21–114

Adis J (1997) Estratégias de sobrevivência de invertebrados terrestres em florestas inundáveis da Amazônia Central: uma reposta à inundação de longo período. Acta Amaz 27:43–54

Adis J, Junk WJ (2002) Terrestrial invertebrates inhabiting lowland river floodplains of Central Amazonia and Central Europe: a review. Freshw Biol 47:711–731

Adis J, Mahnert V (1985) On the natural history and ecology of Pseudoscorpiones (Arachnida) fron na Amazonian black water inundation forest. Amazoniana 9(3):297–314

Adis J, Mahnert V (1990) Vertical distribution and abundance of Pseudoscorpion species (Arachnida) in the soil of a Neotropical secondary forest during the dry and the rainy season. Acta Zool 190:11–16

Adis J, Mahnert V (1993) Vertical distribution and abundance of Pseudoscorpions (Arachnida) in the soil two different Neotropical primary forest during the dry and rainy seasons. Mem Queensl Mus 33(2):431–440

Adis J, Messner B (1997) Adaptations to life under water: Tiger beetles and millipedes. In: Junk WJ (ed) The Central Amazon floodplain, 318-330. Ecology of a pulsing system. Ecological Studies 126, Springer, Berlin

Adis J, Schubart HOR (1984) Ecological research on arthropods in central Amazonian forest ecosystems with recommendations for study procedures. In: Cooley JH, Golley FB (eds) Trends in ecological research for the 1980s. NATO Conference Series, Series I: Ecology. Plenum Press, New York/London, pp 11–144

Adis J, Lubin YD, Montgomery GG (1984) Arthropods from the canopy inundated and terra firme forests near Manaus, Brazil, with critical considerations on the pyrethrum-fogging technique. Stud Neotropical Fauna Environ 19(4):223–236

Adis J, Mahnert V, Morais JW, Rodrigues JMG (1988) Adaptation of an Amazonian Pseudoscorpion (Arachnida) from Dryland Forests to Inundation Forests. Ecology 69 (1):287–291

Adis J, Messner B, Hirschel K, Ribeiro MO, Paarmann W (1993) Zum Tauchvermogen eines Sandlaufkäfers (Coleoptera: Carabidae: Cicindelinae) im Überschwemmungsgebiet des Amazonas bei Manaus, Brasilien. Verh. Wesid. Entom. Tag 1992, Lõbbecke-Museum, Düsseldorf 51–62

Aguiar NO (2000) Diversidade e História natural de Pseudoscorpiões (Arachnida), em floresta Primária de terra firme, no alto Rio Urucu, Coari, Amazoniaas. Tese de Doutorado, Manaus, INPA/FUA. p 225

Aguiar NO, BÜhrnheim PF (1998) Ppseudoscorpions (Arachnida) of the Ilha de Maracá. pp. 381–389. In: Mlliken W, Ratter J (ed). Maracá. The Biodiversity and Environment of an Amazoniaian Rainforest. Jon Wiley & sons ltd, England

Aguiar NO, Bührnheim PF (2003) Pseudoscorpiões (Arachnida) da vegetação de sub-bosque da floresta primária tropical de terra firme (Coari, Amazoniaas, Brasil). Acta Amazon 33(3):515–526

Amorim MA, Adis J, Paarmann W (1997a) Life cycie adaptations of a diurnal tiger beetle (Coleoptera, Carabidae, Cicindelinae) to conditions on Central Amazonian floodplains. In: Ulrich H (ed) Tropical Biodiversity and Systematics, pp. 233–239. Proceedings Int. Symp. on Biodiversity and Systematics in Tropical Ecosystems, Bonn, 1994. Zoologisches Forschungsinstitut und Museum Koenig. pp 357

Amorim MA, Adis J, Paarmann W (1997b) Ecology and adaptations of the tiger beetle Pentacomia egrégia (Chaudoir) (Cicindelinae: Carabidae) to Central Amazonian floodplains. Ecotropica 3(2):71–82

Arrieira RL, Schwind LTF, Joko CY, Alves GM, Velho LFM, Lansac-Tôha FA (2016) Relationships between environmental conditions and the morphological variability of planktonic testate amoeba in four neotropical floodplains. Eur J Protistol 56:180–190. https://doi.org/10.1016/j.ejop.2016.08.006

Ayres JM (1995) As Matas de Várzea do Mamirauá. MCT-CNPq-Sociedade Civil Mamirauá, Rio de Janeiro 3–36

Battirola LD, Marques MI, Adis J (2006). The importance of organic material for arthropods on Attalea phalerata (Arecaceae) in the Pantanal of Mato Grosso, Brazil. What's up? ICAN, 12:1–3

Beck L (1967) Beiträge zur Kenntnis der neotropischen Oribatidenfauna. 5. Archegozetes (Arach., Acari). Senckenb boil 48(5–6):407–414

Beck L (1968) Zum jahreszeitlichen Massenwechsel zweier Oribatidenarten (Acari) im neotropischen Überschwemmungswald. Verhandlungen der Deutschen Zoologischen Gesellschaft in Innsbruck 535–540

Beck L (1969) Zum jahreszeitlichen Massenwechsel zweier Oribatidenarten (Acari) im neotropischen Überschwemmungswald. Verh Dtsch Zool Ges (Innsbruck 1968):535–540

Beck L (1971) Bodenzoologische Gliederung und Charakterisierung des amazonischen Regenwaldes Amazoniana 3(1):69–132

Beck L (1972) Der EinfluB der jahresperiodischen Überflutungen auf den Massenwechsel der Bodenarthropoden im zentralamazonischen Regenwaldgebiet. Pedobiologia 12:133–148

Benchimol M, Peres CA (2015) Widespread forest vertebrate extinctions induced by a mega hydroelectric dam in lowland Amazonia. PLoS One 10:e0129818

Besch W (1969) South American Arachnida. In: Fittkau EJ, Lilies J, Klinge H, Schwabe GH, Sioli H (eds) Biogeography and ecology in South America. Junk, The Hague. pp 723–740

Bonaldo AB, Carvalho LS, Pinto-da-Rocha R, Tourinho A, Miglio LT, Candiani DF, Lo-Man-Hung NF, Abrahim N, BVB R, Brescovit AD, Saturnino R, Bastos NC, Dias SC, Silva BJF, Pereira-Filho JMB, Rheims CA, Lucas SM, Polotow D, Ruiz G, Indicatti RP (2009) Inventário e história natural dos aracnídeos da Floresta Nacional de Caxiuanã. In: PLB L (ed) Caxiuanã: Desafios para a conservação de uma Floresta Nacional na Amazônia. Museu Paraense Emílio Goeldi, Belém, pp 577–621

Bonecker CC, Simões NR, Minte-Vera CV, Lansac-Tôha FA, Velho LFM, Agostinho AA (2013) Temporal changes in zooplankton species diversity in response to environmental changes in alluvial valley. Limnologica 43:114–121
Borges SH, Brescovit AD (1996) Inventário preliminar da aracnofauna (Araneae) de duas localidades na Amazônia Ocidental. Boletim do Museu Paraense Emílio Goeldi, série zoológica 12:9–21
Botero-Trujillo R, Flórez E (2017) Two new ricinuleid species from Ecuador and Colombia belonging to the *peckorum* species-group of *Cryptocellus* Westwood (Arachnida, Ricinulei). Zootaxa 4286(4):483–498. https://doi.org/10.11646/zootaxa.4286.4.2
Brescovit AD, Bonaldo AB, Bertani R, Rheims CA (2003) Araneae. In: Adis J (ed) Amazonian Arachnida and Myriapoda. Pensoft Publishers, Sofia/Moscow, pp 303–343
Cardoso P, Arnedo MA, Triantis KA, Borges PAV (2010) Drivers of diversity in Macaronesian spiders and the role of species extinctions. J Biogeogr 37:1034–1046
Castilho ACC, Marques MI, Adis J, Brescovit AD (2005) Distribuição sazonal e vertical de Araneae em área com predomínio de Attalea phalerata Mart. (Arecaceae), no Pantanal de Poconé, Mato Grosso, Brasil. Amazoniana, 18:215–239
Chapin FS, Bret-Harte MS, Hobbie SR, Zhong HL (1996) Plant functional types as predictors of transient responses of arctic vegetation to global change. J Veg Sci 7:347–358
Chobot K, Řezáč M, Boháč J (2005) Epigeic groups of invertebrates and its indicative possibilities. In: Vačkář D (ed) Indicators of biodiversity changes. Academia, Praha, pp 239–248
Coddington JA, Levi HW (1991) Systematics and evolution of spiders (Araneae). Annu Rev Ecol Syst 22:565–592
Cokendolpher JC, Reddell JR (2000) New and rare Schizomida (Arachnida: Hubbardiidae) from South America. Amazoniana 16(1/2):187–212
Condé B (1993) Description du mâle de deux espèces de Palpigrades. Rev Suisse Zool 100:279–287
Cook OF (1899) *Hubbardia*, a new genus of Pedipalpi. Proc Entomol Soc Wash 4:249–261
Cooke JAL (1967) Observations on the biology of Ricinulei (Arachnida) with descriptions of two new species of *Cryptocellus*. J Zool 151:31–42. https://doi.org/10.1111/j.1469-7998.1967.tb02864.x
Diaz S (1995) Elevated CO2 responsiveness, interactions at the community level and plant functional types. J Biogeogr 22:289–295
Erwin TL (1983) Beetles and other insects of the tropical rainforest canopies at Manaus, Brazil, sampled with insecticidal fogging. In: Suttoj SL, Whitmore TC, Chadwick AC (eds) Tropical rain Forest: ecology and management. Special Pub. No. 2 of the British Ecological Society. Blackwell Scientific Publications, Oxford, pp 59–75
Erwin TL, Adis J (1982) Amazonian inundation forests: their role as short-term refuges and generators of species richness and taxon pulses. In: International symposium of the association for tropical biology, 5. 1979, Caracas. Proceedings, vol 5. Columbia University Press, New York, pp 358–371
Ferreira LV, Parolin P (2011) Effects of flooding duration on plant demography in a black-water floodplain forest in Central Amazonia. Invited paper. Pesquisas Botânica 62:323–332
Flores BM, Holmgren M, Xu C, van Nes EH, Jakovac CC, Mesquita RCG, Scheffer M (2017) Floodplains as an Achilles' heel of Amazonian forest resilience. Proc Natl Acad Sci U S A 114(17):4442–4446. https://doi.org/10.1073/pnas.1617988114
Franklin E (1994) Ecologia de oribatídeos (Acari: Oribatida) em florestas inundáveis da Amazônia Central. Tese de doutorado, Inpa/Universidade do Amazonas, Manaus 266
Franklin EN, Woas S, Schubart HOR, Adis J (1980) Ácaros oribatídeos (Acari:Oribatida) arborícolas de duas florestas inundáveis da Amazônia Central. Rev Bras Biol 58(2):317–335
Franklin EN, Schubart HOR, Adis JU (1996) Ácaros (Acari: Oribatida) edáficos de duas florestas inundáveis da Amazônia Central: distribuição vertical, abundância e recolonização do solo após a inundação. Rev Bras Biol 57(3):501 520
Franklin EN, Adis J, Woas S (1997) The Oribatid Mites. In: Junk WJ (ed) Central Amazonian river floodplains: ecology of a pulsing system. Springer-Verlag, Berlin/Heidelberg, pp 331–349

Franklin EN, Woas S, Schubart HOR, Adis J (1998) Ácaros Oribatídeos (Acari: Oribatida) arborícolas de duas florestas inundáveis da Amazônia Central. Rev Bras Biol 58(2):317–335
Franklin EN, Guimarães RL, Adis J, Schubart HOR (2001) Resistência á submersão de ácaros (Acari: Oibatida) terrestres de florestas inundáveis e de terra firme na Amazônia central em condições experimentais de laboratório. Acta Amaz 31(2):285. https://doi.org/10.1590/1809-43922001312298
Franklin E, Hayek T, Fagundes E, Silva L (2004) Oribatid mite (Acari: Oribatida) contribution to decomposition dynamic of leaf litter in primary forest, second growth, and polyculture in the Central Amazon. Braz J Biol 64(1):59–72. https://doi.org/10.1590/S1519-69842004000100008
Franklin E, Santos EMR, Albuquerque MIC (2006) Diversity and distribution of oribatid mites (Acari: Oribatida) in a lowland rain forest in Peru and in several environments of the Brazilians states of Amazonas, Rondônia, Roraima and Pará. Braz J Biol 66(4):999–1020. https://doi.org/10.1590/S1519-69842006000600007
Friebe B, Adis J (1983) Entwicklungszyklen von Opiliones (Arachnida) im Schwarzwasser-Uberschwemmungswald (Igapo) des Rio Tarumã Mirim (Zentralamazonien, Brasilien). Amazoniana 8:101–110
Furch K, Junk WJ (1997) Physicochemical conditions in the floodplains. In: Junk WJ (ed) The Central Amazon Floodplain: Ecology of a Pulsing System. Springer-Verlag, Berlin, 69–108.
Gervais PM (1843) Les principaux résultats d'un travail sur la famille des Scorpions. C R Hebd Seances Acad Sci. 5(7):129–131
Hammen van der L (1982) Comparative studies in Chelicerata II. Epimerata (Palpigradi and Actinotrichida). Zool Verhand 196:1–70
Harvey MS (2002) The neglected cousins: what do we know about the smaller arachnid orders? J Arachnol 30:357–372. https://doi.org/10.1636/0161-8202(2002)030[0357:TNCWDW]2.0.CO;2
Harvey MS (2003) Catalogue of the smaller arachnid orders of the world: Amblypygi, Uropygi, Schizomida, Palpigradi, Ricinulei and Solifugae. CSIRO Publishing, p 400
Harvey MS (2013) Whip Spiders of the World, version 1.0. Western Australian Museum, Perth. Accessed 28 April 2017, online at http://museum.wa.gov.au/catalogues/whip-spiders
Hayek T (2000) Ácaros do solo (Acari: Oribatida): Diversidade, abundância e biomassa na decomposição de serapilheira em parcelas de floresta primária, capoeiras e policultivo da Amazônia Central. MSc. Thesis. Manaus, INPA, 93
Hofer H (1990) The spider community (Araneae) of a Central Amazonian blackwater inundation forest (igapó). Acta Zool Fenn 190:173–179
Höfer H (1997) The spider communities. In: Junk W (ed) The central Amazonian river floodplains. Ecology of a pulsing system. Springer, Berlin, pp 570–576
Höfer H, Brescovit AD (1996) On the genus *Xyccarph* in Central Amazonia (Araneae: Oonopidae). Bull Br Arachnol Soc 10:149–155
Höfer H, Brescovit AD, Gasnier T (1994a) The wandering spiders of the genus *Ctenus* (Ctenidae, Araneae) of Reserva Ducke, a rainforest reserve in Central Amazonia. Andrias 13:81–98
Höfer H, Brescovit AD, Adis J, Paarmann W (1994b) The spider fauna of Neotropical tree canopies in Central Amazonia: first results. Stud Neotropical Fauna Environ 29(1):23–32
Irion G, Junk WJ, Mello JASN (1997) The large central Amazonian river floodplains near and geomorphological aspects. In: Junk WJ (ed) The Central Amazon Floodplain: Ecology of a Pulsing System. Springer-Verlag, Berlin, 23–46
Irmler U (1981) Überlebensstrategien von Tieren im saisonal überfluteten amazonischen Überschwemmungswald. Zool Anzeiger, 206(1/2): 26–38
Irmler U (1985) Temperature dependent generation cycle for the cicindelid beetle Pentacomia egregia Chaud. (Coleoptera, Carabidae, Cicindelinae) of the Amazon valley. Amazoniana 9:431–439
Junk WJ (1983) As águas da região Amazônica. In: Salati E, Schubart H, Junk WJ, Oliveira AR (eds). Amazônia: Desenvolvimento, Integração e Ecologia. Ed Brasiliense/CNPq, São Paulo, 45–100

Junk WJ (1989) Flood tolerance and tree distribution in central Amazonia. In: Holm-Nielsen LB, Nielsen IC, Balslev H (eds) Tropical Forest Botanical Dynamics. Speciation and Diversity. Academic Press, London, p 47–64

Junk WJ (2000) The Central Amazon floodplain: ecology of a pulsing system. Springer, Berlin, p 525

Junk WJ, Furch K (1985) The physical and chemical properties of Amazonian waters and their relationships with the biota. In: Prance GT, Lovejoy TE (eds) Key environments: Amazonia. Pergamon Press, New York, pp 3–17

Kropf C (1995) *Pseudanapis hoeferi*, n. sp. from central Amazonia, Brazil (Araneae, Anapidae). Bull Br Arachnol Soc 10:19–22

Kury AB (1995) A review of *Huralvioides* (Opiliones, Gonyleptidae, Pachylinae). Amazoniana 13:315–323

Kury AB (2003) Annotated catalogue of the Laniatores of the new world (Arachnida, Opiliones). Rev Ibéric Aracnol 1:1–337

Kury AB (2010) Opilionological record – a chronicle of harvestman taxonomy. Part 1: 1758–1804. J Arachnol 38(3):521–529

Kury AB, Pinto-Da-Rocha R (2002) Opiliones. In: Adis J (ed) Amazonian Arachnida and Myriapoda. Pensoft, Sofia, p 590

Lourenço WR (1988) Synopsis de la faune scorpionique de la région de Manaus, Etat d'Amazonas, Brésil, avec description de deux nouvelles espèces. Amazoniana 10(3):327–337

Lourenço WR (1997) A propos de deux nouvelles espèces de Tityus Koch du Brésil (Scorpiones, Buthidae). Rev Arachnol, 12(5):53–59.

Lourenço WR (2002) Scorpiones. In: Adis J (ed) Amazonian Arachnida and Myriapoda: identification keys to all classes, orders, families, some genera and lists of known terrestrial species. Pensoft Publisher, Moscow

Lourenço WR (2005) Scorpion diversity and endemism in the Rio Negro region of Brazilian Amazonia, with the description of two new species of Tityus C. L. Koch (Scorpiones, Buthidae). Amazoniana, 18(3/4):203–213

Lourenço WR, Cuellar O (1999) A new all-female scorpion and the first probable case of arrhenotoky in scorpions. J Arachnol 27:149–153

Lourenço WR, Eickstedt VR (2003) Escorpiões de Importância Médica. In: Cardoso JLC, França FOS, Wen FH, Málaque CMS, Haddad V (eds) Animais Peçonhentos no Brasil: Biologia, Clínica e Terapêutica dos Acidentes. Sarvier, Fapesp, São Paulo

Lourenço WR, Francke OF (1986) A new species of *Chactopsis* from Brazil (Scorpiones, Chactidae). Amazoniana 9(4):549–558

Lourenço WR, Pézier A (2002) Addition to the scorpion fauna of the Manaus region (Brazil), with a description of two species of Tityus from the canopy. Amazoniana, 17(1):177–186

Lourenço WR, Pinto-da-Rocha R (2000) Two new species of *Tityus* from Brazilian Amazonia (Scorpiones, Buthidae). Revue Arachnologique 13(13):187–195

Lourenço WR, Adis J, Araújo JS (2005) A new synopsis of the scorpion fauna of the Manaus region in Brazilian Amazonia, with special reference to an inundation forest at the Tarumã Mirím river. Amazoniana 18(3/4):241–249

Maelfai TJP (1996) Spiders as bioindicators. In: van Straalen NM, Krivolutsky DM (eds) Bioindicator systems for soil pollution. Kluwer Academic Publishers, Dordrecht, pp 165–178

Maelfai TJP, Baer TL (1988) Les araignées sont-elles de bons indicateurs écologiques? Bull Soc Scient Bretagne 59:155–160

Mahnert V (1979) Pseudoskorpione (Arachnida) aus dem Amazonas-Gebiet (Brasilien). Rev Suisse Zool 86(3):719–810

Mahnert V (1985) Pseudoscorpions (Aracnhida) from the lower Amazon region. Revta bras Ent 29(1):75–80

Mahnert V, Adis J (1985) On the occurrence and habitat of Pseudoscorpions (Arachnida) from Amazoniaian forest of Brazil. Studi Neotropical Fauna Environ 20(4):211–215

Mahnert V, Adis J (2002) Pseudoscorpiones. In: Adis J (ed) Amazonian Arachnida and Miriapoda. Pensoft Publishers, Sofia/Moscow

Marques MI, Santos GB, Battirola LD, Tissiani ASO (2009) Entomofauna associada à matéria orgânica em bainhas foliares de Attalea phalerata Mart. (Arecaceae) na região norte do Pantanal de Mato Grosso. Acta Biol Parana 38:93–112

Martius C, Höfer H, Verhaagh M, Adis J, Mahnert V (1994) Terrestrial arthropods colonizing an abandoned termite nest in a floodplain forest of the Amazon river during the flood. Andrias 13:17–22

Mayoral JG (2015) Clase Arachnida. Orden Palpigradi Revista electronica IDE@ – SEA, 10:1–9

Mertes LAK, Dunne T, Martinelli LA (1996) Channel-floodplain geomorphology along the Solimões-Amazon River, Brazil. Geol Soc Am Bull 108:1089–1107

Morais JW (1985) Abundância e distribuição vertical de Arthropoda do solo numa floresta primária não inundada. Dissertação de Mestrado, Manaus, INPA/FUA, 92p

Morais JW, Adis J, Mahnert V, Berti-Filho E (1997) Abundance and phenology of pseudoscorpiones (Arachnida) from a mixed water inundation forest in central Amazonia, brazil. Revue Suisse Zool 104(3):475–483

Muchmore WB (1975) Two Miratemnid Pseudoscorpions from the western hemisphere (Pseudoscorpionida, Miratemnidae). Southwest Nat 20(2):231–239

Norton RA, Behan–Pelletier VM (2009) Suborder Oribatida. In: Krantz GW, Walter DE (eds) A manual of acarology, 3rd edn. Texas Tech University Press, Lubbock, pp 430–564

Oliveira AR, Argolo PS, De GJ, Norton RA, Schatz H (2017) A checklist of the oribatid mite species (Acari: Oribatida) of Brazil. Zootaxa 4245(1):1–89. https://doi.org/10.5281/zenodo.437584

Paarmann W, Irmler U, Adis J (1982) Pentacomia egregia Chaud. (Carabidae, Cicindelinae), an univoltine species in the Amazonian inundation forest. Coleopts Bull, 36(2):183–188

Paoletti MG (1999) Using bioindicators based on biodiversity to assess landscape sustainability. Agric Ecosyst Environ 74(1–3):1–18

Petchey OL, Gaston KJ (2006) Functional diversity: back to basics and looking forward. Ecol Lett 9:741–758

Pinto-da-Rocha R, Kury AB (2003) Third species of Guasiniidae (Opiliones, Laniatores) with comments on familial relationships. J Arachnol 31(3):394–399

Platnick NI, Hofer H (1990) Systematics and ecology of ground spiders (Araneae, Gnaphosidae) from central Amazonian inundation forests. Am Mus Novit 2971:1–16

Pocock RI (1897) Descriptions of some new species of scorpions of the genus *Tityus*, with notes upon some forms allied to *T. americanus* (Linn.). Ann Mag Nat Hist 19(6):510–521

Prendini L (2011a) Order Palpigradi Thorell, 1888. In: Zhang ZQ (ed) Animal biodiversity: an outline of higher-level classification and survey of taxonomic richness, vol 3148. Zootaxa, New Zealand, p 121

Prendini L (2011b) Order Solifugae Sundevall, 1833. In: Zhang ZQ (ed) Animal biodiversity: an outline of higher-level classification and survey of taxonomic richness, vol 3148. Zootaxa, New Zealand, p 118

Raub F, Höfer H, Scheuermann L, Brandl R (2014) The conservation value of secondary forests in the southern Brazilian Mata Atlântica from a spider perspective. J Arachnol 42:52–73

Reddell JR, Cokendolpher JC (1995) Catalogue, bibliography, and generic revision of the order Schizomida (Arachnida). Texas Memorial Museum Speleol Monogr 4:1–170

Reddell JR, Cokendolpher J (2002) Schizomida. In: Adis J (ed) Amazonian Arachnida and Myriapoda – keys for the identification to classes, orders, families, some genera, and lists of known species. Pensoft, Sofia, pp 387–398 590p

Rego FNAA, Venticinque EM, Brescovit AD, Rheims CA, Albernaz ALKM (2009) A contribution to the knowledge of the spider fauna (Arachnida: Araneae) of the floodplain forests of the main Amazon River channel. Rev Ibé Aracnol 97:85–96

Resende AF, Nelson BW, Flores BM, Almeida DR (2014) Fire damage in seasonally flooded and upland forests of the Central Amazon. Biotropica 46:643–646

Ribeiro MOA, Fonseca CRV, Foronda EH (1996) Bionomi a de Megacephaia sobrina punctata Laporte 1835 e inventário das espécies do gênero Megacephaia Latreille (Coleoptera:

Cicindelidae) em áreas alagáveis na Amazônia Central. Rev UA Série: Ciências Biológicas 1(1):31–54

Rocha LS (2002) Solifugae. In: Adis J (ed) Amazonian Arachnida and Myriapoda. Pensoft Publishers, Sofia

Rocha LS, Cancello EM (2002) South American Solifugae: new records, occurrence in humid forests and concurrence with termites. News Lett Br Arachnol Soc 93:4–5

Rowland JM (1975) Classification, phylogeny and zoogeography of the American arachnids of the order Schizomida. PhD dissertation. Texas Tech University, Lubbock

Silva D (1996) Species composition and community structure of Peruvian rainforest spiders: a case study from a seasonally inundated forest along the Samiria river. Revue Suisse de Zoologie Hors Série 2:597–610

Sioli H (1984) The Amazon and its main affluents: hydrography, morphology of the river courses, and river types. In: Sioli H (ed) Amazon: Limnology and Landscape Ecology of a Mighty River and its Basin. Dr. W. Junk Publishers, Dordrecht, 127–165

Smrž J, Kováč Ĺ, Mikeš J, Lukešová A (2013) Microwhip scorpions (Palpigradi) feed on heterotrophic cyanobacteria in slovak caves – a curiosity among arachnida. PLoS One 8(10):e75989. https://doi.org/10.1371/journal.pone.0075989

Souza MFVR, Ferreira RL (2010) *Eukoenenia* (Palpigradi: Eukoeneniidae) in Brazilian caves with the first troglobiotic palpigrade from South America. J Arachnol 38:415–424. https://doi.org/10.1636/Ha09-112.1

Souza MFVR, Ferreira RL (2013) Two new species of the enigmatic Leptokoenenia (Eukoeneniidae: Palpigradi) from Brazil: first record of the genus outside intertidal environments. PLoS ONE 8(11):e77840

Tilman D, Reich PB, Knops J, Wedin D, Mielke T (2001) Diversity and productivity in a long-term grassland experiment. Science 294:843–845

Tizo-Pedroso, Del-Claro K (2007) Natural history and social behavior in Neotropical Pseudoscorpions. In: Encyclopedia of life support systems (UNESCOEOLSS, 2007). UNESCO: Eolss Publishers, Oxford, pp 1–12

Tourinho AL, Lo Man-Hung NF, Bonaldo AB (2010) A new species of Ricinulei of the genus *Cryptocellus* Westwood (Arachnida) from northern Brazil. Zootaxa 2684:63–68

Uehara-Prado M, Fernandes JO, Bello AM, Machado G, Santos AJ, Vaz-de-Mello FZ, Freitas AVL (2009) Selecting terrestrial arthropods as indicators of small-scale disturbance: a first approach in the Brazilian Atlantic Forest. Biol Conserv 142:1220–1228

Valdecasas AG, Camacho AI (2003) Conservation to the rescue of taxonomy. Biodivers Conserv 12:1113–1117

Vieira RS, Höfer H (1994) Prey spectrum of two army ant species in Central Amazonia, with special attention on their effect on spider populations. Andrias 13:189–198

Voigt W, Perner J, Jones TH (2007) Using functional groups to investigate community response to environmental changes: two grassland case studies. Glob Chang Biol 13:1710–1721

Ward JV, Tockner K (2001) Biodiversity: toward a unifying theme for river ecology. Freshw Biol 46:807–819

Werner F (1939) Neu-Eingänge von Skorpionen im Zoologischen Museum in Hamburg. II. Teil. Festschrift zum 60. Geburstage von Professor Dr. Embrik Strand 5:351–360

Weygoldt P (1969) Biology of Pseudoscorpions. Harvard University Press, Cambridge, 145p

Weygoldt P (2000) Biology of whip spiders (Chelicerata: Amblypygi): their biology, morphology and systematics. Apollo Books, Stenstrup

Woas S (2002) Acari: Oribatida. In: Adis J (ed) Amazonian Arachnida and Myriapoda. Pensoft, Sofia/Moscow, pp 21–291

World Spider Catalog (2017) World Spider Catalog. Natural History Museum Bern, online at http://wsc.nmbe.ch, version 17.5, Accessed on 01.20.2017

Chapter 7
Influence of Flood Levels on the Richness and Abundance of Galling Insects Associated with Trees from Seasonally Flooded Forests of Central Amazonia, Brazil

Genimar R. Julião, Eduardo M. Venticinque, and G. Wilson Fernandes

7.1 Introduction

The plant species from the Amazonian igapó and várzea forests have yearly their roots and stems submerged for longer or shorter periods, which requires a combination of physiological adaptations and reproductive and growth strategies (Junk et al. 2010; Parolin and Wittmann 2010). These traits can also aid in their survival in an environment with high levels of stress (Junk 1993; Ayres 1993). Likewise, several organisms associated with these plants also develop strategies that allow their maintenance in this biphasic (aquatic and terrestrial) system. In well-documented cases, categories of behavioral adaptations of invertebrates have been described, such as vertical and horizontal migrations, temporal flight, dormant stages, and active stages in shelters (Adis 1997a, b).

In the case of sessile invertebrates such as gall-inducing insects, the environmental stresses extended to their host plants seem to favor this guild (Fernandes and Price 1988, 1992; Julião et al. 2014). These insects usually spent the majority of their life cycle – immature phase – in the gall structure, derived from the hyperplasia

G. R. Julião (✉)
Coordenação de Ecologia, Instituto Nacional de Pesquisa da Amazonia (INPA), Manaus, Amazonas, Brazil

Fiocruz Rondônia, Laboratório de Entomologia, Porto Velho, Rondônia, Brazil

E. M. Venticinque
Departamento de Ecologia, CB/Universidade Federal do Rio Grande do Norte, Campus Universitário, Lagoa Nova, Brazil

G. W. Fernandes
Ecologia Evolutiva e Biodiversidade/DBG, CP 486, ICB/Universidade Federal de Minas Gerais (UFMG), Belo Horizonte, Minas Gerais, Brazil

R. W. Myster (ed.), *Igapó (Black-water flooded forests) of the Amazon Basin*, https://doi.org/10.1007/978-3-319-90122-0_7

and/or hypertrophy of plant cells, tissues, and organs (Isaias et al. 2014). Higher numbers of galling insects have been found in vegetation growing on conditions of soil nutrient shortage, such as phosphorus (e.g., Blanche and Westoby 1995; Cuevas-Reyes et al. 2004), magnesium, potassium, and iron (Gonçalves-Alvim and Fernandes 2001), or hydric stresses such as Mediterranean-type vegetation (Price et al. 1998). For instance, *Calophyllum brasiliense* Cambess. plants subjected to a submersion period were more frequently attacked by *Lopesia* spp. gallers than non-flooded plants (Ribeiro et al. 1998).

Besides, gall-inducing insects can notably support water inundation during the flood pulses. Cogni et al. (2003) compared the survival of gall-inducing insect larvae associated with submerged and non-submerged leaves. The host plant *Symmeria paniculata* Benth was sampled in a igapó vegetation in the Anavilhanas region, Amazonas, Brazil. Plant leaves stayed submerged approximately for 114 days and the survival rate of larvae inside the submerged galls did not differ from that observed for the non-submerged leaves.

In the present study, we reexamined three hypotheses which addressed gall abundance and richness in the canopies of igapó and várzea forests, considering different flooding levels. Based on the "nutritional stress hypothesis" (Fernandes and Price 1991, 1992; Blanche and Westoby 1995), igapó forests, flooded by nutrient-poor waters and growing in low-fertility soils, should support a greater richness and abundance of galling insects compared to the várzea forests, which occur in areas with nutrient- and sediment-rich waters (see Junk et al. 2011). In addition, host plants located in these lowland floodplains are expected to exhibit differential values of gall abundance and richness in function of terrain elevation, which in turn determines the flood height and duration. This natural gradient results in plant species zonation, and some extremes such as low and high várzea/igapó can be recognized by its forest types with intrinsic individual densities, height levels (higher or lower than 3 meters), and flood duration lower and higher than 50 days *per* year (Wittmann et al. 2002; Wittmann et al. 2010). In this context, plants that experience greater flooding extent should undergo greater physiological stress ("hydric stress hypothesis," Fernandes and Price 1988, 1992). We also tested the "plant species richness hypothesis" – that predicts that the highest richness of galling insects should be positively related to a more species rich flora, due to the greater diversity of niches and resource availability (Southwood 1960; Fernandes 1992; Lara et al. 2002).

To address these questions, samples of galling insects were done in the Mamirauá (1,124,000 hectares, Fig. 7.1) and Amanã (2,350,000 hectares, Fig. 7.2) Sustainable Development Reserves (Mamirauá Institute for Sustainable Development – MISD 2018), which present various research facilities, including funding programs, infrastructure, and logistical support, via Mamirauá Institute for Sustainable Development (MISD) and Mamirauá Civil Society.

Fig. 7.1 Floating research base at the Mamirauá Sustainable Development Reserve, Amazonas, Brazil. This structure gives support to multidisciplinary research teams

Fig. 7.2 Floating research base at the Amanã Sustainable Development Reserve, Amazonas, Brazil

7.2 Material and Methods

7.2.1 *Study Sites*

The collections were made at the Mamirauá (MSDR) and at Amanã (ASDR) Sustainable Development Reserves, located near the city of Tefé, Amazonas, Brazil. Overall 28 sites were sampled, 14 sampling points were performed in várzea and 14 in igapó forest areas (Table 7.1, Fig. 7.3). All igapó points were sampled at the Amanã Reserve. Várzea sampling points (n = 14) were divided in the two reserves:

Table 7.1 Number of individuals and species of host plants and richness and abundance of galling insects in the sampling points at the Mamirauá and Amanã Sustainable Development Reserves, Amazonia, Brazil. Sampling points were also characterized as low/high várzea and low/high igapó forests and with its geographical coordinates

			Host plant		Galling insect			Coordinates	
Point	Forest	Habitat	Individuals	Species	Richness	Abundance	Reserve	S	W
1	Várzea	Low	31	25	53	3502	Mamirauá	−2.98358	−64.9266
2	Várzea	Low	23	18	36	2005	Mamirauá	−3.0581	−64.8489
3	Várzea	Low	14	12	22	1032	Mamirauá	−2.8248	−64.9534
4	Várzea	Low	18	15	30	1430	Mamirauá	−2.79353	−65.1044
5	Várzea	Low	26	16	35	3392	Mamirauá	−3.02557	−65.0063
6	Várzea	High	11	6	19	744	Mamirauá	−2.98275	−65.0948
7	Várzea	High	29	25	50	4566	Mamirauá	−2.39093	−65.3365
8	Várzea	Low	22	16	33	2964	Mamirauá	−2.7354	−65.2241
9	Igapó	High	18	9	25	1805	Amanã	−2.72483	−64.379
10	Igapó	Low	15	6	16	2822	Amanã	−2.72093	−64.3787
11	Igapó	High	18	14	35	1717	Amanã	−2.69972	−64.3497
12	Igapó	Low	18	7	17	2190	Amanã	−2.4727	−64.612
13	Igapó	High	18	13	36	1025	Amanã	−2.48463	−64.6403
14	Igapó	Low	17	11	34	633	Amanã	−2.48253	−64.6301
15	Igapó	High	16	10	30	1681	Amanã	−2.71222	−64.3569
16	Igapó	Low	12	10	22	1573	Amanã	−2.70878	−64.346
17	Várzea	High	8	7	13	355	Amanã	−2.94583	−64.5357
18	Várzea	High	9	8	13	823	Amanã	−2.93943	−64.5411
19	Várzea	High	13	12	26	999	Amanã	−2.76262	−64.7379
20	Várzea	High	12	11	23	562	Amanã	−2.76782	−64.7283
21	Várzea	Low	17	12	20	3187	Amanã	−2.71022	−64.8242
22	Várzea	High	13	11	19	683	Amanã	−2.71095	−64.7954
23	Igapó	Low	16	10	18	720	Amanã	−2.55717	−64.6954
24	Igapó	Low	16	11	24	1827	Amanã	−2.56098	−64.7091
25	Igapó	High	17	15	41	2671	Amanã	−2.64748	−64.6857
26	Igapó	Low	17	11	24	3497	Amanã	−2.64323	−64.6669
27	Igapó	High	14	9	22	1254	Amanã	−2.63205	−64.6974
28	Igapó	Low	17	11	21	579	Amanã	−2.62582	−64.6803

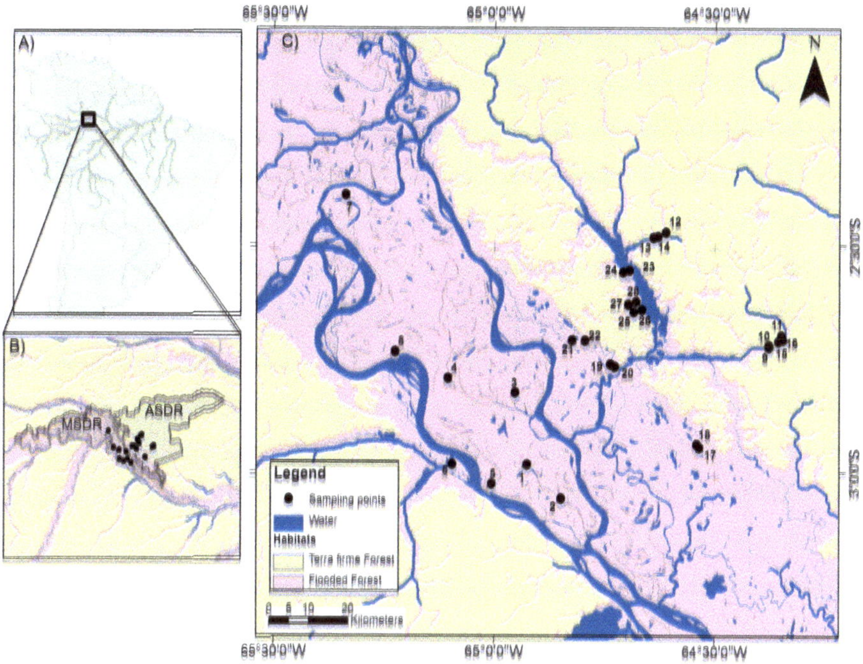

Fig. 7.3 Maps showing the localization of study area (**A**), the Mamirauá (MSDR) and Amanã (ASDR) Sustainable Development Reserves (**B**), and the sampling points ($n = 28$) (**C**). Decimal geographical coordinates of points are given in Table 7.1

eight sites were located at the Mamirauá and the other six sites were located at the Amanã Reserves.

Additionally, the sampling points were categorized in high várzea, low várzea, high igapó, and low igapó, according to their flooding quotas and local vegetation. This classification was based on the terminology proposed for the zonation study of várzea vegetation described by Wittmann et al. (2002, 2010). The same criteria were assumed for the igapó areas. Besides, the height of watermarks was observed in the tree trunks to confirm the local flooding levels (Fig. 7.4). Low várzea usually shows the highest amplitude of canopy height, from 15 to 35 meters. In the high várzea, the upper forest canopy ranges between 35 and 45 meters, with a few trees reaching 58 m (Wittmann et al. 2002, 2004). Canopy height of low igapó forest reaches 16 meters, while in high igapó the upper canopy varies from 25 to 30 meters (Maia and Piedade 2000).

7.2.2 *Insect Gall Sampling*

Sampling was carried out between May and September 2004 at the Mamirauá Reserve. At the Amanã, collections were done in November and December 2005. Each sampling point was composed of eight plots of 5 × 20 meters (total area of

Fig. 7.4 Watermarks in the tree trunks observed during insect gall sampling in a low várzea site, at the Mamirauá Reserve, Brazil

800 m^2), which were separated 20 meters from each other. The plots were established in the understory, using a 50 m measuring tape. The plot lines were then visually projected into the canopy, delimiting tree crowns to be collected. Only the upper canopy crowns were sampled, i.e., crowns localized at the air-canopy interface, which have no shading from other trees. At some sites, and depending on flood level, tree crowns were easily sampled with a telescoping aluminum pole (10 m) inside a boat. The crowns were also accessed by climbing with a "peconha device" – apparatus made of plant fiber or fabric – and using the telescopic pole. Then, 10 crown terminal units were randomly clipped ("branch unit" sensu Bell et al. 1999; Julião et al. 2005), terminal unit length ranged from 30 to 50 cm, and it comprised leaves, stems, branches, and rarely flowers and fruits. Each set of plant units was individually enclosed in a plastic bag and identified with date, local, and plant code. At the floating research bases (Mamirauá and Amanã Reserves, Figs. 7.1 and 7.2), the insect gall morphotypes were described and recorded as well as their abundance for each tree individual. Gall morphotype description was based on gall external morphology (shape, color, trichomes presence/absence, single/grouped occurrence), and the host plant organ attacked. Each gall morphotype was considered a galling insect species since there is an inherent specificity in this insect-plant interaction: usually one galling insect species induces a morphologically distinct gall in its host plant and host organ (Carneiro et al. 2009).

7.2.3 *Data Analysis*

In our study, we considered as dependent variables galling insect richness (GIR), the gall abundance (GIA), the number of tree species (richness) and individuals, and a proportion between GIR and tree richness (GIR/TR ratio). GIR was considered the number of gall morphotypes found in the tree crowns of each sampling point. GIA was given by the sum of the number of gall structures sampled for each gall morphotype. Tree richness and number of tree individuals encompassed host and nonhost plants. GIR/TR ratio was established due to the great variability in the number of individuals/species of trees sampled in each point. Forest types (igapó and várzea) and habitats (flooding quota categories, low and high) were computed as explanatory variables, as well as the landscape categories high várzea (HV), low várzea (LV), high igapó (HI), and low igapó (LI). Analyses of covariance (ANCOVA) were employed to compare the GIR, GIA, and GIR/TR ratio among forest types and habitats, considering the possible interaction between these predictors. The effects of landscape categories (HV, LV, HI, LI) in GIR, GIA, and number of tree species and individuals were explored by means of ANOVA. For significant analyses of variance, multiple pairwise comparisons were employed by means of the Tukey's HSD test. Simple linear regressions were used in the relationship between galling insect richness (response variable) and tree richness (predictor), considering forest types and habitats separately. Except by the hypothetical scheme in the Fig. 7.10, all analyses and figures were performed in the R environment (R Core Team 2017).

7.3 Results

A relatively similar abundance of galling insects was recorded in the two forest types. In the várzea 26,244 galls were collected, corresponding to 297 species of galling insects; 23,994 galls were sampled in the igapó and identified as 235 galling species (Table 7.1). The number of nonhost trees was approximately 3 times greater in várzea forests (64 individuals) than in igapó, with 21 nonhosts recorded. Overall, 205 tree host species were sampled in the várzea and igapó forests, 22 of which occurred in both forest types. Host richness was higher in várzea, with 127 tree species attacked by gallers, and there were 78 host species exclusive to igapó forest. The number of host tree individuals was very variable among low/high igapó and várzea; however, similar numbers of host trees were observed between flooding categories, considering each forest type (Table 7.2).

Table 7.2 Number of individuals and species of host plants and the galling insect richness and abundance sampled in the high/low várzea and igapó forests of the Mamirauá and Amanã Sustainable Development Reserves, Brazil

		Host plant		Nonhost plant		Galling insect	
Forest	Habitat	Individuals	Species	Individuals	Species	Richness	Abundance
Igapó	High	101	55	9	9	136	10,153
	Low	128	60	12	10	136	13,841
Várzea	High	95	72	21	18	155	8732
	Low	151	74	43	28	178	17,512

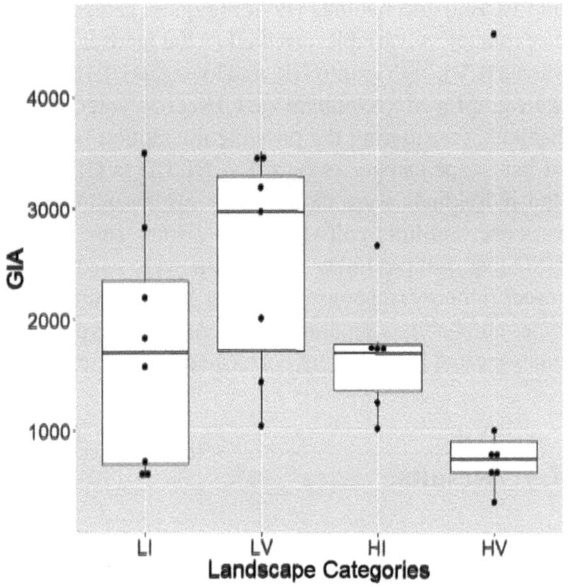

Fig. 7.5 Variation of galling insect abundance (GIA) according to the landscape categories low igapó (LI), low várzea (LV), high igapó (HI), and high várzea (HV) at the Mamirauá and Amanã Sustainable Development Reserves, Brazil

7.3.1 *Galling Insect Abundance × Forest Type and Habitat*

The forest type ($F_{1,24} = 0.150$, $p = 0.702$) and habitat ($F_{1,24} = 2.454$, $p = 0.130$) had no significant effect on GIA, as well as the interaction between these variables ($F_{1,24} = 2.132$, $p = 0.157$). Considering the categories high várzea (HV), low várzea (LV), high igapó (HI), and low igapó (LI), ANOVA also showed that variance in GIA was not predicted by forest nor by habitat ($F_{3,24} = 1.579$, $p = 0.220$, Fig. 7.5).

7.3.2 *Galling Insect Richness × Forest Type and Habitat*

In contrast to GIA, the interaction between habitat and forest type accounted for the variance ($F_{1,24} = 6.868$, $p = 0.015$) observed in the richness of galling insect (GIR), indicating a differential effect of habitat on GIR between várzea and igapó. However,

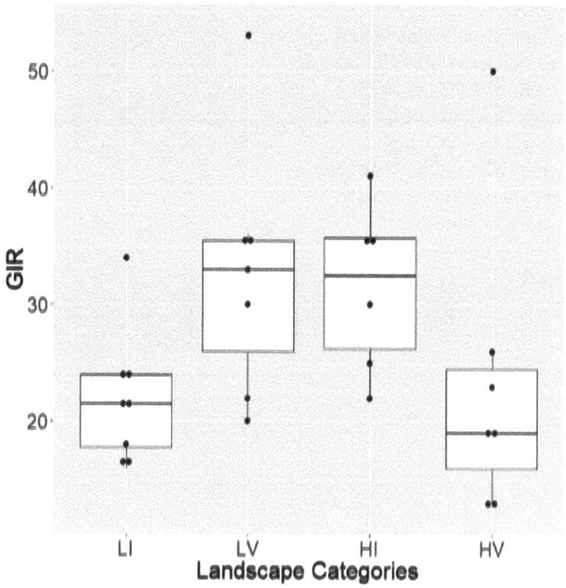

Fig. 7.6 Variation of galling insect richness (GIR) according to the landscape categories low igapó (LI), low várzea (LV), high igapó (HI), and high várzea (HV) at the Mamirauá and Amanã Sustainable Development Reserves, Brazil

these variables alone had no significant effect in GIR ($p > 0.05$). The richness of galling insects also was not statistically different among the landscape categories (HV, LV, HI, LI) ($F_{3,24} = 2.386$, $p = 0.094$, Fig. 7.6).

7.3.3 *Tree Species and Individuals × Forest Type and Habitat*

There was a great amplitude in the number of tree species and individuals, both for host or nonhost plants of landscape categories (Table 7.2), and all analyses of variance were significant ($p < 0.05$). However, significant pairwise comparisons (Tukey's HSD test, adjusted P value <0.05) were observed mainly in the low várzea-low igapó contrast, the number of nonhost individuals, and the number of host and nonhost species differed statistically between these categories. Besides, the number of host individuals was significantly higher in low várzea compared with high várzea (adjusted P value = 0.0205) and showed no difference for the other categories.

7.3.4 *GIR/Tree Richness × Forest Type and Habitat*

Higher values in the ratio between galling insect richness and tree richness [GIR/TR (host + nonhost species)] were observed in high/low igapó categories compared to high/low várzea (mean ± standard deviation: HI = 2.4 ± 0.3, LI = 2.0 ± 0.5,

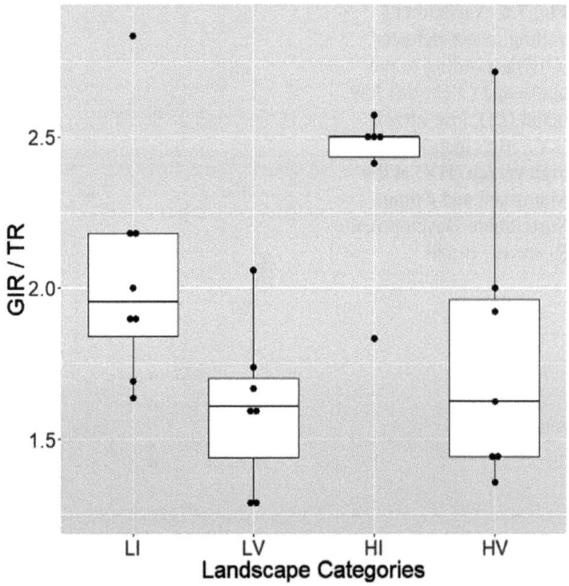

Fig. 7.7 The ratio between galling insect richness and tree richness (GIR/TR) in function of the landscape categories low igapó (LI), low várzea (LV), high igapó (HI), and high várzea (HV) at the Mamirauá and Amanã Sustainable Development Reserves, Brazil

Table 7.3 Simple regression values to the relationships among galling insect richness (GIR) and tree richness in the different forest types and habitats (quotas), at the Mamirauá and Amanã Sustainable Development Reserves, Brazil

Forest/habitat	Intercept	Coefficient	Adj R^2	N	t	P
Várzea	2.9016	1.4764	0.8199	14	7.759	5.14e-06
Igapó	−8.6006	2.9419	0.6974	14	5.564	0.000123
Low quota	8.1670	1.2229	0.7479	15	2.585	0.0227
High quota	2.2370	1.8774	0.6657	13	4.989	0.000409

Adj R^2 adjusted R-squared; *N* number of samples; *t*, *t*-test values

HV = 1.8 ± 0.4, LV = 1.6 ± 0.3, Fig. 7.7). The results of the analysis of covariance showed that the forest type (várzea and igapó forests) influenced the GIR/TR ratio ($F_{1,24} = 12.896$, $p = 0.001$). However, habitat (low and high quota) and interaction between habitat and forest type did not affect this variable, despite the visual difference between flood quotas (habitat, $F_{1,24} = 3.622$, $p = 0.069$; interaction, $F_{1,24} = 0.350$, $p = 0.559$).

7.3.5 *Plant Species Richness Hypothesis*

Galling insect richness – GIR – showed a significant strong positive relationship with the number of tree species sampled in the two types of flooded forests. The number of tree species was responsible for 82% of the variation in várzea GIR, while in the igapó a smaller value – 70% – was obtained (Table 7.3, Fig. 7.8a).

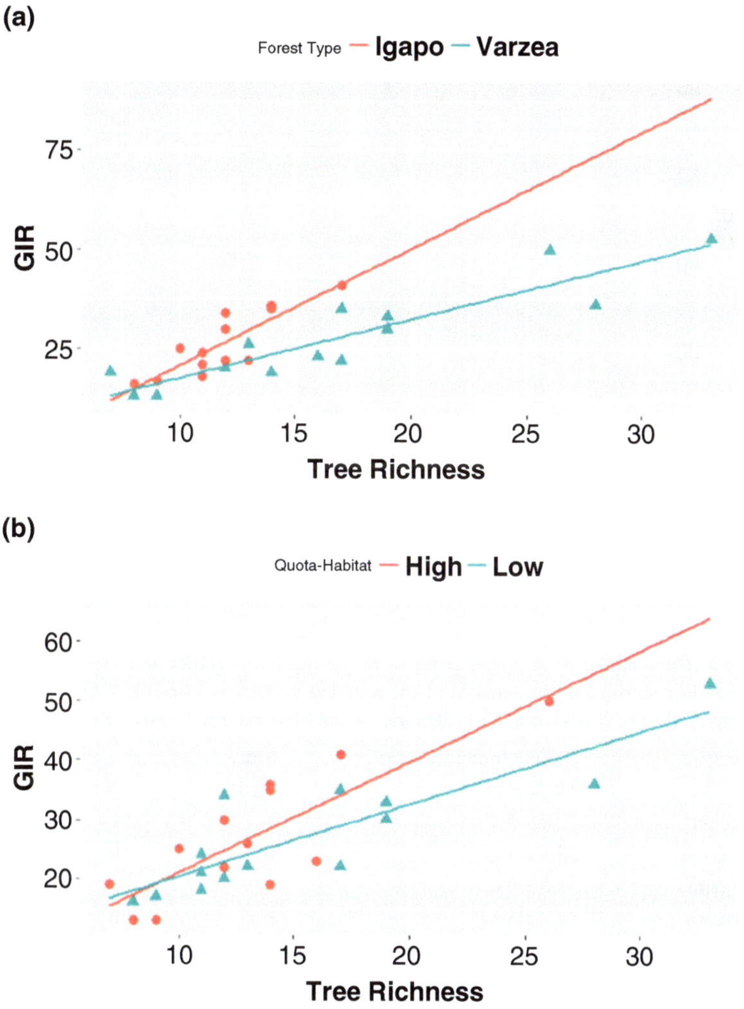

Fig. 7.8 Relationships between galling insect richness (GIR) and the richness of trees sampled (**a**) in the várzea and igapó forests and (**b**) in the high and low quota habitats at the Mamirauá and Amanã Reserves, Brazil

Considering only quota habitats, a similar pattern was verified regardless of the forest type (igapó or várzea). Approximately 75% of the variation in GIR was explained by the variation in the tree richness at the low-level quota; at the high-level habitat, the percentage of explanation was 67% (Table 7.3, Fig. 7.8b). The relationship – galling insect richness × tree richness – was maintained when considering the landscape categories (Fig. 7.9), as predicted by our hypothesis.

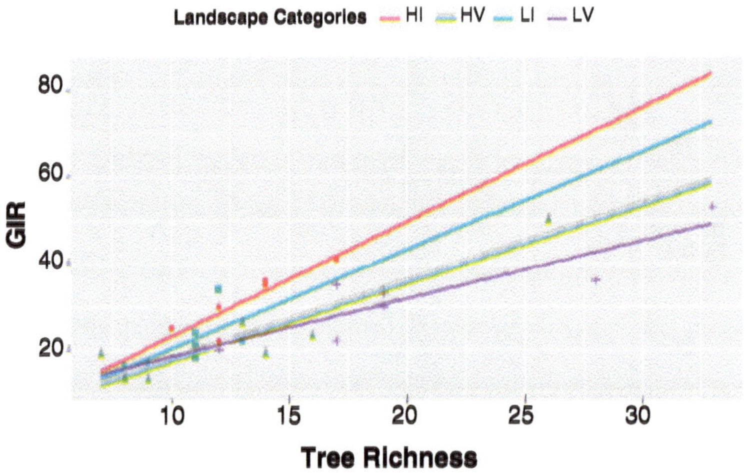

Fig. 7.9 Relationship between galling insect richness (GIR) and tree richness according to the landscape categories low igapó (LI), low várzea (LV), high igapó (HI), and high várzea (HV) at the Mamirauá and Amanã Reserves, Brazil

Table 7.4 Comparison of averaged richness of galling insect (GIR) and hypothetical stresses (hydric and/or nutritional) expected in plants sampled in the low igapó (LI), high várzea (HV), high igapó (HI), and low várzea (LV) categories at the Mamirauá and Amanã Reserves, Brazil

Landscape	Galling insect richness	Stress	
Categories	Mean ± sd	Hydric	Nutritional
Low igapó	22.0 ± 5.4	+	+
High várzea	23.3 ± 11.8	0	0
High igapó	31.5 ± 6.6	0	+
Low várzea	32.7 ± 10.1	+	0

7.4 Discussion

Our findings indicated that local nutrient availability (igapó and várzea), flood levels (high and low quota), and their interaction affect both insects and plants, as well as their ecological relationships, resulting in different richness patterns in the floodplain forests of Central Amazonia. However, most of the significant outcomes showed that nutritional stress is the main factor driving the variation in the galling insect richness associated with upper canopy in these forests. In its turn, the abundance of gall-inducing insects was not affected by flood quotas or even by the forest type.

It was also observed that the hydric and nutritional stresses do not act independently, and their influence on the host plants and galling insects was more remarkable when considering the average of galling insect richness *per* landscape category (Table 7.4, Fig. 7.10). Higher GIR (averaged for landscapes categories) were ascer-

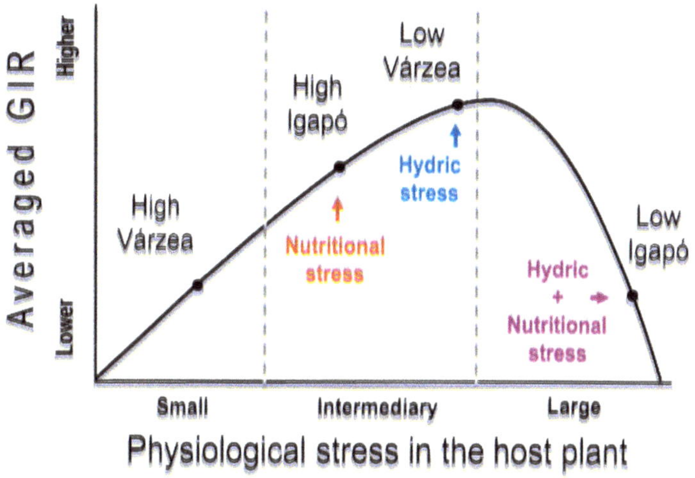

Fig. 7.10 Hypothetical scheme of the relationship between the real average values of galling insect richness (averaged GIR) to the supposed physiological stress levels in the host plants sampled at floodplain forests of Central Amazonia. (This figure was adapted from Schwartz et al. 2003)

tained with the intermediary range of hypothetical conditions of physiological stress in the host plants. The convergence of lower levels of hydric stress and higher levels of nutritional stress was associated with the second highest value of galling insect richness (High igapó, HI, Table 7.4). On the other hand, the highest averaged GIR was computed in the higher hydric stress and lower nutritional stress condition, in the low várzea (LV). These results indicate that galling insect assemblages can be affected by the amalgamation of environmental stress conditions, in which the responses and adaptations of host and also nonhost plants are the cornerstone of these relationships (Fig. 7.10).

As pointed by Schwartz et al. (2003), there is an optimal level of physiological stress of the host plant in which the establishment of galling insects would be favored. Adapted from the intermediate disturbance hypothesis (Begon et al. 1996), plants under severe stress conditions would not be prone to galling insect attack since their organs and tissues would have accelerated the process of senescence (Munne-Bosch and Alegre 2004). Moreover, since galling insects act as resource drains to the host plant (Larson and Whitham 1997; Isaias et al. 2014; Oliveira et al. 2016), it would not be advantageous for the insect to spend much of its life cycle, the larval stage, under nutritional limitations, which could affect its development and survival. On the other hand, host plants submitted to low levels of hydric and nutritional stress would be able to develop efficient defenses against gall-inducing insects (Fernandes 1990; Barbosa and Fernandes 2014). In virtue of great nutritional content, the tissues of these plants would be very attractive to other herbivorous insects, which could act as competitors and predators of galling insects (Fernandes et al. 2005; Barbosa and Fernandes 2014).

Thus, galling insects that live in host plants under intermediate stress levels could guarantee tissues with high nutritional contents and weakly protected against herbivores. De Bruyn et al. (2002) observed that leaf miner larvae showed higher developmental rates when their host plants grew in moist soils with intermediate levels of nutrients. Furthermore, the nutrient excess could trigger physiological stress in host plants and reduce the leaf miner performance (Scheirs and De Bruyn 2004).

Based on the data in Table 7.4, a hypothetical scheme illustrates the relationship between the levels of physiological stress in host plants and average richness of galling insects in the Amazonian ecosystem studied. In this scenario, low/high quotas in the várzea and igapó forests represent two hypotheses tested in our study (Fig. 7.10). It was assumed that there are no growth and development limitations for the várzea host plants. Since these forests are flooded by sediment- and nutrient-rich waters (Prance 1979; Junk 1984), and considering the nutritional stress hypothesis, we expected lower GIR in this forest type, compared to the igapó. Moreover, variability in the galling insect richness can be attributed to the flooding levels to which the forest plants are subjected (hydric stress hypothesis). Thus, host plants would not be affected by hydric stress in the high quotas of várzea forest ("small" stress on the *x*-axis in the Fig. 7.10), while tree roots remain submerged for a longer period in the low quota (Wittmann et al. 2004), which would lead to intermediary levels of stress.

On the other hand, low igapó forests are subject to both hydric stress (see Wittmann et al. 2004) and nutritional stress (Prance 1979; Junk 1984) and would represent the most inhospitable environment for host plants ("large" stress on the x-axis in the Fig. 7.10). In this range of the stress gradient, galling insects would not be favored, and consequently the host plants would present lower values of GIR. At high igapó, host plants would be subject only to nutritional stress, representing an intermediate level of stress. In this habitat, the second highest average value of galling insect richness was found. The diversity of species observed in vegetal communities of flooded forests is also supported by the intermediate disturbance hypothesis (Junk and Piedade 1997). The predictability of flood pulses would be responsible for the high number of species (Junk et al. 1989).

A great variability in the number of tree individuals and species was found among the landscape categories, low/high várzea and low/high várzea. The distribution and diversity of plant species in flooded forests have been related to three main factors: flood gradient, causing a floristic distinction between high and low levels, the process of natural succession (initial stages: poorest in number of species, higher densities, and monospecific stands), and the geographical distance between the sampled areas (Ferreira 2000; Wittmann et al. 2002; Haugaasen and Peres 2006; Wittmann et al. 2006). Furthermore, a higher degree of endemism can be found in the low and high quotas, as observed to várzea forests (Wittmann et al. 2010).

In our analyses, the effects of interaction between high and low habitat and forest type (igapó and várzea) evidently regulated the number of galling insect species (GIR). Subsequently, the results reveal the importance of the number of plant species sampled – the tree richness (TR) – and the forest type, indicating that the nutrient availability factor in the soil + water system is also preponderant. The ratio between GIR and TR reiterated the relevance of the number of plant species in a given area/habitat. In spite of the smaller number of tree species *per* sampling site, the igapó plants present a greater richness of galling insects, corroborating the hypothesis of nutritional stress.

In addition, two other factors confirm the tendency of a speciose fauna of galling insects in the igapó forests: (i) the relatively fewer occurrence of nonhost trees (plants not attacked by this insects) and (ii) the presence of "superhosts," host plants that shelter a diverse fauna of gallers (Veldtman and McGeoch 2003; Julião et al. 2014). Forty-three plants did not present insect galls in the low várzea, while in the high várzea only 21 tree individuals were not attacked. The number of nonhosts was smaller in the igapó forest and similar between quotas: 9 individuals in the high igapó and 12 in the low igapó. In the low igapó areas, at least three plant species were attacked by several species of galling insects (Julião et al. 2014), which were frequently sampled in the collection sites. Among them, the most predominant was the superhost species *Licania micrantha* Miq. (Chrysobalanaceae), attacked by ten galling insect species. Six gall morphotypes of *L. micrantha* were described in Fig. 7.11. For the purpose of showing the great morphological variability of insect galls, another two host plants were included (Fig. 7.11).

The hydric stress hypothesis was partially corroborated due to the interaction between the environmental variables and their correlation with plant richness, as a potential confounding factor. Such linkages suggest that the floristic composition should be incorporated in the analysis in order to understand the diversity patterns of gall-inducing insects. The environmental factors, pillars of our predictive hypotheses, simultaneously modulate the distribution and diversity patterns of these insects, as well as determine the composition of the local and regional flora. The availability of nutrients (clear and black waters) and flood levels determines the species-specific distribution and plant community zonation (Parolin et al. 2004; Wittmann et al. 2010). Therefore, high values of GIR and GIA may reflect only the presence of "superhosts" in a peculiar vegetation and/or mask the cryptic mechanisms responsible for the observed patterns (Julião et al. 2014). In any case, galling insect diversity in the trees from flooded forests such as várzea and igapó represents a large frontier to be examined; vertical distribution of gallers, seasonality, adaptations of gall structure to longer flooding, and absence of natural enemies remain obscure. Future studies should also address and quantify the chemical traits and sclerophylly of tree leaves, contrasting host and nonhost plants. These issues may contribute to the understanding of differential mechanisms of plant resistance and mortality of gall-inducing insects in the igapó and várzea forests of Central Amazonia.

Fig. 7.11 Insect galls found in the Amazonian igapó forest, Amanã Reserve, Brazil. On *Licania micrantha*: (**a**) stem gall induced by Coleoptera, (**b**) spherical/pubescent leaf galls induced by Cecidomyiidae (abaxial surface), (**c**) discoid leaf galls induced by Cecidomyiidae (abaxial surface), (**d**) conical leaf galls induced by Cecidomyiidae (abaxial surface), (**e**) spherical leaf galls induced by Cecidomyiidae (adaxial surface), (**f**) conical leaf galls induced by Cecidomyiidae (adaxial surface). On *Caraipa heterocarpa* (Calophyllaceae), (**g**) spherical leaf gall induced by Cecidomyiidae, and on *Brosimum lactescens* (Moraceae), (**h**) conical leaf galls induced by Cecidomyiidae

Acknowledgments We thank M.T. Piedade, S. Ribeiro, C. Fonseca, J. Almeida-Cortez, and H. Vasconcelos for helpful comments on the initial versions of this manuscript. J. Guedes, J. Ramos, E. Pereira, P. Assunção, and J. Inuma provided valuable technical assistance in several phases of this project. Funds were provided by the Program FEPIM 2003 from Mamirauá Sustainable Development Institute and Mamirauá Civil Society, which also provided all logistical support to the field work. G.R. Julião received a PhD grant from CNPq (Process Number: 141415/2003-7). GWF thanks CNPq and FAPEMIG for grant supports.

References

Adis J (1997a) Estratégias de sobrevivência de invertebrados terrestres em florestas inundáveis da Amazônia Central: uma resposta à inundação de longo período. Acta Amaz 27(1):43–54

Adis J (1997b) Terrestrial invertebrates: survival strategies, group Spectrum, dominance and activity patterns. In: Junk WJ (ed) The Central Amazon Floodplain. Ecological studies (analysis and synthesis), vol 126. Springer, Berlin, Heidelberg

Ayres JM (1993) As Matas de Várzea do Mamirauá. Estudos de Mamirauá, vol 1. Sociedade Civil Mamirauá, Brasilia

Barbosa M, Fernandes GW (2014) Bottom-up effects on gall distribution. In: Fernandes GW, Santos JC (eds) Neotropical insect galls. Springer, New York, pp 99–113

Begon M, Harper JL, Townsend CR (1996) Ecology, 3rd edn. Osney Mead, Oxford

Bell AD, Bell A, Dines TD (1999) Branch construction and bud defense status at canopy surface of a West African rainforest. Biol J Linn Soc 66:481–499

Blanche KR, Westoby M (1995) Gall-forming insect diversity is linked to soil fertility via host plant taxon. Ecology 76:2334–2337

Carneiro MAA, Branco CSA, Braga CED, Almada ED, Costa MBM et al (2009) Are gall midge species (Diptera: Cecidomyiidae) host plant specialists? Revista Brasileira de Entomologia 53:365–378

Cogni R, Fernandes GW, Vieira DLM, Marinelli CE, Jurinitz CF, Guerra BR, Zuanon J, Venticinque EM (2003) Galling insects (Diptera: Cecidomyiidae) survive inundation during host plant flooding in Central Amazonia. Biotropica 35:155–119

Cuevas-Reyes P, Quesada M, Siebe C, Oyama K (2004) Spatial patterns of herbivory by gall-forming insects: a test of the soil fertility hypothesis in a Mexican tropical dry forest. Oikos 107:181–189

De Bruyn L, Scheirs J, Verhagen R (2002) Nutrient stress, host plant quality and herbivore performance of a leaf-mining fly on grass. Oecologia 130:594–599

Fernandes GW (1990) Hypersensitivity: a neglected plant resistance mechanism against insect herbivores. Environ Entomol 19:1173–1182

Fernandes GW (1992) Plant historical and biogeographical effects on insular gall-forming species richness. Lett Glob Ecol Biogeogr 2:71–74

Fernandes GW, Price PW (1988) Biogeographical gradients in galling species richness: test of hypotheses. Oecologia 76:161–167

Fernandes GW, Price PW (1991) Comparisons of tropical and temperate galling species richness: the role of environmental harshness and plant nutrient status. In: Price PW, Lewinsohn TM, Fernandes GW, Benson WW (eds) Plant-animal interactions: evolutionary ecology in tropical and temperate regions. John Wiley, New York, pp 91–116

Fernandes GW, Price PW (1992) The adaptive significance of insect gall distribution: survivorship of species in xeric and mesic habitats. Oecologia 90:14–20

Fernandes GW, Gonçalves-Alvim SJ, Carneiro MAA (2005) Habitat-driven effects on the diversity of gall-inducing insects in the Brazilian cerrado. In: Raman A, Schaefer CW, Withers TM (eds) Biology, ecology, and evolution of gall-inducing arthropods. Science Publishers, Enfield, NH, pp 693–708

Ferreira LV (2000) Effects of flooding duration on species richness, floristic composition and forest structure in river margin habitat in Amazonian Blackwater floodplain forest: implications for future design of protected areas. Biodivers Conserv 9:1–14

Gonçalves-Alvim SJ, Fernandes GW (2001) Biodiversity of galling insects: historical, community, and habitat effects in four tropical savannas. Biodivers Conserv 10:79–98

Haugaasen T, Peres CA (2006) Floristic, edaphic and structural characteristics of flooded and unflooded forests in the lower Rio Purús region of Central Amazonia, Brazil. Acta Amazon 36:25–36

Isaias RMS, Oliveira DC, Carneiro RGS, Kraus JE (2014) Developmental anatomy of galls in the neotropics, arthropods stimuli versus host plant constraints. In: Fernandes GW, Santos JC (eds) Neotropical insect galls. Springer Netherlands, Dordrecht, pp 15–34

Julião GR, Venticinque EM, Fernandes GW (2005) Richness and abundance of gall-forming insects in the Mamirauá Várzea, a flooded Amazonian forest. Uakari 1:39–42 http://www.mamiraua.org.br/uakari/home.htm

Julião GR, Venticinque EM, Fernandes GW, Price PW (2014) Unexpected high diversity of galling insects in the Amazonian upper canopy: the savanna out there. PLoS One 9:e114986. https://doi.org/10.1371/journal.pone.0114986

Junk WJ (1984) Ecology of the várzea, floodplain of Amazonian white water rivers. In: Sioli H (ed) The Amazon: Limnology and landscape ecology of a mighty tropical river and its basin. W. Junk, Dordrecht, pp 215–243

Junk WJ (1993) Wetlands of tropical South America. In: Whigham DF, Dykyjova D, Hejny S (eds) Wetlands of of the World I. Kluwer Academic Publishers, Dordrecht, pp 679–739

Junk WJ, Piedade MTF (1997) Plant life in the floodplain with special reference to herbaceous plants. In: Ecological Studies, Junk (ed) The Central Amazon floodplain. Springer – Verlag, Berlen/Heidelberg, vol. 126, pp. 147–185

Junk WJ, Bayley PB, Sparks R (1989) The flood pulse concept in river-floodplain systems. In: Dodge DP (ed). Proceedings of the international large river Symposium.Can.Spec.Publ.Fish.Aquat.Sci., vol. 106, pp. 110–127

Junk WJ, Piedade MTF, Wittmann F, Schöngart J, Parolin P (2010) Amazonian floodplain forests: ecophysiology, biodiversity and sustainable management. Springer, Dordrecht

Junk WJ, Piedade MTF, Schöngart J, Cohn-Haft M, Adeney JM, Wittmann F (2011) A classification of major naturally-occurring Amazonian lowland wetlands. Wetlands 31:623–640. https://doi.org/10.1007/s13157-011-0190-7.

Lara ACF, Fernandes GW, Gonçalves-Alvim SJ (2002) Tests of hypotheses on patterns of gall distribution along an altitudinal gradient. Trop Zool 15:219–232

Larson KC, Whitham TG (1997) Competition between gall aphids and natural plant sinks: plant architecture affects resistance to galling. Oecologia 109:575–581

Maia LA, Piedade MTF (2000) Phenology of *Eschweilera tenuifolia* (Lecythidaceae) in flooded forest of the Central Amazonia – Brazil. Neotropical ecosystems. Proceedings of the German-Brazilian Workshop, Hamburg

Mamirauá Institute for Sustainable Development – MISD (2018) Reserves: Mamirauá and Amanã. www.mamiraua.org/en-us

Munne-Bosch S, Alegre L (2004) Die and let live: leaf senescence contributes to plant survival under drought stress. Funct Plant Biol 31:203–216

Oliveira DC, Isaias RMS, Fernandes GW, Ferreira BG, Carneiro RGS, Fuzaro L (2016) Manipulation of host plant cells and tissues by gall-inducing insects and adaptive strategies used by different feeding guilds. J Insect Physiol 84:103–113. https://doi.org/10.1016/j.jinsphys.2015.11.012

Parolin P, Wittmann F (2010) Struggle in the flood: tree responses to flooding stress in four tropical floodplain systems. AoB plants, 2010, plq003. https://doi.org/10.1093/aobpla/plq003

Parolin P, De Simone O, Haase K, Waldhoff D, Rottenberger S, Kuhn U, Kesselmeier J, Kleiss B, Schmidt W, Piedade MTF, Junk WJ (2004) Central Amazonian floodplain forests: tree adaptations in a pulsing system. Bot Rev 70:357–380

Prance GT (1979) Notes on the vegetation of Amazonia III. The terminology of Amazonian forest types subject to inundation. Brittonia 3:26–38
Price PW, Fernandes GW, Lara ACF, Brawn J, Barrios H, Wright MG, Ribeiro SP, Rothcliff N (1998) Global patterns in local number of insect galling species. J Biogeogr 25:581–591
R Core Team (2017) R: a language and environment for statistical computing. R Foundation for Statistical Computing, Vienna, Austria. https://www.R-project.org/
Ribeiro KT, Madeira JA, Monteiro RF (1998) Does flooding favour galling insects? Ecol Entomol 23:491–494
Scheirs J, De Bruyn L (2004) Excess of nutrients results in plant stress and decreased grass miner performance. Entomol Exp Appl 113:109–116
Schwartz G, Hanazaki N, Silva MB, Izzo TJ, Bejar MEP, Mesquita MR, Fernandes GW (2003) Evidence for stress hypothesis: hemiparasitism effect on the colonization of *Alchornea castaneaefolia* A.Juss. (Euphorbiaceae) by galling insects. Acta Amazon 33:275–279
Southwood TRE (1960) The abundance of Hawaiian trees and the number of their associated insect species. Proc Hawaiian Entomol Soc 17:229–303
Veldtman R, McGeoch MA (2003) Gall-forming insect species richness along a non-scleromorphic vegetation rainfall gradient in South Africa: the importance of plant community composition. Austral Ecol 28:1–13
Wittmann F, Anhuf D, Junk WJ (2002) Tree species distribution and community structure of central Amazonian várzea forests by remote sensing techniques. J Trop Ecol 18:805–820
Wittmann F, Junk WJ, Piedade MTF (2004) The várzea forests in Amazonia: flooding and the highly dynamic geomorphology interact with natural forest succession. For Ecol Manag 196:199–212
Wittmann F, Schöngart J, Montero JC, Motzer T, Junk WJ, Piedade MTF, Queiroz HL, Worbes M (2006) Tree species composition and diversity gradients in white-water forest across the Amazon basin. J Biogeogr 33:11334–11347
Wittmann F, Schöngart J, Junk WJ (2010) Phytogeography, species diversity, community structure and dynamics of Amazonian floodplain forests. In: Junk WJ, Piedade MTF, Wittmann F, Schöngart J, Parolin P (eds) Amazonian floodplain forests: ecophysiology, biodiversity and sustainable management. Springer Verlag, Berlin, pp 61–102

Part V
Vertebrates

Chapter 8
Primates of Igapó Forests

Adrian A. Barnett and Thays Jucá

8.1 Introduction

Igapó is a long-overlooked habitat whose importance, as shown in other chapters in this book, is only now being fully appreciated. For primates, the number and extent of field studies are far less in igapó than in either never-flooded terra firme or in várzea, igapó's much more productive hydrogeographic twin that borders Amazonia's whitewater rivers (e.g., the Solimões and affluents: Goulding et al. 2003). Although igapó is present on the edges of some of the neotropics largest rivers, such as the blackwater Negro and the clearwater Xingu and Tapajós (Goulding et al. 2003; Prance 1979), and covers an area of some 100,000 km^2, there have been relatively few studies of primates in igapó (as indeed, is true for mammals in general). The studies concerned are listed in Table 8.1. Their geographical distribution is shown in Fig. 8.1.

The reasons for the paucity of studies appear to be linked primarily to the methodological difficulties associated with studies in flooded forests, especially the canoe-based studies that the flooded season inundations necessitate. The modifications to standard canonical methods that are consequently required appear to be off-putting, both to field investigators and to the agencies that fund them (Pinto et al. 2013; Barnett et al. 2018a).

A. A. Barnett (✉)
Department of Zoology, Universidade Federal do Amazonas, Manaus, AM, Brazil

Amazon Mammal Research Group, Biodiversity Studies,
Instituto Nacional de Pesquisas da Amazonia, Manaus, AM, Brazil

Centre for Evolutionary Anthropology, University of Roehampton, London, England

T. Jucá
Amazon Mammal Research Group, Biodiversity Studies, Instituto Nacional de Pesquisas da Amazonia, Manaus, AM, Brazil

R. W. Myster (ed.), *Igapó (Black-water flooded forests) of the Amazon Basin*,
https://doi.org/10.1007/978-3-319-90122-0_8

Table 8.1 Summary of primate studies in Amazonian igapó forests[a, b]

Authors	Location	Study type	Species reported from igapó
Antunes et al. (in preperation)	Jaú National Park, Central Amazonas state, Brazil (site 1 on map)	Ecology, use of forest floor when unflooded	*Cebus albifrons*, *Sapajus apella*
Ayres and Milton (1981)	Tapajós National Forest, Tapajós River, western Pará state, Brazil (site 2 on map)	Species survey	
Barnett (2005); Barnett and Bezerra (submitted)	Jaú National Park, Central Amazonas state, Brazil (site 1 on map)	Survey and species dietary and habitat use ecology	*Alouatta juara*, *Aotus vociferans*, *Cacajao ouakary*[c], *Cebus megacephalus*, *Pithecia chrysocephala*, *Saimiri sciureus*, *Sapajus apella*
Barnett (2010) and subsequent papers	Jaú National Park, Central Amazonas state, Brazil (site 1 on map)	Ecology: diet, foraging, and habitat use	*Cacajao ouakary*[c]
Barnett and da Cunha (1991)	Uaupes(3a) & Curicuriari (3b) rivers, western Amazonas state, Brazil	Species surveys, ecology	*Cacajao ouakary*
Barnett et al. (2017); Barnett, Stone et al. (in press).	Tapajós River, western Pará state, Brazil (site 4 on map)	Predation studies	*Chiropotes albinasus*
Barnett, Cavalcanti et al. (submitted)	Tapajós River, western Pará state, Brazil (site 4 on map)	Species surveys, ecology	*Alouatta nigerrima*, *A. ululata*, *Ateles chamek*, *A. marginatus*, *Callicebus hoffmansii*, *Cebus albifrons*, *Chiropotes albinasus*, *Pithecia irrorata*, *Saimiri ustus*, *Sapajus apella*
Benchimol and Peres (2015)	Islands in the reservoir of the Balbina Dam, North-Central Amazonas state, Brazil (site 5 on map)	Ecology, impact of habitat fragmentation on population survival on recently created islands in dam reservoir	*Alouatta macconnelli*, *Ateles paniscus*, *Chiropotes sagulatus*, *Pithecia chrysocephala*, *Saguinus midas*, *Sapajus apella*
Bezerra (2010) and subsequent papers	Jaú National Park, Central Amazonas state, Brazil (site 1 on map)	Species social ecology	*Cacajao ouakary*[d]

(continued)

Table 8.1 (continued)

Authors	Location	Study type	Species reported from igapó
Boubli and de Lima (2009)[e]	Rio Araça, north-Central Amazonas state, Brazil (site 6 on map)	Species surveys reported	*Cacajao melanocephalus*[g]
Defler (1999) and other papers	Apaporis River, Vaupes department, eastern Amazonian Colombia (site 7 on map)	Ecology and social biology	*Cacajao ouakary*
Dia da Silva RHP (2017) and subsequent papers	Jaú National Park, Central Amazonas state, Brazil (site 1 on map)	Foraging ecology	*Cacajao ouakary*[d]
Haugaasen and Peres (2005)	Purus River, western Pará state, Brazil (site 8 on map)	Community-level interactions	*Alouatta seniculus*, *Ateles chamek*, *Cheracebus torquatus*[h], *Cebus albifrons*, *Lagothrix lagotricha*, *Pithecia albicans*, *Sapajus apella*, *Saimiri ustus*
Lehman and Robertson (1994)	Southern Amazonas state, southern Venezuela (site 9 on map)	Species survey	*Cacajao melanocephalus*
Negreiros (2017) and subsequent papers	Jaú National Park, Central Amazonas state, Brazil (site 1 on map)	Multi-species ecological interactions	*Cacajao ouakary*[d]
Rocha (2016) and subsequent papers	Açutuba forest, Rio Negro, Central Amazonas state, Brazil (site 10 on map)	Community-level responses to anthropic change, species diet, and habitat use	*Alouatta juara*, *Aotus vociferans*, *Cacajao ouakary*[d], *Cebus megacephalus*, *Saimiri sciureus*, *Sapajus apella*

Notes: [a]Studies such as those of Ayres (1986), Bowler (2006), and Paim et al. (2013) occurred in várzea and not igapó

[b]The table does not include studies such as those of Spironello (2001) that studied the use by primates on flooded vegetation on the margins of blackwater streams within terra firme forest, as this does not accord with the definition of igapó given by Prance (1979)

[c]Taxonomy of the genus *Cacajao* follows Ferrari et al. (2014)

[d]Reported as *C. melanocephalus*

[e]Thesis work on *Cacajao melanocephalus* conducted by JP Boubli (e.g., Boubli 1993, 1997, 1999; Boubli and Tokuda 2008[f]) occurred in unflooded white-sand forests (campina) and not igapó and so strictly lies outside the purview of this chapter

[f]Reported as *Cacajao hosomi* (but see Ferrari et al. (2014) regarding the validity of this name)

[g]Reported as *Cacajao ayresi* (but see Ferrari et al. (2014) and Bertuol (2015) regarding the validity of this name)

[h]*Callicebus torquatus*

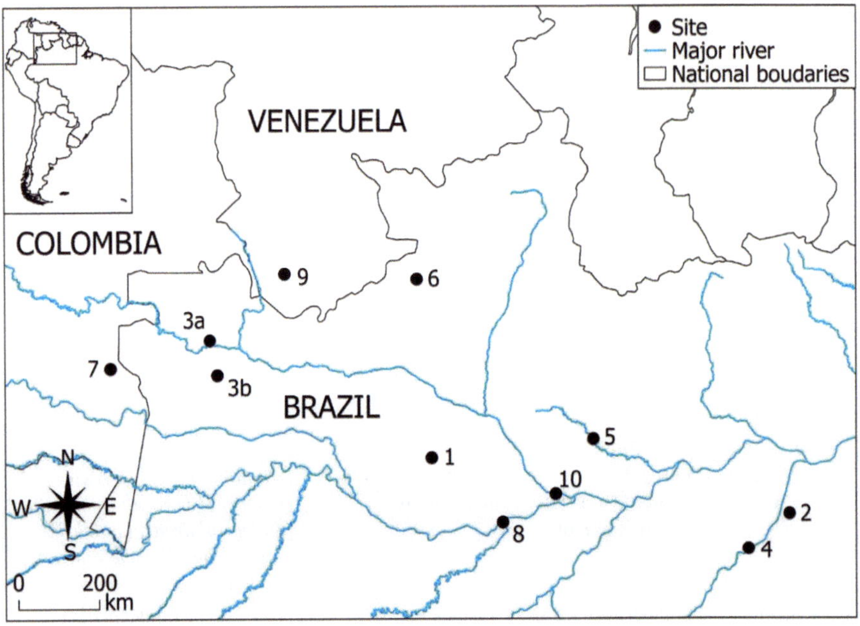

Fig. 8.1 Distribution of primate studies in igapó forests in Brazil, Colombia, and Venezuela where 1 = Jaú National Park, 2 = Tapajós National Forest, 3a = Uaupes River, 3b = Curicuriari River, 4 = Tapajós River, 5 = Balbina dam reservoir, 6 = Rio Araça, 7 = Apaporis River, 8 = Purus River, 9 = southern Amazonas State and Venezuela, and 10 = Açutuba forest

8.2 Summary of Existing Knowledge

Of the many species of Amazonian primate that use igapó (Table 8.1), only *Cacajao ouakary* appears to have this habitat as its primary habitat (Barnett et al. 2018 a, b). Others, such as *Cebus albifrons* and *Chiropotes albinasus*, may have populations that use igapó extensively, while other populations of the same species rarely, if ever, visit flooded forests, remaining strictly in terra firme (Barnett et al. 2018c). Some, such as species of *Alouatta*, appear to have occasional populations that are restricted entirely to igapó on isolated riverine islands (e.g., Rocha 2016 for *A. juruensis*). Other taxa, such as *Pithecia* spp. and *Sapajus* spp., appear to visit only to exploit specific resources that are available briefly as a result of the marked seasonal pulsing of the igapó resource profile (see Ferreira 2002; Parolin et al. 2002; Scudeller 2018).

The extent of our understanding of the ecology and natural history of igapó primates varies greatly. Some taxa, such as species of night monkey (*Aotus*), are known to occur in igapó but have not been the subject of any form of ecological study. Others, such as members of the genera *Alouatta*, *Ateles*, *Cebus*, *Pithecia*, and *Sapajus*, have been the subject of general surveys and community-level studies (e.g., Haugaasen and Peres 2005; Rocha 2016), or general natural history treatments (Barnett et al. 2002; Barnett and Bezerra submitted; Barnett et al. submitted). Very few have been the focus of specific autecological and synecological studies. Of these, *Chiropotes albinasus* has been the subject of studies of predation by raptors (Barnett et al. 2017), and of its response to predators and pseudopredators (Barnett et al. 2018d), while predation of wasp nests by *Cebus albifrons* in igapó forests has been analyzed by Barnett et al. (submitted - a), and Jucá et al. (in review) have analyzed nocturnal and diurnal resting sites of *Alouatta nigerrima* and *A. ululata* in igapó in terms of the different predators likely to attack them. The species of igapó-based primate to have received the most intensive studies to date is the golden-backed uacari, *Cacajao ouakary*, studies on which are summarized in Table 8.2.

8.2.1 Seasonal Use

Terra firme and igapó forests are congruent habitats, with little ecotone and few tree species in common (Barnett et al. 2015b). However, their phenologies have two asynchronous peaks of leaf flush and fruit production (e.g., Haugaasen and Peres 2006; Barnett 2010, 2012a, b), and, consequently, seasonal movements of mammal species occur between the two forest types, as individuals track these resources. This not only occurs with primates (Haugaasen and Peres 2005), but also with other arboreal mammals (e.g., Haugaasen and Peres 2007), as well as bats (Bobrowiec et al. 2014; Marques et al. 2012) and terrestrial rodents (Antunes et al. 2017). Such movement most commonly follows the flooded forest fruit production peaks. These follow seasonal changes in river levels, since the majority of trees, palms, and lianas are either fish- or water-dispersed (Anderson et al. 2009; Correa et al. 2007). As a result, canopies of unflooded igapó contain little fruit, and many primate species move to the adjacent terra firme. The move is not a simple all-or-nothing affair. Species that remain switch diets and eat higher proportions of insects and leaves and fewer fruits, as resource availability changes (e.g., Barnett et al. 2013b for *C. ouakary*, a species where group sizes also change with resource availability profile shifts (Barnett 2010)), and visit terra firme only briefly during the year. In contrast, on the Purus River, Haugaasen (2004) reported *Ateles chamek*, *Cebus albifrons*, *Lagothrix lagotricha*, and *Pithecia albicans* visited igapó sporadically to exploit pulses of fruit availability. Such primates appear to be primarily terra firme species that make occasional use of flooded habitats. Species that appear to remain in igapó year-round tend to have small home ranges and reliable food sources (e.g., *Aotus* sp., insects – AA. Barnett, unpublished data).

In addition, the extent of movement varies between sites; while *C. ouakary* on the Apaporis River, Colombia, seasonally travel many km into terra firme, the same species in Jaú National Park, Brazil (some 900 km east), have never been recorded more than 500 m from igapó. Other differences between sites also exist: at Jaú less than 4% of *C. ouakary* group encounters recorded associations with other primate, while on the Apaporis such encounters were very common, sometimes lasting several days (Barnett et al. 2013a). It is unknown whether such differences reflect different resource availability patterns or regional traditions (Barnett et al. 2013a).

Table 8.2 Studies of the aut- and synecology of *Cacajao ouakary*, the golden-backed uacari

Author(s)	Topic summary
Barnett (2010)	Diet, seasonal habitat use, feeding and social ecology, conservation biology
Barnett et al. (2011)	Predation by harpy eagles on *C. ouakary* – Event reported, plus reactions during and after detailed. Contrasted with reactions of *Cebus albifrons* to same events in igapó
Barnett et al. (2012a)	Terrestrial foraging in season when igapó forest floor is unflooded shows sensitivity to possibility of attack by predators
Barnett et al. (2012b)	Primary seed dispersal occurs despite the species being considered a seed predator, as many eaten fruits multi-seeded and discarded seeds subsequently germinate
Barnett et al. (2012c)	Sleeping site trees have features indicating they are chosen to minimize risk of predator attack
Barnett et al. (2013a)	Ecological summary and between-site variation in diet and social organization of *C. ouakary* populations in Brazil and Colombia
Barnett et al. (2013b)	Insectivory in *C. ouakary* was reported and noted as the most common in parts of the year when fruits availability was lowest
Barnett and Shaw (2014)	Effect of presence of *C. ouakary* on foraging success of two guilds of insectivorous birds: sit-and-wait predators and bark-crevice foragers
Barnett et al. (2015a)	Presence of *Pseudomyrmex* ants on *Macrolobium acaciifolium* (Fabaceae) plants reduced incidence of *C. ouakary* seeds predation
Barnett et al. (2016)	*C. ouakary* selectively bites hard fruits at their weakest points, facilitating entry, saving energy, and minimizing risk of canine tip damage
Bezerra (2010)	Vocal repertoire, social interactions, and group structure in *C. ouakary*[a]
Bezerra et al. (2010a)	*C. ouakary*[a] response to playback of conspecific, aggressive, and contact calls
Bezerra et al. (2010b)	Vocal repertoire of *C. ouakary*[a] defined nine call types, including the *tchó* contact and alarm call and play-specific calls
Bezerra et al. (2011a)	Ethogram of *C. ouakary*[a] defined nine behavioral categories and noted low levels of aggression and social behaviors such as grooming. Also noted fission-fusion sociality and extensive home range
Bezerra et al. (2011b)	Complexity of *C. ouakary* vocal repertoire and its confounding of theoretical predictions of relationships between call length and frequency of use
Deler (1991)	Observations on the diet and habitat use of *C. ouakary* in Colombia
Defler (1999)	Group sizes change seasonally in respond to resource availability
Defler (2001)	Report on *C. ouakary* densities and seasonal movement between igapó and terra firme habitats
Dias da Silva et al. (submitted)	*C. ouakary* selects size-weight combinations of *Aldina latifolia* (Fabaceae) pods to optimize processing time
Mourthé and Barnett (2014)	Response of *C. ouakary* to predators and pseudopredators (harmless species that resemble predators) and the reason why some category confusion exists
Negreiros et al. (2018)	When eating young leaves of *Buchenavia ochroprumna* (Combretaceae), *C. ouakary* avoids those canopies infested with caterpillars of Heliconiid butterflies

(continued)

Table 8.2 (continued)

Author(s)	Topic summary
Negreiros et al. (submitted)	Canopies infested with Heliconiid caterpillars are, on average, larger than those of uninfested trees. Consequently, individual *C. ouakary* is forced to forage from smaller canopies in order to avoid the toxic caterpillars and the phytochemical changes they induce in the *B. ochroprumna* leaves
Rocha (2016)	Tolerance of human impacts on flooded forest ecology; diet, density, seasonal ranging patterns

[a]As *Cacajao melanocephalus*

One complicating element is the use of terrestrial resources by primates when igapó is unflooded. At such times there may be substantial volumes of fruits and germinating seeds on the dried-out forest floor (Antunes et al. 2017; Barnett et al. 2012a). Such resources attract a variety of graminivores (rodents) and herbivores (tapir, deer) (Antunes et al. submitted), as well as *Cebus albifrons* (Antunes et al. in pre-peration) and *Cacajao ouakary* (Barnett et al. 2012a), which descend to the ground specifically to forage on fallen fruits, so such movements might be delayed or not occur equally in all populations depending on whether this pattern of resource exploitation is part of the local repertoire (Barnett et al. 2013a; Perry et al. 2003).

Because of the relative sediment loads of the rivers which flood them, the fertility of igapó soils is much lower than those of várzea (Furch 1997; Irion 1978), leading to less productive forests (Cintra et al. 2013; Junk et al. 2011). This has been observed to have effects on plant community structure and composition (Haugaasen and Peres 2006), growth rates of individual trees and species (da Fonseca et al. 2009), leaf nitrogen levels (Kreibich et al. 2002), and seed mass (Parolin 2000). On the Purus River, igapó and várzea are in close proximity, allowing Haugaasen and Peres (2005) to make direct comparisons between the two habitat types for primate species density and biomass, and with terra firme. They found the várzea primate assemblage biomass to be more than twice that of terra firme and that both were much higher than igapó.

8.2.2 Species Use of Igapó

Studies of use of flooded and non-flooded forest types by Amazonian primate assemblages reveal that few primate species completely avoid such habitats; there is widespread use of both habitats on a seasonal, near-permanent, or permanent basis. Studying a 12-species strong primate community on the Purus River, central Brazilian Amazon, Haugaasen and Peres (2005) found that only 1 species (*Callicebus cupreus*) did not visit igapó. At Jaú, Barnett et al. (2002) and Barnett (2010) recorded only three of eight primate species did not visit igapó (*Pithecia chrysocephala, Saguinus inustus, Sapajus apella*). As might be expected from a habitat that floods so profoundly, species absent from flooded forests are often those with preferences for the understory and mid-story, notably members of the genus *Saguinus* (Peres 1997). In flooded forests their role as insectivores may be occupied by species of *Saimiri.* Members of this genus have a slightly broader diet, and are highly flexible over which parts of the vegetation column they use, and are often at their most abundant densities in riverine forests (Peres 1993, 1997b; Peres et al. 1996).

However, there are between-site differences for such aspects; so, while *Sapajus apella* was one of the commonest species in flooded forests on the Rio Purus (Haugaasen and Peres 2005), it was never recorded entering igapó at Jaú (Barnett et al. 2002; Barnett and Bezerra submitted). The reasons for such differences are currently unexplored. However, in some cases the physical nature of the habitat may be involved. Thus, while southern red howlers (*Alouatta juara*) were rarely seen in igapó at Jaú, they had resident populations at Açutuba, some 220 km to the east. But the latter were resident on large isolated islands of igapó, which may have made it difficult for any seasonal migration to occur. On the Tapajós, Jucá et al. (submitted) report year-round use of igapó by two species of *Alouatta.* However, this was largely a result of the positioning of nocturnal sleeping sites in igapó trees; feeding generally occurred in the adjacent terra firme. Such quotidian bi-habitat use may be facilitated by very narrow nature of igapó on clearwater rivers such as the Tapajós, where such forest is rarely more than 10 m wide (Ferreira and Prance 1998), in contrast to the 100–300 m for igapó on the Rio Negro (Rocha 2016), and several kilometers extension for the várzea floodplains (Paim et al. 2013).

In other cases, species patterns of habitat use may be structured by competition: noting the predominance of unripe seeds from hard-husked fruits in the diets of the two Pitheciinae genera, *Cacajao* and *Chiropotes*, Ayres (1986, 1989) proposed that members of the two genera were ecological competitors, but where their ranges overlapped, they coexisted via habitat choice. However, it is now known that members of another Pitheciinae genus, *Pithecia*, may also be canopy-dwelling specialists in such resources (Norconk 2007). Where these genera overlap geographically, one species will often occur in terra firme, while another will use mostly igapó (e.g., *P. chrysocephala* and *C. ouakary* at Jaú), whereas on the Purus, where it is the only pitheciid, *P. albicans* uses flooded forest extensively (Haugaasen and Peres 2005). On the Tapajós River, *Chiropotes albinasus* occurs with *Pithecia mittermeieri* on the west and *Pithecia* sp. on the eastern bank; in 19 months only *C. albinasus* was ever seen in the igapó (A. Barnett unpublished data). This mirrors the observation by Boubli (1999) and Boubli and de Lima (2009) of *Cacajao* in north-west Amazonia occurring in non-flooded forests only in areas from which *Chiropotes* were absent, but being restricted to them where the two general were geographically congruent. On the Solimões River, the recently described *Pithecia* species (*P. cazuzai* and *P. vanzolinii*) have the same broad habitat preference (várzea) as *Cacajao calvus* (Marsh 2014). Though it has not yet been studied, they would be expected either to be using the cavascal and retsinga components of this habitat differentially or sharing the várzea but using different species of this highly productive habitat.

8.3 Recommendations for Future Studies

Topics:

- Comparisons of fruit and leaf crop volumes (not just phenology) between igapó and adjacent terra firme
- Comparisons of concentrations of secondary compounds (especially tannins and phenols) in leaves and fruits in terra firme and igapó congeneric plants
- Studies of group size and range size of folivorous and frugivorous primate species using the above information as key causal drivers
- Comparisons of primate communities in igapó forests with low and high levels of human impact (including and excluding hunting)
- Ecology of primates on mid-river islands
- Multisite estimates of range sizes and densities for all primate taxa inhabiting igapó, to allow population density estimations to be made
- Studies of seasonal migration of primates into and out of igapó, to distinguish between primarily igapó-based taxa that migrate seasonally to terra firme and terra firme-based species (or populations) that use igapó seasonally
- Overlap in diet between primates and other vertebrates (including birds and fish) in igapó
- Studies of the influence of influence of W-E fertility gradients on igapó tree species diversity and ecology and the knock-on effects this has on primate ecology (including diet breadth, seasonality, range size)

References

Anderson JT, Rojas JS, Flecker AS (2009) High-quality seed dispersal by fruit-eating fishes in Amazonian floodplain habitats. Oecologia 161:279–290

Antunes AC, Baccaro F, Barnett AA (2017) What bite marks can tell us: use of on-fruit tooth impressions to study seed consumer identity and consumption patterns within a rodent assemblage. Mamm Biol 82:74–79

Antunes AC, Bezerra BM, Baccaro F, Barnett AA (In preparation) Terrestrial use of central Amazonian seasonally-flooded forest by *Cebus albifrons* and *Sapajus apella*. For J. Trop Ecol

Antunes AC, Bacarro F, Andrade V, Ramos J, Barnett AA (Submitted) Igapó seed patches: An overlooked, but potentially key resource for a vertebrate assemblage in a seasonally flooded forest of Amazonian Brazil? At J. Zool

Ayres JM (1986) Uakaris and Amazonian flooded Forest. PhD dissertation, University of Cambridge

Ayres JM (1989) Comparative feeding ecology of the uakari and bearded saki, *Cacajao* and *Chiropotes*. J Hum Evol 18:697–716

Ayres JM, Milton K (1981) Levantamento de primatas e habitat no Rio Tapajós. Boletim do Museu Paraense Emilio Goeldi (ciências biológicas) 111:1–11

Barnett AA (2005) Cacajao melanocephalus. Mamm Spec 776:1–6

Barnett AA (2010) Diet, habitat use and conservation ecology of the golden-backed uacari, *Cacajao melanocephalus ouakary*, in Jaú National Park, Amazonian Brazil. PhD Dissertation, Roehampton University, London

Barnett AA, Bezerra BM (Submitted) Primate assemblage in blackwater flooded forests in the Negro-Solimões riverine interfluve; diversity, ecology, and natural history'. At J Trop Ecol

Barnett AA, da Cunha AC (1991) The golden-backed uacari on the upper Rio Negro, Brazil. Oryx 25:80–88

Barnett AA, Shaw P (2014) More food or fewer predators? The benefits to birds of associating with a Neotropical primate varies with their foraging strategy. J Zool (Lond) 294:224–233

Barnett AA, Borges S, de Castilho CV, Neri F, Shapley RL (2002) Primates of Jaú National Park, Amazonas, Brazil. Neotropical Primates 10:65–70

Barnett AA, Schiel V, Deveny A, Valsko J, Spironello W, Ross C (2011) Predation on *Cacajao ouakary* and *Cebus albifrons* (Primates: Platyrrhini) by harpy eagles. Mammalia 75:169–172

Barnett AA, Almeida T, Spironello WR, Sousa Silva W, MacLarnon A, Ross C (2012a) Terrestrial foraging by *Cacajao melanocephalus ouakary* (Primates) in Amazonian Brazil: is choice of seed patch size and position related to predation-risk? Folia Primatol 83:126–139

Barnett AA, Boyle S, Pinto L, Lourenço WC, Almeida T, Sousa Silva W, Ronchi-Teles B, Bezerra B, Ross C, MacLarnon A, Spironello WR (2012b) Primary seed dispersal by three Neotropical seed-predators (*Cacajao melanocephalus, Chiropotes chiropotes* and *Chiropotes albinasus*). J Trop Ecol 28:543–555

Barnett AA, Shaw P, Spironello WR, MacLarnon A, Ross C (2012c) Sleeping site selection by golden-backed uacaris, *Cacajao melanocephalus ouakary* (Pitheciidae), in Amazonian flooded forests. Primates 53:273–285

Barnett AA, Bezerra BM, Oliveira M, Queiroz H, Defler TR (2013a) *Cacajao ouakary* in Brazil and Colombia: patterns, puzzles and predictions. In: Veiga L, Barnett A, Ferrari S, MNorconk M (eds) *Evolutionary Biology and Conservation of Titis, Sakis and Uacaris*. Cambridge University Press, Cambridge, pp 179–195

Barnett AA, Ronchi-Teles B, Almeida T, Sousa Silva W, Bezerra B, Deveny A, Schiel-Baracuhy V, Spironello W, Ross C, MacLarnon A (2013b) Arthropod predation by the golden-backed uacari, *Cacajao melanocephalus ouakary* (Pitheciidae), in Jaú National Park, Brazilian Amazonia. Int J Primatol 34:470–485

Barnett AA, Almeida T, Andrade R, Boyle S, Gonçales-Lima M, Sousa Silva W, Spironello WR, Ronchi-Teles B (2015a) Ants in their plants: *Pseudomyrmex* ants reduce primate, parrot and squirrel predation on *Macrolobium acaciifolium* (Fabaceae) seeds in Brazilian Amazonia. Biol J Linn Soc 114:260–273

Barnett AA, Silva WS, Shaw PJA, Ramsay RM (2015b) Inundation duration and vertical vegetation stratification: a preliminary description of the vegetation and structuring factors in Borokotóh (hummock igapó), an overlooked, high-diversity, Amazonian habitat. Nord J Bot 33:601–614

Barnett AA, Bezerra BM, Spironello WR, Shaw P, Ross C, MacLarnon A (2016) Foraging with finesse: a hard-fruit-eating primate selects weakest areas as bite sites. Am J Phys Anthropol 160:113–125

Barnett AA, Silla JM, de Oliveira T, Boyle SA, Bezerra BM, Spironello WR, Setz EZF, Soares R, de Albuquerque Teixeira S, Todd LM, Pinto LP (2017) Run, hide or fight: anti-predation strategies in endangered red-nosed cúxiu (*Chiropotes albinasus*, Pitheciidae) in South-Eastern Amazonia. Primates 58:353–360

Barnett AA, Hawes JE, Mendes Pontes AR, Guedes Layme VM, Chism J, Wallace R, de Alcântara Cardoso N, Ferrari SF, Beltrão-Mendes R, Wright B, Haugaasen T, Cheyne SM, Bezerra BM, Matsuda I, dos Santos RR (2018a) Survey and study methods for flooded habitats. In: Barnett AA, Matsuda I, Nowak K (eds) *Primates in flooded habitats: ecology and conservation*. Cambridge University Press, Cambridge pp. 33–44

Barnett AA, Tománek P, Todd LM (2018b) A ecologia do uacari-de-costas-douradas (*Cacajao ouakary*) (Pitheciidae) na bacia Amazônica. In: Urbani B, Kowalewski M, da RGT C, de la Torre S, Cortés-Ortiz L (eds) *La primatología en Latinoamérica 2 – A primatologia na* America Latina 2. Instituto Venezolano de Investigacoes Cientificas, Caracas, pp. 219–228

Barnett AA, McGoogan KM, Mendes Pontes AR, Guedes Layme VM, Lehman SM (2018c) Primates of riverine and gallery forests: a worldwide overview. In: Barnett AA, Matsuda I,

Nowak K (eds) *Primates in flooded habitats: ecology and conservation*. Cambridge University Press, Cambridge pp. 259–262

Barnett AA, de Oliveira T, Soares da Silva FM, de Albuquerque Teixeira S, Todd LM, Boyle SA (2018d) Honest error, precaution or alertness advertisement? Reactions to vertebrate pseudopredators in red-nosed cuxiús (*Chiropotes albinasus*), a high-canopy neotropical primate. Ethology 124:177–187

Barnett AA, Stone AI, Shaw P, Ronchi-Teles B, Pimenta NC, Spironello WR, Ross C, Wenzel JW (Submitted-a) When food fights back: at nest predation of larval and adult paper wasps (Hymenoptera, Vespidae, Polistinae), by *Cebus albifrons* and *Saimiri collinsi* (Primates, Cebidae) and the high-energy yield of three risk-sensitive foraging techniques. To Biotropica

Barnett AA, Cavalcanti G, Kasper BH, de Oliveira (Submitted-b) Primates of the middle Rio Tapajós, eastern Brazilian Amazon, and the importance of flooded forests. To Primates

Benchimol M & Peres CA (2015) Predicting local extinctions of Amazonian vertebrates in forest islands created by a mega dam. Biol Cons 187:61–72

Bertuol F (2015) Testando limites especificos dos uacaris pretos, sensu Hershkovitz 1987 (Pitheciidae, Primates). MSc theses, Amazonas Federal University, Manaus, Brazil

Bezerra BM (2010) Behavior and communication in golden-backed uacaris, *Cacajao melanocephalus*. PhD Diss., University of Bristol

Bezerra BM, Souto AS, Jones G (2010a) Responses of golden-backed uakaris, *Cacajao melanocephalus*, to call playback: implications for surveys in the flooded igapó forest. Primates 51:327–336

Bezerra BM, Souto AS, Jones G (2010b) Vocal repertoire of golden-backed uakaris (*Cacajao melanocephalus*): call structure and context. Int J Primatol 31:759–778

Bezerra BM, Barnett AA, Souto A, Jones G (2011a) Ethogram and natural history of golden-backed uakaris (*Cacajao melanocephalus*). Int J Primatol 32:46–68

Bezerra BM, Souto AS, Radford AN, Jones G (2011b) Brevity is not always a virtue in primate communication. Biol Lett 7:23–25

Bobrowiec PED, Rosa LDS, Gazarini J, Haugaasen T (2014) Phyllostomid bat assemblage structure in Amazonian flooded and unflooded forests. Biotropica 46:312–321

Boubli JP (1993) Southern expansion of the geographical distribution of *Cacajao melanocephalus melanocephalus*. Int J Primatol 14:933–937

Boubli JP (1997) Ecology of the Black Uakari monkey, *Cacajao melanocephalus melanocephalus*, in Pico de Neblina National Park, Brazil. PhD thesis, University of California, Berkeley, CA

Boubli JP (1999) Feeding ecology of black-headed uacaris (*Cacajao melanocephalus melanocephalus*) in Pico da Neblina National Park, Brazil. Int J Primatol 20:719–749

Boubli JP, Tokuda M (2008) Socioecology of black uakari monkeys, *Cacajao hosomi*, Pico da Neblina National Park, Brazil: the role of the peculiar spatial-temporal distribution of resources in the Neblina forests. Primate Report 75:3–10

Boubli JP, de Lima MG (2009) Modeling the geographical distribution and fundamental niches of *Cacajao* spp. and *Chiropotes isrealita* in northwestern Amazonia via a maximum entropy algorith, Int J Primatol 30:217–288

Bowler M (2006) The ecology and conservation of the red uakari monkey on the Yavari River, *Peru*. PhD dissertation, Canterbury, University of Kent, Kent

Cintra BBL, Schietti J, Emillio T, Martins D, Moulatlet G, Souza P, Levis C, Quesada CA, Schöngart J (2013) Productivity of aboveground coarse wood biomass and stand age related to soil hydrology of Amazonian forests in the Purus-Madeira interfluvial area. Biogeosci Discuss 10:6417–6459

Correa SB, Winemiller KO, López-Fernández H, Galetti M (2007) Evolutionary perspectives on seed consumption and dispersal by fishes. Bioscience 57:748–756

da Fonseca Júnior SF, Piedade MTF, Schöngart J (2009) Wood growth of *Tabebuia barbata* (E. Mey.) Sandwith (Bignoniaceae) and *Vatairea guianensis* Aubl.(Fabaceae) in Central Amazonian black-water (igapó) and white-water (várzea) floodplain forests. Trees 23:127–134

Defler TR (1999) Fission-fusion behaviour in *Cacajao melanocephalus ouakary*. Neotropical Primates 7:5–8

Defler TR (2001) *Cacajao melanocephalus ouakary* densities on the lower Apaporis River, Colombian Amazon. Primate Report 61:31–36

Deler TR (1991) Preliminary observations of *Cacajao melanocephalus* (Humboldt, 1811) (Primates, Cebidae). Trianea 4:557–559

Dia da Silva RHP (2017) Limitações físicas do tamanho da boca e da mão, e suas relações com a dieta de Cacajão, um gênero de primatas amazônicos especializados em predação de sementes; MSc thesis, Instituto Ncional de Pesquisas da Amazonia, Manaus, Brazil

Dias da Silva RHP, Baccaro F, Todd LM, Tomacek P, Hopkins MG, Barnett AA (Submitted) Juggling options: optimal selection of size-weight combinations of *Aldina latifolia* (Fabaceae) pods by *Cacajao ouakary.* At Proc R Soc B

Ferrari SF, Guedes PG, Figueiredo-Ready WM, Barnett AA (2014) Reconsidering the taxonomy of the Black-Faced Uacaris, *Cacajao melanocephalus* group (Mammalia: Pitheciidae), from the northern Amazon Basin. Zootaxa 3866:353–370

Ferreira LV (2002) A review of tree phenology in central Amazonian floodplains. Ecotropica 52:195–222

Ferreira LV, Prance GT (1998) Structure and species richness of low-diversity floodplain forest on the Rio Tapajós, eastern Amazonia, Brazil. Biodivers Conserv 7:585–596

Furch K (1997) Chemistry of várzea and igapó soils and nutrient inventory of their floodplain forests. In: Junk WJ (ed) *The Central Amazon Floodplain*. Springer, Berlin & Heidelberg, pp 47–67

Goulding M, Barthem R, Ferreira E (2003) *Smithsonian Atlas of the Amazon*. Smithsonian Books, Washington DC

Haugaasen T (2004) Structure, composition and dynamics of a central Amazonian forest landscape: a conservation perspective. PhD dissertation, University of Est Anglia

Haugaasen T, Peres CA (2005) Primate assemblage structure in Amazonian flooded and unflooded forests. Am J Primatol 67:243–258

Haugaasen T, Peres CA (2006) Floristic, edaphic and structural characteristics of flooded and unflooded forests in the lower Rio Purús region of central Amazonia, Brazil. Acta Amazon 36:25–35

Haugaasen T, Peres CA (2007) Vertebrate responses to fruit production in Amazonian flooded and unflooded forests. Biodivers Conserv 16:4165–4190

Irion G (1978) Soil infertility in the Amazon rainforest. Naturwissenschaften 65:515–519

Jucá T, de Oliveira T, Tomacek P, Cavalcanti G, Barnett AA (In review) Being hunted high and low: are differences in howler nocturnal and diurnal sleeping site characteristics linked to most-likely predator attack types? To J Zool

Junk WJ, Piedade MTF, Schöngart J, Cohn-Haft M, Adeney JM, Wittmann F (2011) A classification of major naturally-occurring Amazonian lowland wetlands. Wetlands 31:623–640

Kreibich H, Kern J, Förstel H (2002) Studies on nitrogen fixation in Amazonian floodplain forests. In: Pedrosa FO, Hungria M, Yates G, Newton WE (eds) *Nitrogen Fixation: From Molecules to Crop Productivity*. Springer, Berlin & Heidelberg, pp 544–554

Lehman SM, Robertson KL (1994) Preliminary survey of *Cacajao melanocephalus melanocephalus* in southern Venezuela. Int J Primatol 15:927–934

Marques JT, Pereira MJR, Palmeirim JM (2012) Availability of food for frugivorous bats in lowland Amazonia: the influence of flooding and of river banks. Acta Chiropterologica 14:183–194

Marsh LK (2014) A taxonomic revision of the Saki Monkeys, *Pithecia* Desmarest, 1804. Neotropical Primates 21:1–165

Mourthé I, Barnett AA (2014) Crying Tapir: the functionality of errors and accuracy in predator recognition in two Neotropical high-canopy primates. Folia Primatol 85:379–398

Negreiros AA (2017) Fitoquímica e ecologia da interação de três espécies (planta, lepidoptera e primata) em uma floresta de igapó, Amazonas, Brasil. MSC dissertation, INPA, Manaus, Brazil

Negreiros AA, Pohlit AM, Baccaro F, Koolen HHF, Barnett AA (2018) The bitter end: primate avoidance of caterpillar-infested trees in a central Amazon flooded forest. Can. J. Zool: 96

Negreiros AA, Bacarrio F, Juca T, Barnett AA (in review) Taking safe second best: *Cacajao ouakary* (Pitheciidae) forages in small canopied *Buchenavia ochroprumna* (Combretaceae) trees to avoid caterpillar infestations. At Can J Bot

Norconk MA (2007) Saki, uakaris, and titi monkeys: behavioral diversity in a radiation of primate seed predators. In: Campbell CJ, Fuentes A, KC MK, Panger M, Bearder SK (eds) *Primates in perspective*. Oxford University Press, New York, pp 123–138

Paim FP, Valsecchi J, Harada ML, de Queiroz HL (2013) Diversity, geographic distribution and conservation of squirrel monkeys, *Saimiri* (Primates, Cebidae), in the floodplain forests of Central Amazon. Int J Primatol 34:1055–1076

Parolin P (2000) Seed mass in Amazonian floodplain forests with contrasting nutrient supplies. J Trop Ecol 16:417–428

Parolin P, Armbruester N, Wittmann F, Ferreira L, Piedade MTF, Junk WJ (2002) A review of tree phenology in central Amazonian floodplains. Pesquisas Bôtanica 52:195–222

Peres CA (1993) Structure and spatial organization of an Amazonian terra firme forest primate community. J Trop Ecol 9:259–276

Peres CA (1997) Primate community structure at twenty western Amazonian flooded and unflooded forests. J Trop Ecol 13:381–405

Peres CA, Patton JL, da Silva MNF (1996) Riverine barriers and gene flow in Amazonian saddle-back tamarins. Folia Primatol 67:113–124

Perry S, Panger M, Rose LM, Baker M, Gros-Louis J, Jack K, MacKinnon KC, Manson J, Fedigan L, Pyle K (2003) Traditions in wild white-faced capuchin monkeys. In: Fragaszy DM, Perry S (eds) *The Biology of Traditions: models and evidence*. Cambridge University Press, Cambridge, pp 391–425

Pinto LP, Barnett AA, Bezerra BM, Boubli J-P, Bowler M, Cardoso N, Castelli C, Rodriguez MJO, Santos RR, Setz EZ, Veiga LM (2013) Why we know so little: the challenges of fieldwork on the pithciines. In: Veiga L, Barnett A, Ferrari S, Norconk M (eds) *Evolutionary Biology and Conservation of Titis, Sakis and Uacaris*. Cambridge University Press, Cambridge, pp 145–150

Prance GT (1979) Notes on the vegetation of Amazonia III. The terminology of Amazonian forest types subject to inundation. Brittonia 31:26–38

Rocha A (2016) Disponabilidade de recusros e antropização estructurando a seeembleia de primatas na floresa de igapó. MSc dissertation, Instituto Nacional de Pesquiosas da Amazonia, Manas

Scudeller V (2018) Do Igapo tree species are exclusive to this Phytophysiognomy? In: Myster RM (ed) *Igapo (black-water flooded forests) of the Amazon Basin*. Springer, Cham

Spironello WR (2001) The brown capuchin monkey (*Cebus apella*): ecology and home range requirements in central Amazonia. In: Bierregaard R, Gascon C (eds) *Lessons from Amazonia: the ecology and conservation of a fragmented forest*. Yale University Press, New Haven, pp 271–283

Chapter 9
Influence of Time and Flood on Diurnal Mammal Diversity and Story Level Use in Igapó Forest in the Peruvian Amazon

Rosa R. Palmer and John L. Koprowski

9.1 Introduction

Rainforests represent less than seven percent of the Earth's surface but support the highest biodiversity in the world (Wilson and Peter 1988) containing approximately 50% of all terrestrial species (Myers 1988; Mittermeier et al. 1998). Rainforests are also among the most severely threatened ecosystems (Myers 1988; Phillips 1997; Laurance 2004) mainly due to forest destruction causing changes in species composition, species interactions, ecosystem processes, and microclimate modifications (Newmark 1991; Terborgh 1992; Bierregaard et al. 1992; Hall et al. 1996; Benitez-Malvido 1998; Laurance 1998). Rainforest deforestation is also predicted to affect local and global climate due to an increase in temperature, reduction of evapotranspiration and precipitation, and increase in runoff (Nobre et al. 1991; Laurance and Williamson 2002). Additionally, forest harvest affects water cycles and increases susceptibility to fire, while predicted droughts due to climate change also will increase fire in rainforests (Nobre et al. 1991; Laurance 1998; Laurance and Williamson 2002). Climate change is expected to have a great impact on rainforest ecosystems due to the narrow range of daily and yearly temperature to which species are adapted, as well as their dependence on rainfall (Malhi et al. 2009). Even subtle temperature changes, and, most importantly, changes in precipitation, such as extreme drought events, will affect species distribution, presence, and abundance and are predicted to drive a great percentage of species to extinction (Thomas et al. 2004; Parry et al. 2007; Malhi et al. 2009). Techniques such as rapid inventories, systematic surveys, and monitoring programs are important to measure

R. R. Palmer (✉) · J. L. Koprowski
The University of Arizona, School of Natural Resources and the Environment, Tucson, AZ, USA
e-mail: rjessen@email.arizona.edu; squirrel@ag.arizona.edu

R. W. Myster (ed.), *Igapó (Black-water flooded forests) of the Amazon Basin*,
https://doi.org/10.1007/978-3-319-90122-0_9

and document changes in species diversity and species distribution over time, and even some behavioral patterns (Wilson 1985; Haila and Margules 1996). These tools provide important knowledge for conservation and management of rainforest ecosystems (Haila and Margules 1996).

Rainforests, particularly the Peruvian Amazon, have a more heterogeneous landscape than previously reported (Tuomisto et al. 1995). This has important implications since heterogeneity is positively correlated with species diversity (Pianka 1966). Species diversity is also related to forest structure and differs between canopy and story levels when forests are stratified vertically (Basset et al. 2001; Bernard 2001; Schulze et al. 2001; Viveiros Grelle 2003; Fermon et al. 2005). Each story has unique physical characteristics, microclimate, and resource availability and harbors a unique flora and fauna (Frith 1984; Basset et al. 2001; Bernard 2001; Basset et al. 2003). Consideration of all vertical strata is necessary for accurate estimates of forest biodiversity and abundance since some species use multiple story levels (Bernard 2001; Viveiros Grelle 2003; Stork and Grimbacher 2006).

Rainforests are home to the greatest mammalian species richness in the world, and mammals occupy all vertical strata (Estrada and Coates-Estrada 1985; Bernard 2001; Monteiro Vieira and Monteiro-Filho 2003; Viveiros Grelle 2003); however, little is known about most rainforest species. In the neotropics, mammalian diversity is being discovered at an average of one new genus and eight new species annually (Patterson 2000), but the rate of extinction is unknown. Mammals play important roles in rainforests and can provide ecological services such as seed dispersal and pollination (Gessman and MacMahon 1984; Terborgh 1988; Jansen et al. 2012). Mammals also drive the dynamics and complexity of biological communities by serving as predators and providing a prey base (Gessman and MacMahon 1984; Terborgh 1988). Therefore, changes in mammal communities may have implications for the dynamics and function of rainforest ecosystems with unknown consequences.

We conducted surveys in the Peruvian rainforest to determine what the diurnal mammalian diversity was and if this diversity was influenced by flood and varied by story level. To answer these questions, we stratified the forest vertically into four levels and predicted that the upper canopy would have the highest diversity index compared to the midstory, understory, and ground due to the advantages that a dense canopy offers for protection from predators and high food production and that use of lower levels would decrease in a year of flooding and deposition of silt at the ground level. We also predicted that diversity would be higher during a wet year compared to a dry year because of the influence that rainfall has on resource availability.

9.2 Methods

9.2.1 Study Site

Our study site was a 400 ha research grid located in the Peruvian Amazon at the Amazon Research Center (ARC). The ARC is located in Tamshiyacu-Tahuayo Reserve (TTR) in northeastern Peruvian Amazon between the Tamshiyacu-Tahuayo and Yavarí Miri rivers in the state of Loreto, near the Brazilian border (4°39'S, 73°26'W). This 322,500 ha conservation area is a lowland, evergreen, and seasonally flooded forest that was created by local communities due to overexploitation of natural resources by outside commercial interests (Newing and Bodmer 2003). Hunting without regulation, poaching, large-scale commercial fishing, fishing with chemicals, and large-scale logging were the main disturbances that affected the area (Newing and Bodmer 2003). The plant communities found within the study area include palm swamps (low-lying areas of poor drainage, low tree diversity; the most common species is moriche palm or *Mauritia flexuosa*), bajial (floods with a water level of 5–7 m, low tree diversity, small trees, none to sparse vegetation in understory), high restinga (unflooded forest, clay soils, high tree diversity, and large trees are common), and low restinga (floods with a water level of 2.5–5 m, low tree diversity; Prance 1979; Kvist and Nebel 2001; Myster 2009).

In 2009, the study site experienced above average rainfall, and in 2010 a severe drought affected the area. In 2009, the mean low temperature was 23.3 °C (± 0.05 SE), and the mean high temperature was 28.2 °C (± 0.11 SE). In 2010 the mean low temperature was 23.2 °C (± 0.07 SE), and the mean high temperature was 29.4 °C (± 0.13 SE). Total rainfall was 3914 mm in the wet year of 2009 compared to 3100 mm in the dry year of 2010, a difference of 21%. The lodge at ARC is built on stilts because of seasonal flooding. On 31 May 2009, the depth of water at the lodge was 1.5 m, and on the same day in 2010, the depth of water was 0 m.

9.2.2 Mammalian Survey

In June and July of 2009 and 2010, we conducted mammal surveys in forested habitat to compare species diversity of a wet versus a dry year. Each day from 0600 h to 1700 h, we walked two 2 km transects, a total of 42 transects and 84 km each year. During surveys, we recorded species of mammals, number of individuals, time of day, and story level where individuals were found, categorized as ground, understory (<5 m, some cover and vegetation), midstory (5–15 m, little vegetation, mainly bare trunks from large trees), and canopy (>15 m, usually very dense foliage). We stratified the forest vertically into these story levels to determine whether mammalian diversity varied by level. In the rainforest, these four story levels are distinct and easily recognizable.

9.2.3 Analysis

We used the Shannon-Wiener diversity index to estimate diurnal mammal diversity for wet (2009) and dry (2010) years as well as by story level. We analyzed differences between wet (2009) and dry (2010) years using the Shannon diversity t-test in program PAST (Hammer et al. 2001). We estimated mammalian alpha, beta, and gamma diversity for the four different story levels as different types of habitat for wet, dry, and both years combined (Whittaker 1972). We created a contingency table in program JMP 10 (SAS Institute, INC., Cary, NC) and used a log-likelihood ratio to determine if the frequency of sightings by species varied by time of day or story level.

9.3 Results

The diurnal mammalian diversity index did not differ between wet and dry years ($t_{807.04} = -1.623$, $P = 0.105$; Table 9.1). Species richness was higher in the wet year, at 19 species, compared to 17 species sighted in the dry year, and evenness was low for both years (Fig. 9.1). We sighted a total of 22 different species over 2 years, and 14 species were detected in both years (Table 10.2). In the wet year, we sighted a total of 438 individuals and 829 individuals in the dry year.

The diversity index by story level differed between years, but the diversity index was highest in the canopy for both years (Table 9.1). In the wet year, the diversity index was high in the understory after the canopy, followed by midstory and ground (Table 9.1). Species richness was highest in the canopy, followed by ground, understory, and finally midstory (Fig. 9.2). Evenness was high for ground and canopy and low for understory and midstory (Fig. 9.2). In the dry year, the diversity index was high in the midstory after the canopy, followed by ground and understory (Table 9.1). Species richness was highest in the canopy followed by midstory, ground, and finally understory (Fig. 9.2). Evenness was high for canopy and understory and low for ground and midstory (Fig. 9.2).

Table 9.1 Shannon-Wiener diversity index comparison between wet and dry year and among different story levels and alpha, gamma, and beta diversity by year and combined. Amazon Research Center in Tamshiyacu-Tahuayo Reserve, Loreto, Peru, from June to July 2009 and 2010

	Diversity index	Diversity index by story level				Diversity		
Year	total	Ground	Understory	Midstory	Canopy	Alpha	Gamma	Beta
Wet (2009)	2.01	0.69	1.34	0.95	1.88	7	19	2.7
Dry (2010)	2.08	1.48	0.8	1.64	2.02	8	17	2.1
Combined	–	–	–	–	–	10	22	2.2

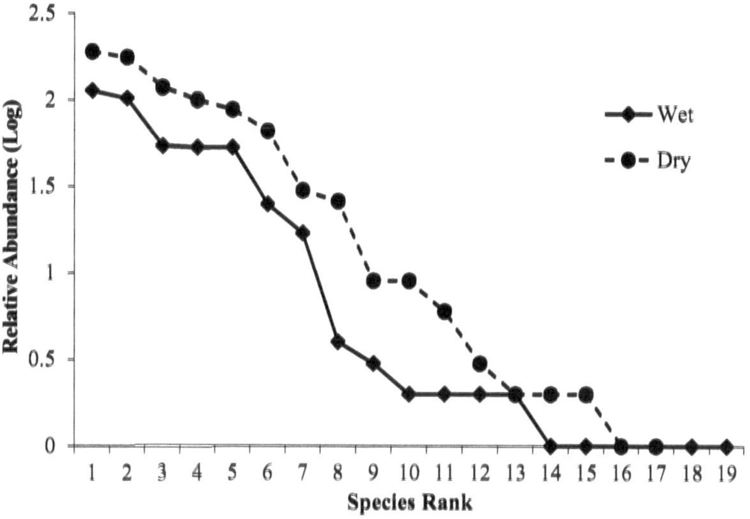

Fig. 9.1 Rank abundance curve: comparison of species richness between wet and dry years, relative species abundance, as well as evenness for both years. On the *x*-axis, each number represents a species and is ranked by the relative abundance. A steep slope indicates low evenness, and a shallow slope indicates high evenness since relative abundance among species is similar. Amazon Research Center in Tamshiyacu-Tahuayo Reserve, Loreto, Peru, from June to July 2009 and 2010

Alpha, beta, and gamma diversity were different for both years separately and combined. Alpha diversity was higher in the dry year compared to the wet year, and gamma and beta diversity were higher in the wet year compared to the dry year (Table 9.1).

Species richness for the wet year and dry year combined was 22, representing 6 orders of mammals (Table 9.2). Primates was the most diverse and most abundant with 8 different species and 967 individuals sighted. Rodentia was represented by 6 species with 171 individuals. We sighted 2 species of the order Carnivora with 116 individuals; the South American coati (*Nasua nasua*) was the most common carnivore species in the study area. Artiodactyla was represented by three species with eight individuals sighted. Pilosa was represented by two species with four individuals, and Didelphimorphia was represented by a single species with one individual sighted (Table 9.1).

During the wet and dry year, the frequency of sightings of species was influenced by time of day ($X^2 = 511.063$, $n = 438$, $P = <0.001$; $X^2 = 473.400$, $n = 829$, $P = <0.001$; Fig. 9.3) and varied by story level ($X^2 = 430.811$, $n = 438$, $P = <0.001$; $X^2 = 403.854$, $n = 829$, $P = <0.001$; Fig. 9.4). We sighted more mammal species during the late morning and early afternoon compared to the early morning and late afternoon, and although there was a difference in the amount of flooding between years, our frequency of sightings of species was similar across time periods for both years (Fig. 9.3). Some mammal species were sighted in only one story level, and several species were sighted in different story levels at different frequencies (Fig. 9.4).

Table 9.2 Mammal abundance and story level sighted (percentage of sightings from both years combined) in 2009 and 2010. Amazon Research Center in Tamshiyacu-Tahuayo Reserve, Loreto, Peru, from June to July 2009 and 2010

			Abundance		Story level			
Order	Common name	Scientific name	Wet (2009)	Dry (2010)	Ground	Understory	Midstory	Canopy
Didelphimorphia	Common opossum	*Didelphis marsupialis*	1	0	100	–	–	–
Pilosa	Two-toed sloth	*Choloepus* sp.	1	0	–	–	–	100
	Southern tamandua	*Tamandua tetradactyla*	1	2	–	–	–	100
Primates	Pygmy marmoset	*Cebuella pygmaea*	0	9	–	–	78	22
	Saddleback tamarin	*Saguinus fuscicollis*	53	190	–	14	51	35
	Black-chested mustached tamarin	*Saguinus mystax*	54	176	–	1	50	49
	White-fronted capuchin	*Cebus albifrons*	0	30	–	–	–	100
	Tufted capuchin	*Cebus apella*	1	66	–	–	30	70
	Common squirrel monkey	*Saimiri sciureus*	102	100	–	15	12	73
	Dusky titi monkey	*Callicebus moloch*	17	26	2	7	47	44
	Monk saki	*Pithecia monachus*	25	118	–	–	12	88
Rodentia	Neotropical pygmy squirrel	*Sciurillus pusillus*	53	88	–	6	15	79
	Northern and southern Amazon red squirrel[a]	*Sciurus igniventris*, *S. spadiceus*	2	6	50	13	–	38
	Black agouti	*Dasyprocta fuliginosa*	4	9	100	–	–	–
	Green acouchi	*Myoprocta pratti*	0	2	100	–	–	–
	Paca	*Cuniculus paca*	2	0	100	–	–	–
	Yellow-crowned brush-tailed rat	*Isothrix bistriata*	3	2	–	80	20	–
Carnivora	Tayra	*Eira barbara*	2	1	67	–	–	33
	South American coati	*Nasua nasua*	113	0	73	–	–	27
Artiodactyla	Collared peccary	*Tayassu tajacu*	2	0	100	–	–	–
	Red brocket deer	*Mazama americana*	1	3	100	–	–	–
	Gray brocket deer	*Mazama gouazoubira*	1	1	100	–	–	–

[a]Northern and southern Amazon red squirrel are too similar to identify in the field due to nearly identical size and color

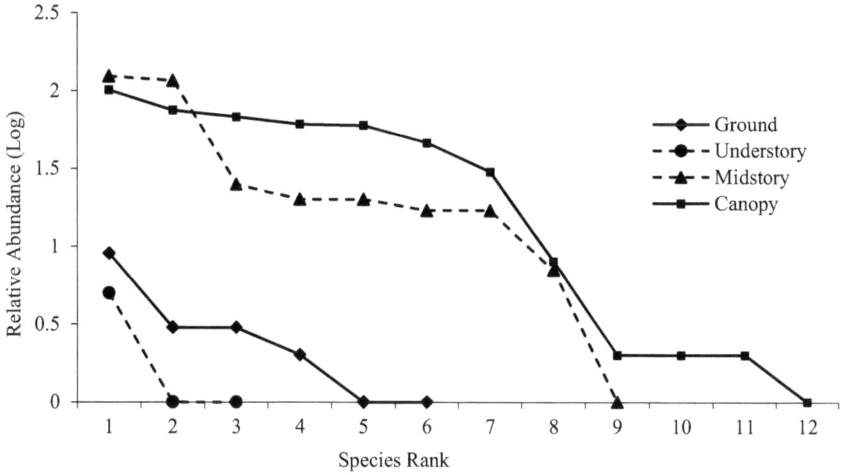

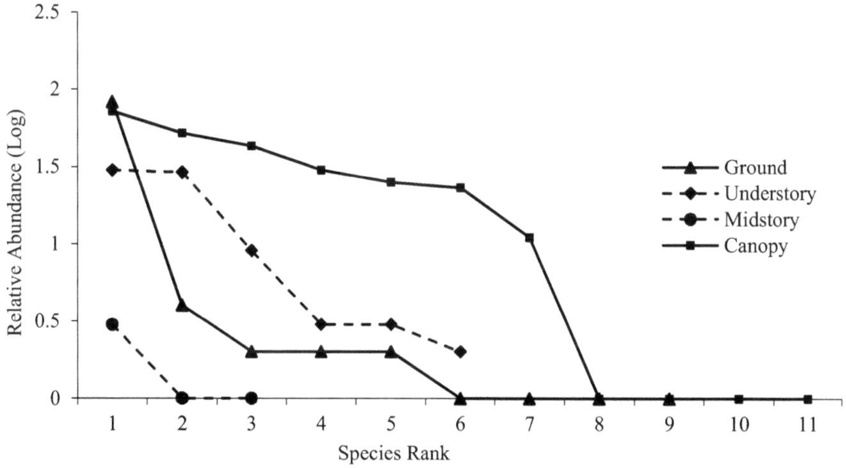

Fig. 9.2 Rank abundance curve: comparison of species richness between the four different story levels in the forest, relative species abundance, and evenness in wet (2009; top) and dry (2010; bottom) year. On the *x*-axis, each number represents a species and is ranked by the relative abundance. A steep slope indicates low evenness, and a shallow slope indicates high evenness since relative abundance among species is similar. Amazon Research Center in Tamshiyacu-Tahuayo Reserve, Loreto, Peru, from June to July 2009 and 2010

9.4 Discussion

During our sampling in the igapó forest, we observed 48 mammal species, including diurnal, nocturnal, terrestrial, arboreal, and aquatic mammals. Diversity index did not differed between wet and dry year, and this is similar to patterns observed in other areas of Amazonia (Emmons 1984). During mammal surveys in a wet and dry

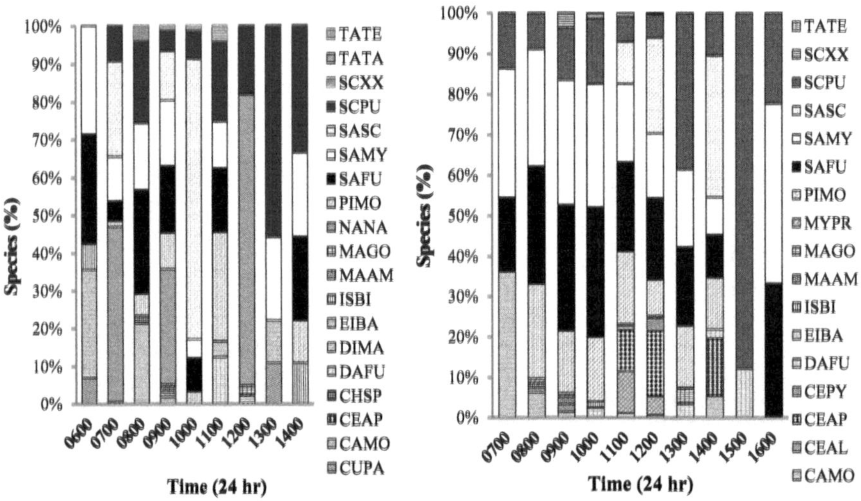

Fig. 9.3 Percentage of mammal species sighted by time in wet year (2009; left) and dry year (2010; right) at the Amazon Research Center in Tamshiyacu-Tahuayo Reserve, Loreto, Peru, from June to July 2009 and 2010. CUPA *Cuniculus paca*; CAMO *Callicebus moloch*; CEAL *Cebus albifrons*; CEAP *Cebus paella*; CEPY *Cebuella pygmaea*; CHSP *Choloepus* sp.; DAFU *Dasyprocta fuliginosa*; DIMA *Didelphis marsupialis*; EIBA *Eira barbara*; ISBI *Isothrix bistriata*; MAAM *Mazama americana*; MAGO *Mazama gouazoubira*; MYPR *Myoprocta pratti*; NANA *Nasua nasua*; PIMO *Pithecia monachus*; SAFU *Saguinus fuscicollis*; SAMY *Saguinus mystax*; SASC *Saimiri sciureus*; SCPU *Sciurillus pusillus*; SCXX *Sciurus igniventris*, *S. spadiceus*; TATA *Tayassu tajacu*; TATE *Tamandua tetradactyla*

year, we sighted 22 diurnal mammalian species with primates representing the more diverse and abundant order. Although Primates are not usually the most common taxon in Peru (Pacheco et al. 2009), sightings have become more frequent at the TTR since the decrease in hunting after creation of the reserve (Pitman et al. 2003). Most of the primates forage in groups, and some groups include different species, making primates more visible compared to solitary species (Chapman 1990; Klein and Klein 2005). We sighted 2 more species of primates and had a total of 463 more primate sightings during the dry year compared to the wet year, which could be explained by the low food availability observed in the dry year, forcing primates to move more while searching for food (Dawson 1979; Vedder 1984; Boinski 1987; Garber 2005; Di Bitetti 2006).

Although most of the mammal species sighted during our surveys were diurnal (Emmons and Feer 1997; Eisenberg and Redford 1999), species sightings varied by time of day and by story level. This is similar to other areas of the tropics, where species were associated with certain story levels when the forest was stratified vertically (Basset et al. 2001; Bernard 2001; Schulze et al. 2001; Viveiros Grelle 2003; Fermon et al. 2005). Each story level offers unique resources and includes different physical and environmental characteristics (Frith 1984; Basset et al. 2001; Bernard

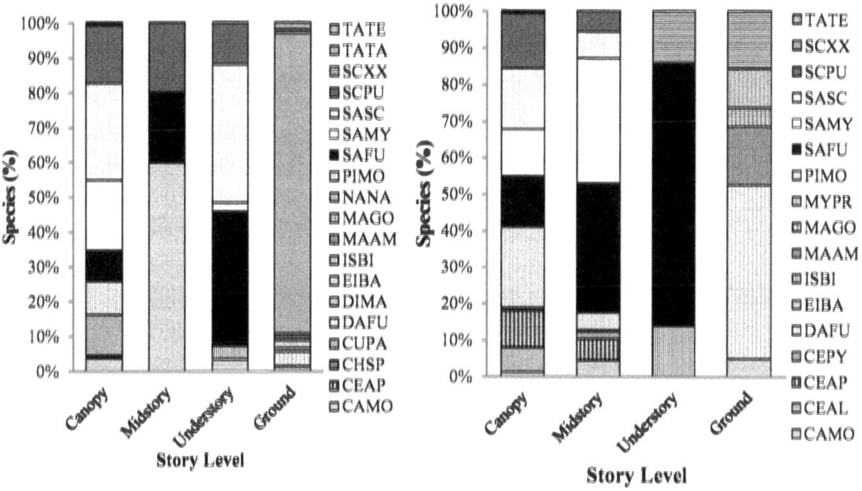

Fig. 9.4 Percentage of mammal species sighted by story level in wet year (2009; left) and dry year (2010; right) at the Amazon Research Center in Tamshiyacu-Tahuayo Reserve, Loreto, Peru, from June to July 2009 and 2010. CUPA *Cuniculus paca*; CAMO *Callicebus moloch*; CEAL *Cebus albifrons*; CEAP *Cebus paella*; CEPY *Cebuella pygmaea*; CHSP *Choloepus sp*; DAFU *Dasyprocta fuliginosa*; DIMA *Didelphis marsupialis*; EIBA *Eira barbara*; ISBI *Isothrix bistriata*; MAAM *Mazama americana*; MAGO *Mazama gouazoubira*; MYPR *Myoprocta pratti*; NANA *Nasua nasua*; PIMO *Pithecia monachus*; SAFU *Saguinus fuscicollis*; SAMY *Saguinus mystax*; SASC *Saimiri sciureus*; SCPU *Sciurillus pusillus*; SCXX *Sciurus igniventris*, *S. spadiceus*; TATA *Tayassu tajacu*; TATE *Tamandua tetradactyla*

2001; Basset et al. 2003), which allows for a complex community with a great diversity of species to occupy the forest (Basset et al. 2001; Bernard 2001; Schulze et al. 2001; Viveiros Grelle 2003; Fermon et al. 2005).

As we predicted, the upper forest canopy had the highest diversity index and species richness for both years among all forest levels, which suggests the importance of protection and management of forests with large trees and closed canopy due to their contribution to biological diversity (Estrada and Coates-Estrada 1985; Dean et al. 1999; Andersson and Östlund 2004; Ishii et al. 2004; Sorensen 2004). Dense, large canopies are important not just for species that move through them (Emmons 1995; Laurance and Laurance 1999; Wilson et al. 2007) but also for species that spend most of their time on the ground, understory, and midstory. Large, mature trees with large and continuous canopies provide protection, offer greater seed and fruit production to species using different story levels, and create different microclimates that support a high biodiversity (Goodrum et al. 1971; Estrada and Coates-Estrada 1985; Burns and Hokala 1990; Bierregaard et al. 1992). Mammal movements are restricted with reduced canopy connectivity, and mammals and bird species will permanently move away from areas where canopy is destroyed due to fire and where food sources are depleted (Kinnaird and O'Brien 1998; Wilson et al. 2007). Gaps in the canopy or lack of canopy connectivity can bring important ecological changes such as altering the plant community in the understory and subsequently altering the animal community (Frumhoff 1995; Kinnaird and O'Brien 1998). Observed and predicted impacts of rainforest alteration on mammal communities are more alarming when considering effects of climate change in rainforests (Laurance 1998).

Species diversity on the ground was approximately twice as high in the dry year than in the wet year. This could be explained by two factors, which are influenced by the above average rainfall observed in 2009. First, water level during the wet year was higher than during the dry year, and most of the study site stayed underwater until mid-June, preventing ground-dwelling mammals to occupy the area. In 2010, due to the extreme drought, the study site stayed almost completely dry. With increases in water level as experienced in wet years, most of the ground-dwelling species move to higher ground, or terra firme, seeking larger areas of dry land (Bodmer 1990b; Alho 2008). Only a few individuals stay in high restinga or small patches of forest surrounded by water that are high enough in elevation to avoid inundation most of the years (Bodmer 1990b; Alho 2008). When the water recedes, terrestrial wildlife move through the forest to areas of lower elevation (Bodmer 1990b; Alho 2008). Alternatively, moriche palm is one of the main sources of food for many terrestrial mammals (Bodmer 1990a, 1991; Tobler 2008; Bowler and Bodmer 2011). Ripe fruit drops to the ground and many mammal species forage around these palm trees. In the wet year, we observed fruit production, but, because of the above average rainfall, the fruit fell in the water and rotted. In the dry year, however, fruit production was very low, and perhaps individuals were actively looking for food, making our ground sightings more common (Vedder 1984; Bonaventura et al. 1992; Dussault et al. 2005). In the wet year, we sighted 113 South American coatis. The reason for the great number of coati sightings in the wet year compared to the dry year is unknown, and food availability alone could not explain this pattern. Coatis are omnivorous, and their diet changes according to availability of insects or plants and fruits between wet and dry seasons (Alves-Costa et al. 2004), and they are known to increase the time spent and use of trees to forage and travel during the dry season (Beisiegel and Mantovani 2006).

Mammals play a crucial role in ecosystem function (Gessman and MacMahon 1984; Terborgh 1988; Jansen et al. 2012). Conservation of mammals as well as their diversity is important for maintaining the complexity of biological communities (Gessman and MacMahon 1984; Terborgh 1988), and this is especially true in the tropics, host to some of the most complex biotic communities on Earth (Covich 1988; Tuomisto et al. 1995; Giller 1996; Scarano 2002). Protection of continuous, mature forests with large canopies has important conservation implications as these areas most likely protect the greatest diversity of mammals while also providing shelter and food for other taxa (Bierregaard et al. 1992; Dean et al. 1999; Andersson and Östlund 2004; Ishii et al. 2004; Sorensen 2004). These continuous forests also facilitate movements of ground and canopy dwellers between lowland and higher elevation areas during wet and dry season as well as years with above average rainfall (Frumhoff 1995; Kinnaird and O'Brien 1998; Wilson et al. 2007). Forest characteristics such as large trees with large canopies and flood regimes need to be taken into consideration when designing management and conservation plans that aim to conserve and protect rainforest biodiversity.

Acknowledgments Observations were made during field research authorized by permit no. 0246-2010-AG-DGFFS-DGEFFS from the Dirección General de Fauna y Flora Silvestre from Peru and by the University of Arizona Institutional Care and Use Committee (AICUC protocol 09-035). Thanks to Dr. Paul Beaver and staff at the Amazonia Expeditions, especially Jeisen Shahuano, Alfredo Do Santos, and Rafael Flores Pinedo for their enthusiasm and assistance in the field. We also wish to thank Tim Jessen, Zachary Koprowski, Melissa Merrick, and Geoffrey Palmer for their help in the field. Financial support was provided by the Scott Neotropical Fund from the Cleveland Metroparks Zoo and the Cleveland Zoological Society, the Exploration Fund from The Explorers Club, Tinker Summer Field Grant Program from the University of Arizona, the Latin American Student Field Research Award from the American Society of Mammalogists, and the University of Arizona.

References

Alho CJR (2008) Biodiversity of the Pantanal: response to seasonal flooding regime and to environmental degradation. Braz J Biol 68:957–966

Alves-Costa CP, Da Fonseca GAB, Christofaro C (2004) Variation in the diet of the brown-nosed coati (*Nasua nasua*) in southeastern Brazil. J Mammal 85:478–482

Andersson R, Östlund L (2004) Spatial patterns, density changes and implications on biodiversity for old trees in the boreal landscape of northern Sweden. Biol Conserv 118:443–453

Basset Y, Aberlenc HP, Barrios H, Curletti G, Bérenger JM, Vesco JP, Causse P, Haug A, Hennion AS, Lesobre L, Marquès F, O'Meara R (2001) Stratification and diel activity of arthropods in a lowland rainforest in Gabon. Biol J Linn Soc 72:585–607

Basset Y, Hammond PM, Barrios H, Holloway JD, Miller SE (2003) Vertical stratification of arthropod assemblages. In: Basset Y, Novotny V, Miller S, Kitching RL (eds) Arthropods of tropical forests: Spatio-temporal dynamics and resource use in the canopy. Cambridge University Press, Cambridge, pp 17–27

Beisiegel BM, Mantovani W (2006) Habitat use, home range and foraging preferences of the coati *Nasua nasua* in a pluvial tropical Atlantic forest area. J Zool 269:77–87

Benitez-Malvido J (1998) Impact of forest fragmentation on seedling abundance in a tropical rain forest. Conserv Biol 12:380–389

Bernard E (2001) Vertical stratification of bat communities in primary forests of Central Amazon, Brazil. J Trop Ecol 17:115–126

Bierregaard RO Jr, Lovejoy TE, Kapos V, Dos Santos AA, Hutchings RH (1992) The biological dynamics of tropical rainforest fragments. Bioscience 42:859–866

Bodmer RE (1990a) Fruit patch size and frugivory in the lowland tapir (*Tapirus terrestris*). J Zool 222:121–128

Bodmer RE (1990b) Responses of ungulates to seasonal inundations in the Amazon floodplain. J Trop Ecol 6:191–201

Bodmer RE (1991) Strategies of seed dispersal and seed predation in Amazonian ungulates. Biotropica 23:255–261

Boinski S (1987) Habitat use by squirrel monkeys (*Saimiri oerstedi*) in Costa Rica. Folia Primatol 49:151–167

Bonaventura SM, Kravetz FO, Suarez OV (1992) The relationship between food availability, space use and territoriality in *Akodon azarae* (*Rodentia, Cricetidae*). Mammalia 56:407–416

Bowler M, Bodmer RE (2011) Diet and food choice in Peruvian red uakaris (*Cacajao calvus ucayalii*): selective or opportunistic seed predation. Int J Primatol 32:1109–1122

Burns RM, Hokala BH (1990) Silvics of North America: 1. Conifers; 2. Hardwoods. United States Forest Service, vol 654. Agriculture Handbook, Washington, DC, pp 1–377

Chapman CA (1990) Ecological constraints on group size in three species of neotropical primates. Folia Primatol 55:1–9

Covich AP (1988) Geographical and historical comparison of neotropical streams: biotic diversity and detrital processing in highly variable habitats. J N Am Benthol Soc 7:361–386
Dawson GA (1979) The use of time and space by the Panamanian tamarin, *Saguinus oedipus*. Folia Primatol 31:253–284
Dean WRJ, Milton SJ, Jeltsch F (1999) Large trees, fertile islands, and birds in arid savanna. J Arid Environ 41:61–78
Di Bitetti MS (2006) Home-range use by the tufted capuchin monkey (*Cebus apella nigritus*) in a subtropical rainforest of Argentina. J Zool 253:33–45
Dussault C, Courtois R, Ouellet JP, Girard I (2005) Space use of moose in relation to food availability. Can J Zool 83:1431–1437
Eisenberg JF, Redford KH (1999) Mammals of the neotropics. The central neotropics. The University of Chicago Press, London
Emmons LH (1984) Geographic variation in densities and diversities of non-flying mammals in Amazonia. Biotropica 16:210–222
Emmons LH (1995) Mammals of rain forest canopies. In: Lowman MD, Nadkarni NM (eds) Forest canopies. Elsevier Academic Press, London, pp 199–223
Emmons LH, Feer F (1997) Neotropical rainforest mammals: a field guide. The University of Chicago Press, London
Estrada A, Coates-Estrada R (1985) A preliminary study of resource overlap between howling monkeys (*Alouatta palliata*) and other arboreal mammals in the tropical rain forest of Los Tuxtlas, Mexico. Am J Primatol 9:27–37
Fermon H, Waltert M, Vane-Wright RI, Mühlenberg M (2005) Forest use and vertical stratification in fruit-feeding butterflies of Sulawesi, Indonesia: impacts for conservation. Biodivers Conserv 14:333–350
Frith DW (1984) Foraging ecology of birds in an upland tropical rainforest in North Queensland. Aust Wildl Res 11:325–347
Frumhoff PC (1995) Conserving wildlife in tropical forests managed for timber. Bioscience 45:456–464
Garber PA (2005) Role of spatial memory in primate foraging patterns: *Saguinus mystax* and *Saguinus fuscicollis*. Am J Primatol 19:203–216
Gessman JA, Macmahon JA (1984) Mammals in ecosystems: their effects on the composition and production of vegetation. Acta Zool Fenn 172:11–18
Giller PS (1996) The diversity of soil communities, the "poor man's tropical rainforest". Biodivers Conserv 5:135–168
Goodrum PD, Reid VH, Boyd CE (1971) Acorn yields, characteristics, and management criteria of oaks for wildlife. J Wildl Manag 35:520–532
Haila Y, Margules CR (1996) Survey research in conservation biology. Ecography 19:323–331
Hall P, Walker S, Bawa K (1996) Effect of forest fragmentation on genetic diversity and mating system in a tropical tree, *Pithecellobium elegans*. Conserv Biol 10:757–768
Hammer O, Harper DAT, Ryan PD (2001) PAST: paleontological statistics software package for education and data analysis. Palaeontol Electron 4:9
Ishii HT, Tanabe S, Hiura T (2004) Exploring relationships among canopy structure, stand productivity, and biodiversity of temperate forest ecosystems. For Sci 50:342–355
Jansen PA, Hirsch BT, Emsens WJ, Zamora-Gutierrez V, Wikelski M, Kays R (2012) Thieving rodents as substitute dispersers of megafaunal seeds. Proc Natl Acad Sci U S A 109:12610–12615
Kinnaird MF, O'Brien TG (1998) Ecological effects of wildfire on lowland rainforest in Sumatra. Conserv Biol 12:954–956
Klein LL, Klein DJ (2005) Observations on two types of neotropical primate intertaxa associations. Am J Phys Anthropol 38:649–653
Kvist LP, Nebel G (2001) A review of Peruvian flood plain forests: ecosystems, inhabitants and resource use. For Ecol Manag 150:3–26
Laurance WF (1998) A crisis in the making: responses of Amazonian forests to land use and climate change. Trends Ecol Evol 13:411–415
Laurance WF (2004) Forest-climate interactions in fragmented tropical landscapes. Philos Trans: Biol Sci 359:345–352

Laurance SG, Laurance WF (1999) Tropical wildlife corridors: use of linear rainforest remnants by arboreal mammals. Biol Conserv 91:231–239

Laurance WF, Williamson GB (2002) Positive feedbacks among forest fragmentation, drought, and climate change in the Amazon. Conserv Biol 15:1529–1535

Malhi Y, Aragao LEOC, Galbraith D, Hintingford C, Fisher R, Zelazowski P, Sitch S, Mcsweeney C, Meir P (2009) Exploring the likelihood and mechanism of a climate-change-induced die-back of the Amazon rainforest. Proc Natl Acad Sci U S A 106:20610–20615

Mittermeier RA, Myers N, Thomsen JB, Da Fonseca GAB, Olivieri S (1998) Biodiversity hotspots and major tropical wilderness areas: approaches to setting conservation priorities. Conserv Biol 12:516–520

Monteiro Vieira E, Monteiro-Filho ELA (2003) Vertical stratification of small mammals in the Atlantic rain forest of South-Eastern Brazil. J Trop Ecol 19:501–507

Myers N (1988) Threatened biotas: "Hot spots" in tropical forests. Environmentalist 8:187–208

Myster RW (2009) Plant communities of Western Amazonia. Bot Rev 75:271–291

Newing H, Bodmer R (2003) Collaborative wildlife management and adaptation to change: the Tamshiyacu Tahuayo communal reserve, Peru. Nomads 7:110–122

Newmark WD (1991) Tropical forest fragmentation and the local extinction of understory birds in the eastern Usambara Mountains, Tanzania. Conserv Biol 5:67–78

Nobre CA, Sellers PJ, Shukla J (1991) Amazonian deforestation and regional climate change. J Clim 4:957–988

Pacheco V, Cadenillas R, Salas E, Tello C, Zeballos H (2009) Diversity and endemism of Peruvian mammals. Revista Peruana de Biologia 16:5–32

Parry ML, Canziani OF, Palutikof JP, Van der Linden PJ, Hanson CE (2007) Contribution of working group II to the fourth assessment report of the intergovernmental panel on climate change. Cambridge University Press, Cambridge/New York

Patterson BD (2000) Patterns and trends in the discovery of new neotropical mammals. Divers Distrib 6:145–151

Phillips OL (1997) The changing ecology of tropical forests. Biodivers Conserv 6:291–311

Pianka ER (1966) Latitudinal gradients in species diversity: a review of concepts. Am Nat 100:33–46

Pitman N, Vriesendorp C, Moskovits D (2003) Peru: Yavari. Rapid Biological Inventories Report 11. The Field Museum, Chicago

Prance GT (1979) Notes on the vegetation of Amazonia III. The terminology of Amazonian forest types subject to inundation. Brittonia 31:26–38

Scarano FB (2002) Structure, function and floristic relationships of plant communities in stressful habitat marginal to the Brazilian Atlantic rainforest. Ann Bot 90:517–524

Schulze CH, Linsenmair KE, Fiedler K (2001) Understorey versus canopy: patterns of vertical stratification and diversity among Lepidoptera in a Bornean rain forest. Plant Ecol 153:133–152

Sorensen LL (2004) Composition and diversity of the spider fauna in the canopy of a montane forest in Tanzania. Biodivers Conserv 13:437–452

Stork NE, Grimbacher PS (2006) Beetle assemblages from an Australian tropical rainforest show that the canopy and the ground strata contribute equally to biodiversity. Proc R Soc: Biol Sci 273:1969–1975

Terborgh J (1988) The big things that run the world: a sequel to E. O. Wilson. Conserv Biol 2:402–403

Terborgh J (1992) Maintenance of diversity in tropical forests. Biotropica 24:283–292

Thomas CD, Cameron A, Green RE, Bakkenes M, Beaumont LJ, Collingham YC, Erasmus BFN, Ferreira de Siqueira M, Grainger A, Hannah L, Hughes L, Huntley B, Van Jaarsveld AS, Midgley JF, Miles L, Ortega-Huerta MA, Peterson AT, Phillips OL, Williams SE (2004) Extinction risk from climate change. Nature 427:145–148

Tobler MW (2008) The ecology of the lowland tapir in Madre de Dios, Peru: using new technologies to study large rainforest mammals. Phd Dissertation, Texas A&M University

Tuomisto H, Ruokolainen K, Kalliola R, Linna A, Danjoy W, Rodriguez Z (1995) Dissecting Amazonian biodiversity. Science 269:63–66

Vedder AL (1984) Movement patterns of a group of free-ranging mountain gorillas (*Gorilla gorilla beringei*) and their relation to food availability. Am J Primatol 7:73–88

Viveiros Grelle CE (2003) Forest structure and vertical stratification of small mammals in a secondary Atlantic forest, Southeastern Brazil. Stud Neotrop Fauna Environ 38:81–85

Wilson EO (1985) The biological diversity crisis. Bioscience 35:700–706

Wilson EO, Peter FM (1988) Biodiversity. National Academy Press, Washington, DC

Wilson RF, Marsh H, Winter J (2007) Importance of canopy connectivity for home range and movements of the rainforest arboreal ringtail possum (*Hemibelideus lemuroides*). Wildl Res 34:177–184

Whittaker RH (1972) Evolution and measurement of species diversity. Taxon 21:213–251

Chapter 10
Primates' Use of Flooded and Unflooded Forests in Peruvian Amazonia

Janice Chism and Richard L. Jackson Jr

10.1 Introduction

10.1.1 Peruvian Amazonia Forest Habitats

The Amazon River basin in Peru contains approximately 600,000 km^2 of tropical lowland rain forest of which approximately 20% is subjected to annual monomodal flooding (Kvist and Nebel 2001). This tropical lowland rain forest can be subdivided into three major types of forest: terra firme forest, which is above the floodplain and does not experience seasonal inundation, and two types of flooded forests, typically characterized based on the type of river they border.

Várzea forests are seasonally flooded by white-water rivers, which are turbid, are close to a neutral pH, and carry a nutrient-rich alluvial suspension from the Andes (Prance 1979; Kvist and Nebel 2001). These white-water rivers dominate the Peruvian Amazon basin. Nutrient-rich várzea forests have the highest floral species richness of any floodplain forest in the world (Whitmann et al. 2010).

Igapó forests are seasonally flooded by black-water rivers, which are clear, dark brown and acidic due to the colloidal suspension of plant compounds, originate in sandy areas, and are nutrient-poor compared to white-water rivers (Prance 1979; Kvist and Nebel 2001). Tree species richness is poorer in igapó forest compared to várzea forest (Whitmann et al. 2010).

Terra firme forest comprises the remaining 80% of tropical lowland rain forest in the Peruvian Amazon basin and is not subject to annual flooding (Kvist and Nebel 2001). As terra firme forest does not benefit from the annual deposition of alluvial sediments, the soil of terra firme is typically nutrient-poor from leaching (Haugaasen and Peres 2005). Even though the soil of terra firme forest is nutrient-poor compared

J. Chism (✉) · R. L. Jackson Jr
Department of Biology, Winthrop University, Rock Hill, SC, USA
e-mail: jacksonr@winthrop.edu; chismj@winthrop.edu

R. W. Myster (ed.), *Igapó (Black-water flooded forests) of the Amazon Basin*,
https://doi.org/10.1007/978-3-319-90122-0_10

to both várzea and igapó forest, tree species richness is higher than in the flooded forests, and this is considered to be because the soil in flooded forest is subjected to periodic lack of oxygen during times of water inundation (Whitman et al. 2010).

Western Amazonia has some of the highest primate species-rich communities found anywhere in the world (Peres and Janson 1999). Primate communities with 14 sympatric species have been found in Brazil on the Juruá River (Peres 1997) and in Peru within the ACRCTT (Aquino and Encarnación 1994; Puertas and Bodmer 1993). Thirteen sympatric primate species have been observed near the headwaters of the Urucu River in Brazil (Peres 1993) and inhabiting the Ponta da Castanha, Brazil (Johns 1985). Twelve sympatric primate species were detected in the Purús region of Brazil (Haugaasen and Peres 2005).

While the Amazon basin is a mosaic of terra firme, igapó, and várzea forests, primate species richness is reported to be higher in terra firme than either of the flooded forests (Peres 1993, 1997; Haugaasen and Peres 2005, 2009). Haugaasen and Peres speculate that lower species richness in várzea and igapó forest is due to their lower floral diversity compared to terra firme as a result of the soil composition and impact of seasonal flooding on the flora (2005).

10.1.2 Primates in Neotropical Flooded Forests

As a radiation, South American (Neotropical) primates, the Platyrrhini, have traditionally been considered to be overwhelmingly forest-dwelling and arboreal in their habitat use (Sussman 2003; Rosenberger et al. 2009). This contrasts markedly with primate radiations in Africa and Asia where species invaded and successfully occupied more open habitats (woodlands, savannahs, and even steppe areas) and which have given rise to many terrestrial species (Lynch Alfaro et al. 2015). Since across their biogeographic range with rare exceptions, Neotropical primates spend relatively little time on the ground, no matter the forest type, until recently, not much attention has been paid to their use of flooded forests, other than as an impediment to studying them (Boyle et al. in press). The assumption appears to have been that, as primarily arboreal species, primates would be relatively unaffected by forest flooding. As recent work on the evolution of Platyrrhines makes clear, however, the Amazon basin with its complex and constantly shifting river drainages, including its flooded forests, has been a major generator of species diversity for South American primates (Lynch Alfaro et al. 2015). The debate as to whether Amazonian rivers should be regarded mainly as barriers to dispersal or as generators of new species (and therefore diversity for monkeys) goes back to Alfred Russel Wallace (1852) and continues to be actively debated. Differences center on the weighting of effects of geographic isolation leading to vicariance as opposed to those of species' adaptation to the complexly changing Amazonian environment which occurred over the late Miocene–Pliocene, including the emergence of forest types discussed here (e.g., Boubli et al. 2015, Jameson Kiesling et al. 2015). A study on the distribution of small Amazonian mammals and frogs found that habitat and geographic distance were better predictors of species distribution than riverbank side supporting the view that rivers were not acting as insurmountable barriers to dispersal (Gascon et al. 2000). This finding is significant considering that the species surveyed by Gascon and colleagues are less able to cross rivers than most primates. Perhaps the most nuanced view of the role of these rivers, their flooding cycles, and their propensity to meander is that they act to separate populations of monkeys at least temporarily and that these episodes of isolation can, if they persist, generate diversity (e.g., Lynch Alfaro et al. 2015). However, all Platyrrhines do not respond identically to these dynamic riverine environments as comprehensive study of diversification of primates in the Rio Negro and Rio Branco regions of Brazil demonstrated (Boubli et al. 2015). Some species' distribution did indeed appear strongly limited by rivers (e.g., titi monkeys, genus *Callicebus*), while others did not including howlers (*Alouatta* spp.) and especially squirrel monkeys (*Saimiri* spp.), which are apparently competent swimmers. Field primatologists are occasionally dramatically reminded that monkeys can and do cross streams that appear to be too wide for them to easily jump. At our study site we have seen a red howler monkey (*Alouatta seniculus*) successfully swim across the Tahuayo River and an adult male saki monkey (*Pithecia monachus*) leap across that same river where it was 15 m wide. However, the monkey actually jumped about 3 m between the overhanging branches of trees on each side (Jackson 2016).

The importance of Amazonia forests to South American primates is indicated by Peres' (1997) calculation that for this taxonomic group, 66% of species and 88% of genera occur in these forests. Worldwide, primate species richness is highest in the Neotropics, and a recent analysis attributes this to vertical structure of the forests, measured using canopy height as a proxy, rather than to rainfall as has been previously supposed (Gouveia et al. 2014).

Among main Amazonian forest habitat types (terra firme, várzea, and igapó), igapó forests have typically been regarded as the least productive due to their nutrient-poor soils and typically show less species diversity in both plants and animals (Wittman et al. 2010; Myster this volume). Myster (2009) reported that in Western Amazonia the number of tree species was lower in flooded than in terra firme forests and trees in flood plains are shorter. While there has been no systematic sampling of primate diversity in these three main forest types across Amazonia, the few comprehensive studies which have been carried out found lower species diversity of primates in flooded than in unflooded forests (e.g., Peres 1997; Palminteri et al. 2011). These findings agreed with the general pattern of lower faunal species diversity in flooded forests (Haugaasen and Peres 2005). This lower primate species diversity in flooded forests has been attributed to two factors. First, the lower floral species diversity may result in fewer potential food sources for primates which are heavily dependent for food on plants (fruit and leaves primarily) or the insects attracted to or living in them. The second factor is tree height, which is reduced in flooded forests (Gouveia et al. 2014). Since Neotropical primate species have long been described as having diversified niches based at least partly on vertical stratification (e.g., Terborgh 1983), if flooded forests are shorter than unflooded ones, they may also provide less complex structures and therefore less opportunity for such niche differentiation based on canopy or understory structure.

However, despite reduced plant species richness in flooded forests compared to unflooded ones, primates do occupy flooded forests, and some species even appear to specialize in them (e.g., howlers, *Alouatta* and squirrel monkeys, *Saimiri* species, in várzea forests in Brazil, Peres 1997). In fact, primate species biomass (as opposed to species richness) is higher in some of these flooded forests (Haugaasen and Peres 2005) suggesting that while niches for primates may be fewer, at least at some times of the year resources can be superabundant. If the seasonal flooding also produces marked seasonality in fruit abundance, some primate species may be based predominantly in terra firme forests and only shift to flooded forests when resources there (presumably fruit) are abundant. Conversely, some flooded forest-specialist species may temporarily abandon these forests for unflooded ones, a pattern suggested for sites in Brazil (Peres 1997; Haugaasen and Peres 2005).

10.2 Primates in Igapó Forests

There have been few studies of primates in igapó habitats. In a recent review of studies conducted at sites in flooded forests, just 7 of 53 sites identified igapó as a habitat (Boyle et al. in press). Further, even fewer studies have compared primates' use of all three forest types, igapó, várzea, and terra firme, in the same area. One of the most comprehensive of these, conducted in Brazil, found that while primate species richness was lower in igapó than in terra firme habitats, the difference was much less than between várzea and terra firme (Haugaasen and Peres 2005). This was somewhat surprising since igapó forests are considered to be nutrient-poor and less floristically diverse than other flooded and unflooded forests (Wittman et al. 2010). However, if these forests are flooded for a limited time (just a few months rather than most of the year) and if the flooding is predictable, trees and the animals which depend on them for food appear to be able to adapt to these conditions. For animals like primates, which do not need to come to the ground to move about, the key to successfully using these habitats may be the ability to move in and out of nearby terra firme habitats when resources in the flooded forest are scarce, as Peres and colleagues described for their sites in Brazil (Peres 1997; Haugaasen and Peres 2005).

10.2.1 Primates' Use of Igapó Forest in Peru

In Peru, species classified in 12 genera of primates have been reported to use igapó forests (Boyle et al. in press). These include species of the genera *Alouatta*, *Ateles*, *Lagothrix*, *Aotus*, *Cebuella*, *Saguinus*, *Cebus*, *Saimiri*, *Sapajus*, *Cacajao*, *Callicebus*, and *Pithecia*. These primates range from the largest New World primates (*Ateles*, spider monkeys) to the smallest (*Cebuella*, pygmy marmosets) and represent the full range of dietary adaptations for South American species including strongly folivorous species such as howler monkeys, gummivores such as the marmosets, and the strongly frugivorous species such as spider monkeys. While some of these species are definitely known to use both igapó and other habitats (terra firme, várzea), so far none have been definitively found to be igapó specialists.

10.2.2 Primates in the Igapó Forest Habitats of the Área de Conservación Regional Comunal Tamshiyacu Tahuayo, Northeastern Peruvian Amazon

The Área de Conservación Regional Comunal Tamshiyacu Tahuayo (ACRCTT) is a communal reserve located in Loreto Department in northeastern Peru covering 420,080 hectares (Penn 2009). The reserve was established in 1991 in response to combined efforts by local inhabitants, conservationists, and researchers to prevent loggers and hunters coming into the area from outside and extracting resources and for the protection of the endangered red uakaris (*Cacajao calvus ucayalii*) (Meyer and Penn 2003; Newing and Wahl 2004). The predominant habitat in the reserve is terra firme, but it includes both várzea and igapó flooded forests along its smaller white and black-water rivers. One of the rivers that defines the reserve, the Tahuayo, is black water at its source and upper reaches and then becomes a mix of black and white water after its confluence with the Río Blanco. This reserve has among the highest primate species diversity reported for any site in South America (Puertas and Bodmer 1993; Peres 1997), with at least 14 species. Indeed due to continuing uncertainty about the taxonomic status of several of these species, there may be as many as 16 or 17 present in the area (Chism and Jackson in prep.).

10.3 Census of Primates in the ACRCTT

Here we summarize information on primate use of habitat use collected during fieldwork conducted in the ACRCTT with several graduate students in 12 years beginning in 2001. We report data for primates seen in igapó habitats and compare these with observations of species in terra firme habitats in the same area. This fieldwork included methodical censuses using transects in terra firme and várzea habitats along the Río Blanco (in 2005, 2006) and systematic but less formal censuses by boat and on foot in igapó habitats along the upper Tahuayo River and its tributary streams during and outside flood periods (in 2001, 2003, 2004, 2008, 2010, 2011, 2012, 2014, 2015, 2016).

Data presented here derive from the formal transect surveys carried out to assess the effects of hunting on primate species richness and abundance (Matthews 2006;

Fig. 10.1 A saki monkey (*Pithecia inusta*) in igapó forest along the Tahuayo River, ÁCRCTT, Peru. Photo by Richard Jackson

Chism, Guinan, and Seidewand unpublished) and from ad libitum observations collected during several field projects which were focused on surveying for red uacaris (*Cacajao calvus ucayalii*) (Ward and Chism 2003) and the behavioral ecology, including habitat use, of saki monkeys (*Pithecia* spp.) (Frisoli and Chism 2009; Chism and Kieran 2008, Chism unpublished data) (see Fig. 10.1). During these later studies, we made systematic surveys along the Tahuayo River covering distances of approximately 5 km up- and downriver from the Tahuayo River Amazon Research Center (ARC) (S 04° 23.334′, W 073° 15.438′), a research station located on the black-water portion of the Tahuayo River on the northwest border of the ACRCTT. The main goal of these surveys was to search for groups of uakaris or saki monkeys, but we systematically recorded the presence of all primates encountered during these searches. In addition, in 2008, 2010, 2012, 2015, and 2016, we also searched for and followed groups of saki monkeys in igapó forest directly behind the ARC on a 2 km × 2 km trail grid consisting of 42 intersecting trails spaced approximately 100 m apart, oriented southeast of the Tahuayo River. Again during this fieldwork, we systematically recorded the identity of all primates encountered in the area. While the majority of this fieldwork occurred in the period just after the peak of the flood season to dry season (mid-May to mid-August), in two years surveys were conducted during February (2011) and late February to mid-April (Lehtonen 2016). Thus observations reported here include periods of maximum flooding, falling water, and when forests were unflooded.

Fig. 10.2 The authors at the Tahuayo River Amazon Research Center, July 2015. In 2015 the annual flood in the igapó was unusually high and prolonged. The thatched building in the background clearly shows the waterline from the flood waters which receded only in late June

Systematic census data in terra firme were collected during two study periods. The first occurred during June–July 2005 at two different sites about 5 km apart in forest along the Río Blanco (S4^0 25.98′, W73^0 7.63′ and W4^0 23.14′, W73^0 9.93′) where we cut four transects ranging in length from 2 to 4 km following a protocol described by Peres (1999). Transects were cut by hand with machetes and marked every 50 m. Transects were walked at a pace of approximately 1.00–1.50 km per hour by teams of one to three researchers. When a primate group was located, we recorded the species, time detected, number of individuals, location, and behavior. During this study we accumulated 279 km of transect effort. In addition, a second study in this same area collected approximately the same amount of census data over 43 days in May–July 2006 (Chism, Guinan, and Seidewand unpublished data).

For the river surveys and for searching for monkeys during periods when the forest was flooded, we used small motorboats or canoes (see Fig. 10.2). When using motorboats, we motored slowly upriver then turned off the motor and floated back down the river. Teams of two to three observers in the boats were able to visually search forest on both sides of the river for signs of primates. While primates were often detected first by their vocalizations, species identity was always confirmed visually except for howler monkeys (*Alouatta seniculus*) whose calls are distinct and definitive for identification. While searching for saki monkeys on the ARC trail grid, teams of one to three observers walked at a pace of approximately 1.00–1.50 km per hour, and visual identification was required for confirmation of a species' presence. Grid location and habitat type were recorded for every primate sighting (Frisoli 2008; Jackson 2016).

10.4 Habitat Use by ACRCTT Primates

Of the minimum of 14 securely identified primate species reported to occur in the ACRCTT (Puertas and Bodmer 1993; Peres 1997), we observed 13 during our field studies in the area. The only species we have never observed in the study area in any habitat is the spider monkey *Ateles chamek*. This species, among the largest Neotropical primates in body size, is classified by IUCN as endangered and hunting, especially market hunting, and is considered its main threat (Matthews 2006; Wallace et al. 2008). After no sightings in the ACRCTT for at least 15 years, this species may be locally extinct. The species we observed and the habitats we found them in are presented in Table 10.1.

Of the 13 species we observed over the period from 2001 to 2016, 11 species were confirmed to occur in terra firme habitats, while 11 were confirmed in igapó habitats. Eight species (red howlers, red uakaris, coppery titis, white-fronted

Table 10.1 Primate species we visually observed in igapó forest within the ACRCTT study site and the habitats in which we observed them based on data from transect studies and our less formal surveys

Scientific name	Common name	Observed in igapó	Observed in terra firme
Alouatta seniculus	Red howler monkey	Yes	Yes
Aotus nancymaae	Owl monkey	No	Yes
Cacajao calvus ucayalii	Red uakari	Yes	Yes
Callicebus cupreus	Coppery titi monkey	Yes	Yes
Cebuella pygmaea	Pygmy marmoset	Yes	No
Cebus albifrons	White-fronted capuchin	Yes	Yes
Lagothrix poeppigii	Poeppig's woolly monkey	No	Yes
Pithecia monachus (cf Marsh 2014)	Monk saki	Yes	Uncertain
Pithecia inusta (cf Marsh 2014)	Burnished saki	Yes	Yes
Saimiri sciureus	Common squirrel monkey	Yes	Yes
Saguinus fuscicollis	Saddleback tamarin	Yes	Yes
Saguinus mystax	Moustached tamarin	Yes	Yes
Sapajus macrocephalus	Large-headed capuchin	Yes	Yes

capuchins, burnished sakis, squirrel monkeys, both species of tamarins and large-headed capuchins) were observed in both habitats and on multiple occasions making it highly probable that these species are habitat generalists able to occupy flooded and unflooded forests. While we did not have any confirmed sightings of the nocturnal owl monkey in igapó, diurnal censuses in any habitat are highly likely to miss this species which is rarely active during the day. Experienced guides who regularly camp in igapó tell us that it is present in this habitat, however. Only one species, the pygmy marmoset, appeared to be a habitat specialist for flooded forests since it was never observed in terra firme. We did have one possible observation of these tiny monkeys in várzea. We had no confirmed observations of the largest surviving primate species in the area, the woolly monkey, in igapó forests during our studies but again experienced guides report seeing it on the research trail grid. Red uacaris enter igapó areas adjacent to terra firme forests at the edge of the ARC trail grid at some times of the year.

10.5 Interpreting These Patterns of Primate Use of Igapó Forests

Our finding that the number of primate species did not differ between the igapó and terra firme habitats we surveyed in the ACRCTT appears to contradict the findings of other studies comparing primates' use of Amazonian habitats that primate species richness was higher in terra firme habitats than in flooded forests (e.g., Peres 1997; Haugaasen and Peres 2005; Palminteri et al. 2011). What factors might explain the species richness of the igapó forest in the ACRCTT of the northeastern Peruvian Amazon, then?

In the ACRCTT, igapó forests visually closely resemble várzea forests which typically grow on younger, more nutrient-rich soils. While igapó forests in other regions of Amazonia are described as having large areas of open, sandy soil and being almost desert-like when they dry out, this is not true of igapó in Peru which grows on richer soil, similar to that of várzea perhaps because of the proximity of this area to the Andes (Prance 1979). Thus, even though igapó typically has fewer tree species (Wittman et al. 2010), it may produce abundant food for primates, especially at certain times of the year. Some igapó tree species pursue a strategy of investing in nutrient-rich seed masses (Parolin 2001), a food source that especially pitheciin primates, such as sakis, seem to be well-adapted to exploit (Kay et al. 2013; Norconk and Veres 2011; Norconk et al. 2013).

Interestingly, the number and identity of species we observed in ACRCTT igapó match that reported by Bennett et al. (2001) for floodplain forest seasonally inundated by black and white-water rivers in the Río Tapiche area of northeastern Peru.

Our observations strongly support the idea that at least some primate species' use of igapó has a seasonal component as suggested by several other authors (e.g., Peres 1997; Haugaasen and Peres 2005; Bowler and Bodmer 2011). Some species seem to be mostly based in igapó but may move into terra firme areas at some times of the year when their preferred foods are scarce (e.g., sakis), while others may make brief seasonal incursions into igapó when fruits there are superabundant. This later scenario may be the case with uacaris and woolly monkeys and may explain their occasional appearances at the edge of the trail grid in igapó in our study area at ARC. It is likely that capuchins may also be moving widely and using both igapó and terra firme habitats in this area.

It has been suggested that small-bodied primates such as tamarins are less abundant or absent from flooded forests because these species occupy lower levels of the forest, descending into the shrub layer to forage on arthropod prey (Peres 1997). Thus, when forests are flooded, their available habitat and an important food source would be curtailed. If these species could not compensate by switching to alternate food sources, they would be disadvantaged in this type of habitat. However, tamarins and other small-bodied primates (pygmy marmosets and titis) were abundant in ACRCTT igapó even when flooding was at its maximum. These primates must be able to find some compensatory food resources during several weeks of flooding they experience each year in our study area, because they do not leave the forest.

An important factor in this unexpected species richness in igapó at least at our site may be the presence of extensive aguajales (palm swamps) dominated by *Mauritia flexuosa* (aguaje palms). These palms fruit abundantly at exactly the time of year when fruit is otherwise scarce in these forests (Bowler and Bodmer 2011). When an aguaje tree is fruiting, it becomes a mecca for most primates in the area as well as for other nonprimate animals, confirming its reputation as a keystone species. Our observations of saki monkeys suggest that groups shift their ranges during these periods to allow them to use this superabundant resource and that the groups within whose range these palms occur furiously attempt to repel encroachment from nonresident saki groups (Frisoli and Chism 2009).

Our observations of primates in igapó habitat in northeastern Peru show that while these flooded forests may be less diverse in tree species, they harbor as much primate richness as nearby terra firme forests. This is true, even though these flooded forests have been subjected to higher human disturbance and hunting pressure because of the easier access to them than to terra firme forests in the same area. (Matthews 2006; Porter et al. 2013). For example, *Alouatta*, which is abundant in flooded forests elsewhere (Haugaasen and Peres 1997), is now rarely encountered at our site in igapó. *Alouatta* is at the top end of the size range for primates surviving in these forests since *Ateles* appears to be locally extinct now and the other large-bodied primates, uacaris and woolly monkeys, are becoming rare. So while flooded forests represent only about 20% of Peruvian Amazonian forests (e.g., Kvist and Nebel 2001), they harbor a significant amount of primate species diversity. If, as many primatologists working in these forests suspect, the ability of at least some of these species to move back and forth between flooded and unflooded forest is critical at some times of the year, these flooded forests may be key to maintaining sustainable populations. This may be particularly true for the endangered and rare species such as spider monkeys, uakaris, and woolly monkeys.

Acknowledgments The authors wish to thank Hope Matthews, Sean Guinan, Amy Seidewand, Lauren Frisoli Brasington, Troy Kieran, and Candace Stenzel for their work and companionship in the field. We thank Dr. Paul and Dolly Beaver and Alfredo Dosantos for their support and enthusiasm over more years than seems possible. JC thanks Dr. Bill Rogers for years of support and holding down the fort at home while she wandered the Amazon and Will and Ali Rogers for pretending to listen with interest when she insisted on talking about monkeys *again* at dinner. We thank our many Peruvian guides and field assistants without whose help we would never have emerged from the igapó.

References

Aquino R, Encarnación F (1994) Primates of Peru Primate Report 40:1–127

Bennett C, Leonard S, Carter S (2001) Abundance, diversity, and patterns of distribution of primates on the Tapiche River in Amazonian Peru. Am J Primatol 54:119–126

Boubli J-P, Ribas C, Lynch Alfaro J, Alfaro M, da Silva M, Pinho G, Farias I (2015) Spatial and temporal patterns of diversification on the Amazon: a test of the riverine hypothesis for all diurnal primates of Rio Negro and Rio Branco in Brazil. Mol Phylogenet Evol 82:400–412

Bowler M, Bodmer RE (2011) Diet and food choice in Peruvian red uakaris (*Cacajao calvus ucayalii*): selective or opportunistic seed predators. Int J Primatol 32:1109–1122

Boyle S, Alho C, Chism J, Defler T, Palacios E, Santos R, Wallace R, Wright B, Wright K, Barnett A (in press) Conservation of primates and their flooded habitats in the Neotropics. In: Barnett A et al (eds) Primates in flooded forests. Cambridge University Press, Cambridge

Chism J, Kieran T. (2008) Vocal communication of sympatric equatorial and monk sakis (*Pithecia aequatorialis* and *Pithecia monachus*) in Northeastern Peruvian Amazon. Int. Congress of Primatol XXV, Hanoi, Vietnam

Frisoli L (2008) A behavioral investigation of the *Pithecia* niche in Northeastern Peru. MS thesis Winthrop University, Rock Hill, SC, USA

Frisoli L, Chism J (2009) Habitat and resource use by saki monkeys (*Pithecia* spp.) in Amazonian Peru. Annual Meetings of the American Society of Primatology, San Diego, CA

Gascon C, Malcolm J, Patton J, da Silva M, Boragt J, Loughheed S, Peres C, Neckel S, Boag P (2000) Riverine barriers and the geographic distribution of Amazonian species. PNAS 97:13672–13677

Gouveia S, Villalobos F, Dobrovolski R, Beltrão-Mendes R, Ferrari S (2014) Forest structure drives global diversity of primates. J Anim Ecol 83:1523–1530

Haugaasen T, Peres C (2005) Primate assemblage structure in Amazonian flooded and unflooded forests. Am J Primatol 67:243–258

Haugaasen T, Peres C (2009) Interspecific primate associations in Amazonian flooded and unflooded forests. Primates 50:239–251

Jackson RL Jr (2016) Habitat stratification of *Pithecia* Species in the Área de Concervación Regional Comunal Tamshiyacu Tahuayo in the Northeastern Peruvian Amazon. MS thesis. Winthrop University, Rock Hill SC, USA

Jameson Kiesling N, Yi S, Sperone FG, Wildman D (2015) The tempo and mode of new world monkey evolution and biogeography in the context of phylogenetic analysis. Mol Phylogenet Evol 82:386–399

Johns A (1985) Primates and forest exploitation at Tefé. Brazilian Amazonia Primate Conservation No 6:27–29

Kay RF, Meldrum DJ, Taki M (2013) Pitheciidae and other platyrrhine seed predators. Pp 3-12. In: Viega LM, Barnett AA, Ferrari SF, Norconk MA (eds) Evolutionary biology and conservation of Titis, Sakis and Uakaris. Cambridge University Press, New York

Kvist LP, Nebel G (2001) A review of Peruvian flood plain forests: ecosystems, inhabitants and resource use. For Ecol Manag 150:3–26

Lehtonen E (2016) The behavioural ecology of a potentially undescribed morph of saki monkey (genus *Pithecia)* in a highly diverse primate community. MS thesis. Uppsala University, Sweden
Lynch Alfaro J, Cortés-Ortiz L, Di Fiore A, Boubli J-P (2015) Special issue: comparative biogeography of Neotropical primates. Mol Phylogenet Evol 82:518–529
Matthews, H (2006) The effects of hunting on an primate community in the Peruvian Amazon. MS thesis. Winthrop University, Rock Hill, SC, USA
Meyer D, Penn J (2003) An overview of the Tamshiyacu-Tahuayo communal reserve. In: Pitman N, Vriesendorp C, Moskovits D (eds) Perú: Yavarí: Rapid Biological Inventories: 11. The Field Museum, Chicago, pp 176–177
Myster RW (2009) Plant communities of western Amazonia. Bot Rev 75:271–291
Myster RW (this volume) Introduction. In: Myster RW (ed) Igapó (black-water flooded forests) of the Amazon Basin. Springer, Berlin, p xx
Newing H, Wahl L (2004) Benefiting local populations? Communal reserves in Peru. Cultural Survival. 28.1. www.culturalsurvival.org
Noconck MA, Veres M (2011) Physical properties of fruit and seeds ingested by primate seed predators with emphasis on sakis and bearded sakis. The Anat Rec 294:2092–2111
Norconck MA, Grafton BW, McGraw WS (2013) Morphological and ecological adaptations to seed predation – a primate-wide perspective. In: Viega LM, Barnett AA, Ferrari SF, Norconk MA (eds) Evolutionary biology and conservation of Titis, Sakis and Uakaris. Cambridge University Press, New York, pp 55–71
Palminteri S, Powell G, Peres CA (2011) Regional-scale heterogeneity in primate community structure at multiple undisturbed forest sites across South-Eastern Peru. J Trop Ecol 27:181–194
Parolin P (2001) Seed germination and early establishment of 12 tree species from nutrient-rich and nutrient-poor central Amazonian floodplains. Aquat Bot 70:89–103
Penn J (2009) RCF update. Rainforest conservation fund: July, 2009. http://www.rainforestconservation.org/archives/1
Peres CA (1993) Structure and spatial organization of an Amazonian terra firme forest primate community. J Trop Ecol 9:259–276
Peres CA (1997) Primate community structure at twenty western Amazonian flooded and unflooded forests. J Trop Ecol 13:381–405
Peres CA (1999) General guidelines for standardizing line-transect surveys of tropical forest primates. Neotropical Primates 7:11–16
Peres CA, Janson CH (1999) Species coexistence, distribution, and environmental determinants of neotropical primate richness: a community- level zoogeographic analysis. In: Fleagle JG, Janson C, Reed KE (eds) Primate communities. Cambridge University Press, Cambridge, pp 55–74
Porter L, Chism J, Defler T, Marsh L, Martinez J, Matthews H, McBride W, Tirira D, Velilla M, Wallace R (2013) Pitheciid conservation in Ecuador, Colombia, Peru, Bolivia and Paraguay. In: Veiga LM, Barnett AA, Ferrari SF, Norconk MA (eds) Evolutionary biology and conservation of Titis, Sakis and Uacaris. Cambridge University Press, New York, pp 320–333
Prance GT (1979) Notes on the vegetation of Amazonia III. The terminology of Amazonian forest types subject to inundation. Brittonia 31:26–38
Puertas P, Bodmer RE (1993) Conservation of a high diversity primate assemblage. Biodivers Conserv 2:586–593
Rosenberger A, Tejedo M, Cooke S, Pekar S (2009) Platyrrhine ecophylogenetics in space and time. In: Garber P, Estrada A, Bicca-Marques J, Heymann E, Strier K (eds) South American Primates. Springer, New York, p 113
Sussman R (2003) Primate ecology and social structure, vol 2. Pearson, Boston, MA
Wallace AR (1852) On the monkeys of the Amazon. Proc Zool Soc Lond 20:107–110
Wallace RB, Mittermeier RA, Cornejo F, Boubli J-P (2008) Ateles chamek. The IUCN Red List of Threatened Species 2008
Whitman F, Schöngart J, Junk WJ (2010) Phytogeography, species diversity community structure and dynamics of central Amazonian floodplain forests. In: Junk WJ, Pidade MTF, Wittman F, Schöngart J, Parolin P (eds) Amazonian floodplain forests. Springer, New York, pp 61–102

Chapter 11
Turtles of the Igapó: Their Ecology and Susceptibility to Mercury Uptake

Larissa Schneider and Richard C. Vogt

11.1 Introduction

This chapter focuses on giving an in-depth analysis of turtles living in igapó of the Brazilian Amazon, in a way that has not been covered in more general studies describing the entire Amazonian system. The different water chemistry in the igapó represents important parameters for the study of turtle ecology and for their management. Before going into the ecological particularities of turtles of the igapó and implication on Hg bioaccumulation, we explain the flood-pulse system in the Amazon and define igapó in the context of this chapter and its ecological and chemical characteristics that affect turtle ecology. We then cover the adaptation of turtles to the igapó flood pulses and the ecological role that the igapó chemistry plays on turtle species. This includes explaining how the igapó chemistry and ecology promote the uptake of mercury (Hg) by turtles more efficiently than other environments of the Amazon.

11.2 The Amazonian Flood Pulse and the Igapó

The Amazon system is characterized by floodplain ecosystems which are driven by periodic inundation and oscillation between terrestrial and aquatic phases (Ríos-Villamizar et al. 2013). The fluctuation of river levels throughout the year is a result of heavy seasonal rainfall, concentrated in the eastern Andes and the

L. Schneider (✉)
Archaeology and Natural History, Australian National University, Canberra, ACT., Australia

Instituto Nacional de Pesquisas da Amazônia, Manaus, Brazil
e-mail: Larissa.Schneider@anu.edu.au

R. C. Vogt
Instituto Nacional de Pesquisas da Amazônia, Manaus, Brazil

R. W. Myster (ed.), *Igapó (Black-water flooded forests) of the Amazon Basin*,
https://doi.org/10.1007/978-3-319-90122-0_11

northwest area of the basin (Junk 1997). Flooding radically alters the forest landscape, and as the waters creep over the land, the habitats that are created make it possible for aquatic organisms to navigate gallery forests (found alongside the riverbanks) and to seek food there. This is the key characteristic of these flooded areas to play a crucial role in turtle ecology.

In the Amazon Basin of Brazil, a seasonally water-flooded forest can be called either várzea or igapó. Both environments are different in many regards, but the key difference is in the type of water that floods the forest. Várzea forests are flooded by muddy high nutrient content rivers, while igapó forests are flooded by blackwater and clearwater tributaries (Goulding et al. 2003). These environments typically occur along the lower reaches of rivers and around freshwater lakes.

In this chapter we will discuss two types of igapós in the Brazilian Amazon, the black- and the clearwaters, both influenced by sediment and nutrient-poor water rivers.

(a) *Blackwater igapó*: the blackwater system, represented by the Negro River Basin, originated on the Precambrian shield of the northern region of the Amazon Basin. Its transparent red-brown color originates from a high content of dissolved humic and tannic substances which is about ten times higher than in the Solimões/Amazon River. The water is poor in nutrients and electrolytes with dominance of sodium among the major cations, presenting low alkalinity. The pH and electrical conductivity values are less than 5.0 and 25 μS cm−1, respectively. The black color and acidity of the water are due to the elevated concentrations of dissolved organic material such as humic and fulvic acids (Furch and Junk 1980).

In this chapter we approach the blackwater igapó of the Rio Negro and its tributaries.

(b) *Clearwater igapó*: the rivers of the clearwater type have their upper catchments in the Central Brazilian and Guiana Archaic/Precambrian shields and are characterized by pH values that vary between 5.0 and 7.0 and electrical conductivity is in the range of 10–53.6 μS cm-1. The water transparency can reach up to 355 cm or still higher; but transparency values less than 100 cm are also common in these rivers (Ríos-Villamizar et al. 2013).

In this chapter we approach clearwaters of the Rio Trombetas, Rio Tapajós, and Rio Guapore. The acidity of Rio Trombetas is equivalent to tap water, essentially no electrolytes.

11.3 Igapó Characteristics and Its Implications to Turtle Ecology

Várzea forests have high nutrient contents because they receive a transport of high sediment loads from the whitewater rivers. In contrast, igapó forests do not receive this seasonal influx of sediments which explains the nutrient-poor soils (Goulding et al. 2003). They do have, however, the highest phosphorus concentrations out of comparable várzea and terra firme forest soils (Ríos-Villamizar et al. 2013). This phosphorus is uptaken by turtles when feeding on periphyton.

Igapó forests support comparatively less life, and the environment found within these areas tends to lack species diversity and animal biomass (Ríos-Villamizar et al. 2013). Turtles do not fit the same patterns as other faunal species. The igapó has higher abundance of turtle species (5) than the várzea (3), with lower abundance of individuals per species in the igapó.

The igapó areas are crucial to turtle survival, and their ecology and physiology are highly influenced by flooding. Turtles enter the igapó during flooded seasons to feed on fruits, algae, and seeds, while in the dry season, adults enter the rivers, while juvenile and subadults often remain in lakes and pools left by receding waters (Vogt 2008). It is not known if turtles remain in groups while they are foraging in the flooded forests. As water levels drop, turtles leave the igapó and enter lakes or rivers for the dry season.

During the dry season, *Podocnemis expansa* forms groups of adult females to migrate up- or downstream to the high coarse sand beaches along the river to nest. Males and subadults also migrate 75–400 km with the females. The group is led by mature adult females that have made this journey many times, and they group together using vocalizations. Once at the nesting beaches, females communally bask for 2 weeks to ovulate their eggs and copulate with males in front of the beaches. During this time, they do not feed.

The females communicate vocally to bask together and to emerge onto the beach to nest in large *arribadas* numbering hundreds of females nesting at the same time. Historical descriptions of the *arribada* event (100 years old) record these turtles basking in thousands (Bates 1863). Females remain in the vicinity of the nesting beaches in deep areas in the river until their young hatch in 6–8 weeks, the young vocalize when they enter the water, and the females respond and lead the whole cardume to the feeding grounds in the igapó. When the rainy season starts, the rising water level signals the females that the feeding areas in the igapó will soon be inundated. This increase in water level also stimulates the females to migrate downstream. These migrations are annual, turtles often moving 20 km or more per day (Vogt 2008).

The igapó is not stationary. As the forest is flooded, new territory is covered by water daily, until the water levels reach their peak 7 months later. Thus, the igapó maintains a variety of habitats while the water is rising. Some species, e.g., matamatas, continue moving with the shoreline, remaining in water about 1 m deep, where they forage for small fish. Adult giant South American river turtles and yellow-spotted river turtles forage throughout the forest floor, and when the water is at its height, often 8 m above dry season levels, they are foraging in the treetops.

Igapó tree seeds and fruits are important food for turtles. Tree species adapted to seasonal inundation maximize fruit production during periods of flooding in order to take advantage of newly available seed dispersal methods (Haugaasen and Peres 2006). This is the period when turtles are feeding and storing energy to lay eggs during the dry season. The variety of turtle nesting habitats, secretive behavior, special feeding strategies, and the immense habitat available in the igapós have supported the existence of a few more species of turtles than the várzea.

Table 11.1 Species of turtles in the igapó and their distribution by water type

Species	Blackwater igapó	Clearwater igapó
Podocnemis erythrocephala	X	X
Podocnemis unifilis	X	X
Podocnemis expansa	X	X
Podocnemis sextuberculata		X
Peltocephalus dumerilianus	X	
Chelus fimbriatus	X	
Mesoclemmys raniceps	X	

11.4 Turtles Species of the Igapó

Table 11.1 shows the species of turtles which live on the blackwater and clearwater igapós. Here we describe some ecological characteristics of the most common turtles of the igapós:

*11.4.1 Red-Headed Amazon River Turtle (***Podocnemis erythrocephala***)*

This species is the smallest in the genus *Podocnemis*. Males and hatchlings have a distinctive bright red or reddish orange pattern on the head (Fig. 11.1). In females the reddish coloration begins to fade to a dull brown at carapace lengths between 12 and 15 cm. A broad band of red stretches across the top of the head between the tympanic membranes and the snout.

This species is primarily herbivorous; from hatchling to adult, 80% of its diet is aquatic plants, mostly filamentous algae. The algae colonies of the igapó support the maintenance of *P. erythrocephala* populations, especially in areas where bushfires have left carbonized trees which provide excellent nutrient base for the production of algal colonies. *Podocnemis erythrocephala* feeds on periphyton which is abundant in the igapó, nutrient rich, and easily digestible (Vogt 2001).

The igapó is also important for this species survival as it consumes leaves, fruits, and seeds of terrestrial plants that either fall into the flooding waters or are gleaned from the branches of trees and shrubs when the forest is flooded. Animal parts, particularly bones, scales, and flesh, contribute to less than 2% of the food consumed by irapucas in the Rio Negro Basin (Vogt 2008).

This species is the smallest in the genus *Podocnemis*, yet it is highly predated in the Rio Negro. So far it has been able to maintain high density due to its nesting strategy in the diversity of habitats in the igapó (Vogt 2001). This maintenance has been achieved perhaps due to the high diversity of the igapó, which is explained below.

Fig. 11.1 Turtles of the igapó: (**a**) red-headed Amazon River turtle (*Podocnemis erythrocephala*); (**b**) yellow-spotted Amazon River turtle (*Podocnemis unifilis*); (**c**) giant South American river turtle (*Podocnemis expansa*); (**d**) big-headed sideneck turtle (*Peltocephalus dumerilianus*); (**e**) matamata (*Chelus fimbriata*)

During the nesting season, which stretches over 2 months, the females lay 2–4 nests. *Podocnemis erythrocephala* first nests on the open white sand beaches near the river edge. As the igapó recedes from the campinas (the savannah vegetation above the beach) and the soil dries, the turtles disperse up to 200 m from the rivers to nest within the sparse clumps of vegetation.

Since most of the predators of this species find nests by visual cues, not scent (e.g., monkeys, *Ameiva* lizards, humans), the nests in the campinas are concealed from predators and are almost undetectable. In fact, their nesting site might have been extended from the beach as a result of over-predation by man.

Because of the diversity of nesting habitats, this species has not been extirpated locally in many areas because the human predators cannot find all of the nests easily as they can on a bare sand beach. So even in areas near the cities where they have been exploited for over a 100 years, populations of this species still exist. This is not the case for the two larger species. In Barcelos, a town on the ridges of the Rio

Negro, *Podocnemis expansa* was extirpated over 30 years ago and *Podocnemis unifilis*, more recently in the last 15 years.

Most known localities for *P. erythrocephala* are blackwater rivers and their tributaries. They occur in less abundance in clearwater streams in the Rio Tapajós and Rio Trombetas basins (Vogt 2008). This species is often found in slow-moving water, with the igapó representing an important environment for this species.

*11.4.2 Yellow-Spotted Amazon River Turtle (*Podocnemis unifilis*)*

Adults have a domed carapace with only a slight medial keel; the keel is pronounced in juveniles. The carapace is smooth in adults and flares out over the hind limbs. Carapace length can reach 37 cm in males and 47 in females (Vogt 2008).

The head of juveniles and males is dark green or brown with bright yellow-orange spots. The spotting pattern varies between populations and individuals, but generally there are postorbital, suborbital, interorbital, nasal, and parietal spots, as well as elongated spots above the tympanum and along the lower jaw. The heads of adult females are uniform rust brown. Males have a longer, thicker tail and retain the juvenile head coloration, but it is less brilliant (Fig. 11.1). The pattern may be lost entirely in some males (Schneider et al. 2012).

Podocnemis unifilis thrives in a wide variety of habitats and are less reliant on the igapós. These turtles occur in whitewater, clearwater, and blackwater rivers in Brazil. They are found basking in igapós on any available branch reaching out of the water, basking in large groups on logs, often on top of each other.

Podocnemis unifilis (and other species as well) may find fruiting trees by following the noisy sounds of macaws and parrots when they are feeding in fruiting trees, and the turtle nibbles on the parts of the fruits that fall in the water. We found fruiting trees in the igapó of the Rio Negro by listening for the raucous calls of the parrots and macaws; *Podocnemis unifilis* were often caught in traps set at the base of these fruiting trees (Vogt 2008). Since we now know that turtles vocalize underwater and hear both airborne and subaquatic sounds, it seems plausible that they could follow the calls of birds feeding in the canopy.

In addition to directly feeding on fallen fruits, this species also employs a unique feeding niche, neustophagia, whereby they float at the surface of the water and suck in the film of floating algae, insects, and other plant material that is floating on the surface of the water (Belkin and Gans 1968). Once the gulp of water and food particles has passed by the throat papillae, the water is regurgitated, and the food particles remain trapped by the erected papillae (Vogt et al. 1998). As the igapó fills, the surface film is continuously replenished with a banquet of items for this species to suck in.

As the igapós are nearly dry and the water receding, this species enters the adjacent lakes and rivers. Some individuals within the same population will nest on sandy islands in the lakes or along the clay banks along the lake or river edge. Others will migrate 20–40 km to nest on the large sandy river beaches along with *P. expansa* and *P. sextuberculata* to deposit one or two clutches of eggs during the nesting season.

Although they are solitary nesters along the river banks, groups of 20–40 will often nest in a single night on sandy nesting beaches within lakes or on rivers. They remain foraging in the lakes and rivers until the rainy season begins and the igapós are filled with water again.

11.4.3 *Giant South American River Turtle (***Podocnemis expansa***)*

Podocnemis expansa, the largest pleurodire, is also the largest freshwater turtle in South America, with carapace measuring 70 cm (65–79). The largest female reported measured 109 cm and weighed 90 kg (Vogt 2008). Sizes found nowadays are smaller than in the 1800s due to predation of nesting females by humans (Fig. 11.1).

Males are slightly smaller than females and have a narrower carapace; longer, thicker tail; and much larger anal notch. The broad, low, smooth carapace is slate gray to black and much wider posteriorly than anteriorly, and it has an extremely thick tail (Vogt 2001).

This species occurs in white, black- and clearwater (Vogt 2008). In Rio Negro, however, this species population is low due to over 200 years of exploitation of nesting females in the region, and nothing is known about their ecology in this igapó. In Rio Trombetas, seeds and fruits occurred most commonly in more than 50% of turtles, while leaves and grasses were found in 40%. Parts of mollusk shells were found in seven turtles, and a pebble, clay, turtle shell, and fish bone were found in one turtle each (Vogt 2008).

Female turtles migrate from the *igapós* and migrate hundreds of miles kilometers to nesting beaches, called locally as *tabuleiros*. Males come with the females and they mate in front of the nesting beaches.

Females bask on the beach for hours a day in the hot sun, 36–45 ° C, until they are ready to ovulate their eggs. The females dig a deep body pit, 40–50 cm, and then dig a deeper 90 cm nest cavity to deposit their eggs. After covering the nest, they return to the water, but unlike other turtles, they remain in front of the nesting beach in the deep (20–25 m) areas in the river for 2 months until their hatchlings enter the water. Once the hatchlings enter the water, the females migrate back to the *igapó*, leading their young to the feeding grounds (Vogt 2014).

Adult female giant South American river turtles are vocalizing underwater. They have at least 11 distinct sounds (Ferrara et al. 2014) which are made during migrations—moving out of the water to bask on the beaches, leaving the water to communally nest while nesting, and responding to hatchling vocalizations.

We recorded sounds while the turtles were still in the egg; our experiments suggest that vocalizations in the egg stimulate embryos to speed up development to synchronize hatching. They continue vocalizing in the nest to stimulate cooperative digging out of the nest and mass migration to the water.

Once in the water, they continue vocalizing, eliciting response from the females who lead the hatchlings to their feeding grounds. Most travel downstream, but we have documented through the use of small, 0.02-ounce sonar transmitters and fixed underwater automatic receivers that some hatchlings travel upstream with adults and others make lateral movements into nearby lakes. The females have actually been waiting for their young to hatch for 2 months.

Releasing five to six thousand hatchlings over waiting females with transmitters elicited the females to begin their migrations in 3 different years. In 2012 we documented two hatchlings traveling downstream for 62 km in 16 days; they were recorded at depths of 25 m with adults. In 2017, 35 of our 38 hatchlings with sonar transmitters escaped the gauntlet of predators waiting for them in the water near the nesting beach. The group of females returned back to the same nesting beaches each year. They nest only once a year (Vogt 2014).

*11.4.4 Big-Headed Sideneck Turtle (***Peltocephalus dumerilianus***)*

The high-domed carapace of the big-headed sideneck turtle reaches 50 cm in length. Males are generally larger than females. The head of this turtle is really massive (Fig. 11.1), and it is distinctly triangular in shape and with a strongly hooked upper jaw. The males have longer, thicker tails than females, and the opening between the anal scutes is greater in males (Vogt 2008).

This species is indigenous to blackwater streams and lakes, primarily in forested areas which emphasizes the importance of igapós for their survival. It also occurs to a lesser extent in white- and clearwater rivers.

Big-headed sideneck turtles enter buriti palm igapós (*Mauritia flexuosa*) during the high-water season to feed on the rich pulp of their fruits (Vogt 2008). The high-domed carapace is not designed for fast swimming, and webbing on the feet is not well developed for swimming in fast water. These turtles mostly walk along the bottom and shorelines in search of their preferred food, mollusks, and hard-shelled seeds.

Their very sharp beak and strong jaws make them look like formidable predators; however, stomach contents have revealed that they feed primarily on seeds and fruits (*Mauritia flexuosa*, *Montrichardia arborescens*, *Parinari campestris*, and *Thurnia polycephala*) and aquatic plants. Animal matter, primarily fish and mollusks, make up a smaller portion of their diet (De la Ossa et al. 2011).

Although the jaws are large and sharp, they do not have extremely wide alveolar surfaces, so presumably they are used for shearing rather than crushing. Most of the seeds eaten are found intact in the stomach and intestines (De la Ossa et al. 2011), suggesting that the turtles are absorbing only the nutrients available on the outer layers of the seeds and thus may be important seed dispersers.

It is curious that of all the podocnemid turtles examined, the throat papillae of *Peltocephalus* (also known as Trueb's structures) are considerably more developed than in the four Brazilian species of *Podocnemis* (Magalhaes et al. 2014). Histological examination of the papillae suggested that they function as a filter. Possibly the turtles use these in conjunction with neustophagia, the practice of ingesting floating seeds or particulate vegetation together with a large quantity of water; the papillae would then retain the food particles in the throat as the water is forcibly expelled.

Their fierce nature and high-domed shell probably evolved to protect them from predators, primarily dwarf caimans such as *Paleosuchus* that inhabit the same igarape streams as they do in Brazil.

Unlike all other podocnemid turtles in Amazonia, the big-headed sideneck turtles do not dig their nests communally in open sandy nesting beaches. They lay their eggs within 1–2.5 m of the shoreline of the igarape streams within the forest, in August and September in Reserve Trombetas in Pará, Brazil, and September and October in the Rio Negro, near Barcelos, Amazonas, Brazil.

In Colombia the nesting season is in December. The nests are dug in uplands as water levels diminish and the flooded forests become dry. The shallow (12–24 cm) nest cavities are generally horizontally elliptical with the opening 12–14 cm in diameter; eggs are sometimes packed at both ends, giving the appearance of two nests. This could explain the sizable reported variance in clutch size, resulting from excavating only one side of a nest or counting one nest as two.

The nests are often laid in soil mixed with leaves or in termite mounds; thus it is quite difficult to find the nests, and it is not productive for local hunters and fishermen to exploit the eggs of this species for trade in the markets. Nests in termite mounds may benefit from higher temperatures as do eggs of dwarf caimans (*Paleosuchus*) in the same forests.

Peltocephalus dumerilianus is another highly predated species in the Rio Negro. They maintain themselves by cryptic nesting strategy and omnivorous diet (Vogt 2001).

*11.4.5 Matamata (*Chelus fimbriata*)*

This is a peculiar species, with a triangular flattened head, small eyes, long tubelike proboscis, long neck with numerous papillae, and leaflike flaps of the skin for camouflage and tactile sensing. The carapace is covered with pyramid-like projections, giving a very irregular shape, and also often covered with filamentous algae (Fig. 11.1) (Vogt 2008).

Matamatas reach sexual maturity in 5–7 years, when the carapace is longer than 30 cm, but continue growing up to about 50 cm in carapace length and 15 kg in weight. Females are often slightly larger than males in carapace length, and the tail is short in females and longer and thicker in males. Males also have a concave plastron. They are common in both whitewater and blackwater ecosystems.

Matamatas live in shallow water, along the shoreline of lakes or at the interface of flooded forests and open water or in the region between uplands and flooded forest. Matamatas are among the few turtles that are completely carnivorous, feeding almost exclusively on fish captured by a sit-and-wait strategy. They usually remain in water where they can reach the surface with the tip of their snout without swimming off the bottom. This allows them to breathe without surfacing, attracting attention of predators, or giving away their presence to potential prey, principally fish.

The many loose flaps of skin on the neck serve perhaps to sense movement in the water, allowing them to hunt day and night. Their eyes are extremely small in comparison to their size. When a fish ventures into its vicinity, the turtle lunges its neck forward, striking rapidly like a snake, simultaneously opening the mouth and expanding the well-developed hyoid bones in the neck, sucking the fish and water into its mouth like a powerful vacuum cleaner, and swallowing the fish whole. Their jaws are in fact very weak and not built for macerating or even cutting a fish. Small matamatas in captivity have been observed herding small fish into position for attack using their tail. Although fish make up the greatest proportion of their diet, smaller turtles in particular also consume tadpoles and aquatic invertebrates.

Matamata turtles do not seek out beaches to nest but crawl up rather steep banks to nest along the edge of the forest, near the lake or stream where they are living. There are no concentrated nesting areas; each female nests in a different place; thus there are no concentrations of predators procuring their eggs for consumption. No studies have been conducted regarding sex determination in this species but it is perhaps genetically determined in that all of the other turtles studied in this family have GSD.

The matamata is widely distributed in a number of common habitats throughout the Amazon Basin. Nowhere is it intensively sought for food; although some indigenous tribes consume it when found, most rural people in Brazil refuse to eat this turtle because of its peculiar appearance as well as its musky odor. Unlike other species of turtles, the major edible part of the matamata is the neck, whereas the legs are rather short and thin. Matamata turtles have appeared in the international pet trade for more than 50 years, but not in such numbers as to have an effect on the natural populations of this species today. They appear to be reproducing in captivity in North America, so this production might reduce collection from natural populations.

Matamatas thrive on the dynamics of the igapó, carrying them continually into different shallow habitats deeper into the flooded forest as the water rises. They live camouflaged in the shallow water, among the leaf litter on the submerged forest floor. Since they are moving continually with the water depth, they are continually finding new populations of small fish to exploit. When the rain stops and the igapós begin to recede, they again follow the receding water, and when the igapó is dry, they congregate in isolated pools left by the receding waters until the next rainy season. It is perhaps in these pools when males and females are forced together that courtship and copulation take place (Vogt 2008).

11.5 Conservation

11.5.1 Threats Faced by Turtles of the Igapós

The main threats turtle of the igapós face nowadays are human consumption and hydroeletric dam constructions. Human consumption is a historically known problem as they have been an important commercial and protein resource in the region of the Amazon for centuries. For the last 200 years, turtles have been brought to the point of no return in many areas by uncontrolled collection for commercial interest. This is the case of *Podocnemis expansa* which is already known to have been extirpated from the regions of Barcelos in the Rio Negro, in the Rio Negro, in the Rio Solimoes near Manaus, in the Rio Solimoes near Tefe, in the Rio Purus and in the Rio Acre near Rio Branco, Acre.

Turtle Trade and the Pressure on Turtle Species Consumption of species is higher on the largest species. *Podocnemis expansa*, the largest South American freshwater turtle, has been rapidly declining in Brazil from overharvest of both eggs and adults. Because of this overexploitation of *P. expansa*, more pressure has been put on *P. unifilis* and *Peltocephalus dumerilianus*. Large turtles are usually not consumed in great quantities by the indigenous river people; they eat the smaller ones along with some eggs, and the rest are shipped downstream to Manaus or Belem, where they command higher prices (Schneider et al. 2011).

To give an idea of the profit of turtle black market, Schneider et al. (2011) reported that middlemen were buying *Peltocephalus dumerilianus* turtles in the Rio Negro tributaries for US$ 10 and selling in Manaus for US$ 100. *Podocnemis expansa* was bought for US$ 50 and sold for US$ 450 in Manaus. These prices were recorded in 2007; nowadays these numbers might be even higher. This is a very unfair trade, with indigenous people living in the igapós being exploited by middlemen and the turtles decreasing stocks to alarming numbers.

Hydroeletric Dam Construction as mentioned above in this chapter, the ecological system in the Amazon works according to the annual flood pulses. Hydroeletric dams cut the flood pulses and maintain water level steady. The main negative impact of hydroelectric reservoirs on aquatic life of igapós is that the water flow dynamics are altered by the river dam and its reservoir.

A review on the impacts of dams on turtles has shown that the break of seasonal hydrological flow affects turtle habitats and animal movement and limits the dispersion and migration between flooded forest and river canal and the diversity of habitats decreases dramatically due to the constant existence of lotic conditions (Alho 2017). In addition, sedentary species are benefited at the cost of migratory.

Indirectly, dams decrease water quality. Many floras within seasonally inundated forests are highly adapted to a particular flooding schedule, and alterations in flood patterns and the creation of permanently flooded areas induce higher rates of tree mortality. This means that seeds and fruits will not be available to feed turtles as before. In addition, as a consequence of the infrastructure building, turbidity increases as a result of erosion and sedimentation, with reduction of photosynthesis of submersed plants and plankton algae. The reduced number of plants also reduces primary productivity which decreases feeding habitats, reducing turtle feeding.

There have not been many published studies in the Amazon to fully understand all the effects dams have on turtle populations. That makes it difficult to develop mitigation plans to reduce the impact of dams on turtles. One known study (but not published) is the construction of artificial beaches in Balbina hydropower station, a hydroelectric dam in the Uatumã River in the municipality of Presidente Figueiredo. The construction of artificial beaches has attracted nesting turtles below the dam at Balbina on the Rio Uatuma, where they are not able to pass upstream to the original nesting areas. The genetic maintenance in the upper population, however, is still restricted, and the turtles above the dam are endangered. The turtle population above the dam is endangered in that the 65 females trapped there have only one nesting beach available and it is too low and shaded such that incubation temperature is too low for the eggs to survive.

The study of dam construction effects on turtles needs to be further investigated as it is perhaps the main threat currently for turtles of the igapó. Not only studies, it is necessary to have proactive researchers who publish their study and do not just leave their data sitting in their field book.

11.5.2 Status of Turtle Conservation in the Igapó

Igapós are the least protected areas in the Amazon and are not endangered, because the soils on which the igapós grow are not nutrient rich and the trees that grow in the igapó are short and do not have a high economic value. Therefore, the igapó forests are actually more intact than várzea forests.

The igapó forests along the Rio Negro will likely be there in the long term. The white silica sands that they grow on are so nutrient poor that cutting the forest does not produce any land that can grow grass for producing livestock or even grow cash crops like soybeans. The poor primary production in these forests protects them from being heavily exploited because there is no enough value in the forest products available to sustain large populations of people.

Although the igapó areas have lower density of human beings inhabiting them than the highly fertile soils of whitewater várzeas, the black market of turtles in larger towns, the unprotected areas that lack rangers, and the unregulated harvest of fauna by peoples that move into igapó forests to hunt and fish all directly threaten future stocks. The illegal trade of turtle and the existence of middleman trading turtles from igapó local indigenous to big cities like Manaus and Belem make the turtle population as threatened as várzea areas.

Because of the diverse nesting patterns of *P. erythrocephala* and *Peltocephalus dumerilianus*, even though they have been harvested heavily for over 150 years, the populations remain resilient. Even though there are not any government measures set up to protect the igapós and their fauna on the Rio Negro, they are protected by not being an economic boom. This is not the case for *P. expansa* and *P. unifilis*, which are heavily threatened by illegal trade to supply the main cities in the Amazon (details about the trade in the following section).

The use of *P. erythrocephala* and *P. dumerilianus* by local people in a sustainable manner is not because people want to, but because of the diverse nesting strategies of these two species. Finding their nests and nesting females is not easy and it takes too much work. Therefore, sustainable use of turtles is dictated by the turtles behavior, not by the people who, if given the chance, would eat every last egg available.

11.5.3 The Concern of Traditional Beliefs Hindering Turtle Conservation

Social belief that traditional culture and nature will always be in harmony is a utopic idea for turtle conservation. It is necessary to use exact science to understand population numbers and tackle the problem logically, irrespective of traditional beliefs. The traditional indigenous population and rural populations are growing in an accelerated rate and further are not fishing turtles for sustenance only but also for a cash crop (Schneider et al. 2011).

The law that protects turtles in the Amazon is the Brazilian Law 9.605/98 which allows for the consumption of turtles in the case of starvation. The law is there, but there are two main problems with it: first is the lack of law enforcement related mainly to people's attitude and second is that the law is at odds with reality in the sense that people are allowed to eat turtle in the case of starvation, but this is not the case in the Amazon because fish are more abundant and easier to catch than turtles anywhere. Environmentalists and conservationists who work in the Amazon and are directly committed to turtle conservation work do not agree with this situation because many species of turtles are endangered and should not be exploited (Vogt 2008).

Conservation measures should align with reality: as the traditional life of riverine people has been lost in many places in the Amazon, tradition should not be considered in conservation. Traditional values should be dealt separated from conservation as this type of conservation cannot work in a place where most of these values have been lost. Conservation of turtles in the Amazon should be logical: higher consumption of turtles than the current stock leads to turtle extinction.

Indigenous tribes today have cars, cell phones, industrialized fishing equipment, and motorboats to name a few of modern life acquisitions. There is no way that conservation measures taking traditional beliefs into consideration will work. New conservation strategies should be based in reality, not in traditional beliefs that society only obey when they have benefits out of it, not when they have duties.

Unfortunately, if indigenous populations maintain the right to unlimited catches at the current rate, eventually they will wipe out the food source. This, of course, goes against the belief of social biologists who purport that rural riverine people have the rights to eat and sell all the wildlife in the areas where they live without consideration of the consequences of nonregulated destruction of the populations of these animals.

Social biologists believe that rural people will have the conscience to use these resources sustainably. It has been already shown, however, that turtle populations are the first food resource to go in the Mamiraua Sustainable Development Reserve and other multiple-use reserves. Once turtle populations are destroyed in a river basin, it will take hundreds of years to replace.

The season cannot be opened on wildlife for rural people to harvest willy-nilly; this is nonsensical. Populations of turtles are diminishing even where there is strict law enforcement because rural people are motivated to engage in trafficking turtles for economic gain. Given their low social level, judges release the poachers when apprehended without having to pay nay fines, so there is no stimulus for them to stop poaching.

11.5.4 The Need for Alternative Income Source for Rural People

Most attempts at conservation of turtle species have been directed at the riverine people who consume the smaller species of turtles and capture the larger species for the black market. We need to develop a conservation program involving all of the communities along the rivers and develop other alternative income sources for these riverine people so that they can live comfortably without selling turtles to the black market.

Historically, rural people used to consume any turtle species with no economical expectation; now they do not eat the large *P. expansa* but sell them to traffickers who trade them on illegal black markets in Belem and Manaus. While this lucrative market exists, it will be hard to convince the turtle smugglers to stop. A change can be made in this cycle if something else motivates them to earn more money by less work.

The case of Rio Guapore is a good example that smugglers will stop trafficking turtle if they have an easier and more lucrative activity to pursue. In the Rio Guapore, the main nesting beach of *P. expansa* has increased from 860 turtles in 1989 to 24,000 in 2006. This is not even a protected reserve; only the nesting beaches are protected during the nesting and incubation season. Rural people eat turtles in this region but do not traffic them to the larger cities, because it is much more lucrative to traffic cocaine, along the Bolivian border with Brazil, with much smaller packages and much higher prices for their work.

This is one instance where drug trafficking has protected a turtle population. Give the smugglers something else to do that is easier and more lucrative, and they will stop trafficking turtles: they are not addicted to smuggling turtles. If the turtle black market in the large cities is shut down, rural people and boat owners will not make so much money, what most likely will induce them to abandon this practice.

11.5.5 Conservation Measures Taken and Their Effectiveness

The other issue in conserving turtles in the Amazon is the lack of effective actions taken by the government. Publications on conservation and mitigation measures have been published intensively in the last 15 years (Schneider et al. 2011, 2016; Vogt 2001, 2008). All of them have given several conservation measures to guide the government to protect turtles; large nesting beaches and even smaller nesting beaches are being protected in the igapó areas of the Rio Trombetas and in some of the tributaries of the Rio Negro as well, where communities are guarding the nesting beaches and the nests until the hatchlings emerge.

These measures taken by the government seem to be effective, but populations of turtles continue to decline because there is no incentive to stop the criminals who are poaching turtles. When turtle smugglers are captured, the turtles are released, and they are fined R$5000 for each turtle in their possession, but since these criminals have no money and no possessions to confiscate, they are set free without paying anything. So fines are not a detriment.

The Tuchaua (tribe chief) of the Yanomami indigenous tribe, on the Rio Demini, succeeded in protecting the *Podocnemis expansa* nesting beaches in the area of their territory by a decree of a death penalty to anyone who took nesting females or their eggs from the beaches. Indigenous laws might seem more inhumane and brutal, but it is certainly efficient. Even though the Tuchaua has not executed any offenders, his people respected his order so far because they knew it would be carried out.

Time will perhaps save the turtles of the igapós, not government regulations or education. It is a nice thought that environmental education of poor people will make them prone to not eat or sell turtles on the black market for resale to Manaus. However, there is little evidence over the last 25 years that this works.

Perhaps some of the people developed a conscience and are now protecting turtle nests and releasing hatchlings rather than selling or eating all of the eggs. But it only takes a small portion of the population of people who are hardened criminals to destroy turtle populations. It is unlikely that these criminals will not respond to education since they make a lot of money doing it.

A positive fact is that slowly the generation that had a fetish for eating turtles in the large cities is getting old and dying; young people under 40 in Manaus have no interest in eating turtles or their eggs. When the older generation is gone within another 30 years, the desire to have turtles served for birthday parties and political events will have disappeared. This will perhaps drive down the price of black market turtles, such that the poaching of turtles will not produce enough money to make it economically feasible.

Time and social reorganization will save the turtles, not laws, fines, or government intervention.

11.6 The Issue of Mercury Uptake by Turtles and Why to Concern About Turtles of the Igapó

Mercury is a persistent pollutant with unique chemical and physical characteristics, making this trace element one of the most highly studied chemical of all times (Hintelmann 2010). All mercury compounds are highly toxic, but methylmercury (MeHg, its organic form) is the most harmful form of mercury. Considering that this book intends to reach not only researchers but the general community, this chapter will not venture into analytical aspects which can be sought by the reader from the references throughout the text. Here we will take a simple approach to explain how turtles bioaccumulate mercury and why they are more susceptible to bioaccumulate this toxic element in the igapó.

The igapó deserves special attention to the issue of mercury as the background concentrations in the Rio Negro and Rio Tapajos are naturally high, similar to rivers located in high population density and industrial areas (Fadini and Jardim 2001; Roulet et al. 1998). For a long time, it was believed that high concentrations of mercury in the Amazon were a result of illegal mining only (locally known as garimpo); this came into dispute due to the Rio Negro region being a well-protected area against illegal gold mining and still recording high concentrations of mercury in the soil (Fadini and Jardim 2001). Research has since explained that the high natural mercury background concentrations in the Amazon are more linked to Hg-rich soils, and not as much by garimpos as first thought (Fadini and Jardim 2001; Roulet et al. 2001).

As most of the mercury in the Amazon is of natural origin (Fadini and Jardim 2001; Roulet et al. 2001), the principal concern is with its transformation into MeHg which occurs predominantly in fluvial wetlands like the igapó (Kasper et al. 2017). Methylmercury is the most common contaminant in turtles (Eggins et al. 2015) and most other aquatic organisms throughout the world.

11.6.1 Mercury Forms and Their Importance to Turtle Toxicity

Before talking on how turtles uptake mercury, it is necessary to understand the forms of mercury that exist. Mercury has three chemical forms: (1) elemental mercury, (2) inorganic salts, and (3) organic compounds. These may be present in the natural aquatic environment in three different covalent states and various physical and chemical forms (Morel et al. 1998) which will determine the solubility, mobility, and ultimately the toxicity of mercury in aquatic ecosystems.

The organic forms of mercury are of particular interest because these are most easily taken up by aquatic organisms, including turtles. The organic form of mercury refers to the group of organometallic compounds that contain mercury, such as methylmercury (II) cation, CH_3Hg^+; ethylmercury (II) cation, $C_2H_5Hg^+$; dimethylmercury, $(CH_3)_2Hg$; diethylmercury, $C_4H_{10}Hg$; and merbromin, $C_{20}H_8Br_2HgNa_2O_6$ (Morel et al. 1998).

Organic mercury is not normally released into the environment but formed by natural processes. This transformation of inorganic mercury into organic mercury is called methylation and is carried mainly by bacteria and, to a lesser extent, through abiotic pathways (Hintelmann 2010).

The conditions of the igapó are optimal for the conversion of inorganic to organic mercury by bacteria because (1) igapó has high concentrations of mercury stored and released through flooded soils (Fadini and Jardim 2001; Roulet et al. 1998); (2) igapó has plenty of methylation sites such as floodplain forests and hydromorphic soils (Belger and Forsberg 2006; Branfireun et al. 1996; Jardim et al. 2010); and (3) when the igapó is flooded, it is characterized by frequent periods of anoxia, acidic water, and high concentrations of organic matter, favoring the growth of mercury-methylating bacteria (Junk 1997; Kasper et al. 2017; Ríos-Villamizar et al. 2013). A consequence of this is that the turtles of the region have much greater uptake of mercury due to it being more prevalent in its organic form.

These optimal conditions for methylation of inorganic mercury into its organic form hold true for the Rio Negro, well known for its unique physicochemical characteristics showing a very high content of dissolved organic matter up to 20.7 mg/L, an average pH of 4.86, and average conductivity values approximately 15 μS/cm (Fadini and Jardim 2001; Schneider et al. 2009). In addition, Rio Negro igapó is rich in organic material and the fresh dissolved organic matter (DOM) associated with the increase in the water chemical characteristics that favor mercury methylation by bacteria. MeHg levels in the Rio Negro and other large Amazon rivers have been shown to vary seasonally with the annual flood cycle, peaking at high water when igapó and other floodplain environments are fully inundated and conditions for methylation are optimal.

11.6.2 How Does Mercury Get into Turtles?

Turtles absorb methylmercury from their food as it goes through its digestive system. The organic form of mercury is able to be uptaken by organisms because these organic forms have lipophilic and protein-binding properties (Clarkson 1997). In other words, the organic forms of mercury are able to dissolve in fats, oils, and lipids and are tightly bound to proteins in all turtle tissue, including carapace (Schneider et al. 2013). There is no method of cooking or cleaning turtle that will reduce the amount of mercury in a meal.

Methylmercury has a remarkable bioaccumulation potential, and concentrations in turtles increase with age and are often manifested in the good correlation between mercury concentration and size (age). Methylmercury also has the potential for biomagnification, increasing concentration in the tissues of organisms at successively higher levels in a food chain (Beneditto et al. 2011). Concentrations in water are often near the detection limit (e.g., 0.05 ng/L) but can be biomagnified to over 1 mg/kg in organisms occupying high trophic positions (Hintelmann 2010).

The largest methylmercury biomagnification step occurs at the first step of the food chain, when methylmercury is transferred from water into plankton (Hintelmann 2010). Subsequently, methylmercury is assimilated by planktonic organisms and passed on to its predators (Fig. 11.2), where it retained and concentrated over time as it makes its way up the food web. As explained above, retention is well established because methylmercury is able to be dissolved in fats, oils, and lipids and is tightly bound to proteins.

Mercury bioaccumulation and biomagnification have been studied in an igapó food web system in the Rio Negro. Species of turtles from higher trophic levels presented higher mercury concentration (Fig. 11.2). Note that mercury has increased 24-fold from the bottom (forest leaves with 0.030 μg/g mercury) to the top of the food chain (human beings with 0.750 μg/g mercury), demonstrating the potential mercury has for biomagnification.

11.6.3 Future Directions for Studies Investigating Mercury in Turtles

Although it is clear that the igapó favors the production of methylmercury, studies comparing mercury uptake in the igapó and other areas have been only performed for fish (Belger and Forsberg 2006) and human beings (Silva-Forsberg et al. 1999). Studies on turtles are scarce as they are not as consumed by human beings in the same volume as fish. However, turtles are a significant source of mercury for humans and deserve attention as a source of mercury.

The only comparison possible, at the moment, is the igapós of the Rio Negro and the Rio Purus. The species *Podocnemis expansa* and *P. unifilis* have a lower muscle mercury concentration in the Rio Purus (35.5 μg/g and 21.09 μg/g, respectively) (Eggins et al. 2015) than in the Rio Negro (62.4 μg/g and 34.6 μg/g, respectively) (Schneider et al. 2010). This is expected given that the Rio Negro blackwater has higher dissolved organic carbon (DOC) and more methylation sites and its water is more acidic than the clearwaters of the Rio Purus (Ríos-Villamizar et al. 2013). Studies comparing turtle species mercury concentration in the different environments of the Amazon would largely help to complete an understanding of the influence of the different Amazonian environments and how they contribute to mercury uptake by turtles. This information would allow better action plans for turtle conservations, especially considering the current expansion of mining in the Amazon.

11.7 Conclusion

The igapó areas are crucial to turtle survival since their ecology and physiology are highly influenced by flooding. Turtles enter the igapó during the flooded season to feed on fruits, algae, and seeds, while in the dry season, adults enter the rivers,

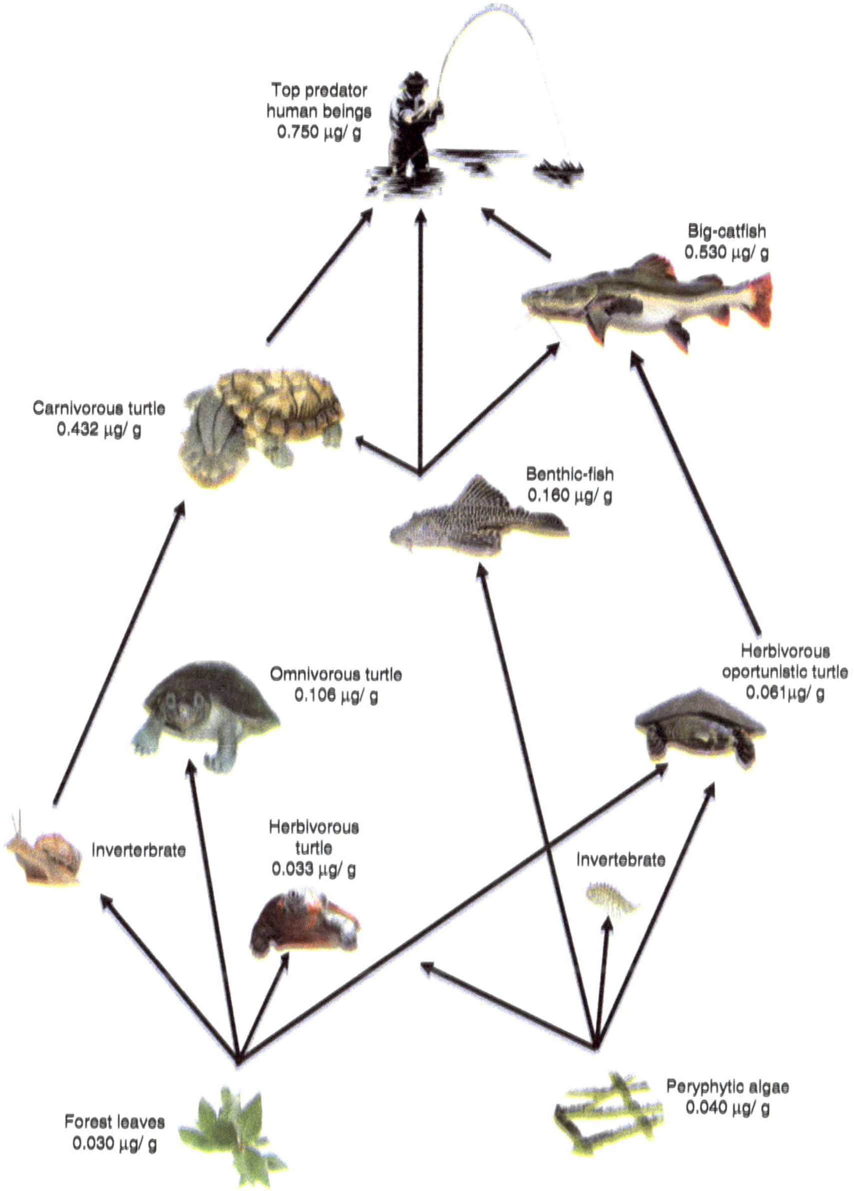

Fig. 11.2 Mean values of total mercury wet mass concentrations for an igapó food web from the Rio Negro. Note the 24-fold biomagnification of mercury from the base to the top of the food chain. (Data extracted from the study of Thome-Souza et al. (2004) and Schneider et al. (2010))

although juvenile and subadults often remain in lakes and pools left by receding waters.

The variety of habitats and peculiar chemical characteristics of the igapó have resulted in higher abundance of turtle species than in the várzea. This is opposite from other fauna species, demonstrating the unique link between turtles and igapós. This is because the variety of turtle nesting habitats, secretive behavior, specific feeding strategies, and the immense habitat available in the igapós have supported the existence of a few more species of turtles than the várzea.

The main threats turtle of the igapós face nowadays are human consumption and hydroelectric dam constructions. Human consumption is a historically known problem as they have been an important commercial and protein resource in the region of the Amazon for centuries. The larger species *Podocnemis expansa*, *P. unifilis*, and *Peltocephalus dumerilianus* face higher pressure from human consumption.

The main issue hydroelectric dams have on turtles is the break from seasonal hydrological flow. There have not been many published studies in the Amazon to fully understand all the effects dams have on turtle populations. This makes it difficult to develop mitigation plans to reduce the impact of dams on turtles.

Most attempts at conservation of turtle species have been directed at the riverine people who consume the smaller species of turtles and capture the larger species for the black market trade. A conservation program, however, should involve all of the communities along the rivers and develop other alternative income sources for these riverine people so that they can live comfortably without selling turtles to the black market.

The igapó deserves special attention to the issue of mercury as the background concentrations in the Rio Negro and Rio Tapajos are naturally high, similar to rivers located in high population density and industrial areas. In addition, turtles of the region have much greater uptake of mercury due to mercury in the igapo being more prevalent in its organic form.

This chapter describes the unique chemical characteristics of the igapo and demonstrates the complexity of these environments on turtle ecology. More research is needed in this area to fully understand the ecological dynamics of the igapo and its importance to the Amazon system as a whole.

Acknowledgments We thank Jeremy Cradock and Bruce Forsberg for their valuable comments on this manuscript.

References

Alho CJR (2017) Environmental effects of hydropower reservoirs on wild mammals and freshwater turtles in Amazonia: a review. Oecologia Aust 15:593–604

Bates HW (1863) The naturalist on the river Amazons: a record of adventures, habits of animals, sketches of Brazilian and Indian life, and aspects of nature under the Equator, during eleven years of travel. London: John Murray. 364 pp.

Belger L, Forsberg BR (2006) Factors controlling Hg levels in two predatory fish species in the negro river basin, Brazilian Amazon. Sci Total Environ 367:451–459 https://doi.org/10.1016/j.scitotenv.2006.03.033

Belkin DA, Gans C (1968) An unusual chelonian feeding niche. Ecology 49:768–769

Beneditto APM, Bittar VT, Camargo PB, Rezende CE, Kehrig HA (2011) Mercury and nitrogen isotope in a marine species from a tropical coastal food web. Arch Environ Contam Toxicol 62:264–271 https://doi.org/10.1007/s00244-011-9701-z

Branfireun BA, Heyes A, Roulet NT (1996) The hydrology and methylmercury dynamics of a Precambrian shield headwater peatland. Water Resour Res 32:1785–1794 https://doi.org/10.1029/96WR00790

Clarkson TW (1997) The toxicology of mercury. Crit Rev Clin Lab Sci 34:369–403 https://doi.org/10.3109/10408369708998098

De La Ossa VJ, Vogt RC, Santos-Júnior LB (2011) Feeding of *Peltocephalus dumerilianus* (Testudines: Podocnemididae) in a natural environment. Actual Biol 33:85–92

Eggins S, Schneider L, Krikowa F, Vogt RC, Da Silveira R, Maher W (2015) Mercury concentrations in different tissues of turtle and caiman species from the Rio Purus, Amazonas, Brazil. Environ Toxicol Chem 34:2771–2781 https://doi.org/10.1002/etc.3151

Fadini PS, Jardim WF (2001) Is the Negro River Basin (Amazon) impacted by naturally occurring mercury? Sci Total Environ 275:71–82 https://doi.org/10.1016/S0048-9697(00)00855-X

Ferrara CR, Vogt RC, Sousa-Lima RS, Bernardes VCD (2014) Sound communication and social behavior in an Amazonian river turtle (*Podocnemis expansa*). Herpetologica 70:149–156

Furch K, Junk WJ (1980) Tropical ecology and development: proceedings of the Vth international symposium of tropical ecology. In: Tropical ecology and development: presented at the proceedings of the Vth international symposium of tropical ecology. International Society of Tropical Ecology, Kuala Lumpur, pp 771–796

Goulding M, Barthem R, Ferreira EJG (2003) Smithsonian atlas of the Amazon. Smithsonian Smithsonian Institution Press, Washington, DC, p 253

Hintelmann H (2010) Organomercurials. Their formation and pathways in the environment. Met Ions Life Sci 7:365–401 https://doi.org/10.1039/BK9781847551771-00365

Jardim WF, Bisinoti MC, Fadini PS, Silva GS (2010) Mercury redox chemistry in the Negro River Basin, Amazon: the role of organic matter and solar light. Aquat Geochem 16:267–278 https://doi.org/10.1007/s10498-009-9086-z

Junk W (1997) The Central Amazon floodplain – ecology of a pulsing system. Springer, Berlin

Kasper D, Forsberg BR, Amaral JHF, Py-Daniel SS, Bastos WR, Malm O (2017) Methylmercury modulation in Amazon rivers linked to basin characteristics and seasonal flood-pulse. Environ Sci Technol 51:14182–14191 https://doi.org/10.1021/acs.est.7b04374

Magalhães MS, Vogt RC, Barcellos JFM, Moura CEB, Da Silveira R (2014) Morphologyof the digestive tube of the Podocnemididae in the Brazilian Amazon. Herpetologica 70(4):449–463

Morel FMM, Kraepiel AML, Amyot M (1998) The chemical cycle and bioaccumulation of mercury. Annu Rev Ecol Syst 29:543–566 https://doi.org/10.1146/annurev.ecolsys.29.1.543

Ríos-Villamizar EA, Piedade MTF, Da Costa JG, Adeney JM, Junk WJ (2013) Chemistry of different Amazonian water types for river classification: a preliminary review. In: WIT transactions on ecology and the environment, vol 178. WIT Press, Ashurst, United Kingdom. pp 17–28 https://doi.org/10.2495/WS130021

Roulet M, Lucotte M, Saint-Aubin A, Tran S, Rhéault I, Farella N, De Jesus Da silva E, Dezencourt J, Sousa Passos CJ, Santos Soares G, Guimarães JR, Mergler D, Amorim M (1998) The geochemistry of mercury in central Amazonian soils developed on the Alter-do-Chão formation of the lower Tapajós River Valley, Pará state, Brazil. Sci Total Environ 223:1–24

Roulet M, Lucotte M, Canuel R, Farella N, Goch YG, Peleja JR, Guimarães JRD, Mergler D, Amorim M (2001) Spatio-temporal geochemistry of mercury in waters of the Tapajós and Amazon rivers, Brazil. Limnol Oceanogr 46:141–1157

Schneider L, Belger L, Burger J, Vogt R (2009) Mercury bioaccumulation in four tissues of *Podocnemis erythrocephala* (Podocnemididae: Testudines) as a function of water parameters. Sci Total Environ 407:1048–1054 https://doi.org/10.1016/j.scitotenv.2008.09.049

Schneider L, Belger L, Burger J, Vogt RC, Ferrara CR (2010) Mercury levels in muscle of six species of turtles eaten by people along the Rio Negro of the Amazon basin. Arch Environ Contam Toxicol 58:444–450 https://doi.org/10.1007/s00244-009-9358-z

Schneider L, Ferrara CR, Vogt RC, Burger J (2011) History of turtle exploitation and management techniques to conserve turtles in the Rio Negro Basin of the Brazilian Amazon. Chelonian Conserv Biol 10:149–157 https://doi.org/10.2744/CCB-0848.1

Schneider L, Iverson JB, Vogt RC (2012) *Podocnemis unifilis* Troscheil 1848, Yellow-Spotted River turtle. Tracaja Catalogue Am Amphibians Reptiles 890:1–33

Schneider L, Maher WA, Green AD, Vogt RC (2013) Mercury contamination in reptiles: an emerging problem with consequences for wild life and human health. In: Kim KH, Brown RJC (eds) Mercury: sources, applications and health impacts. Nova Science Publishers, New York, pp 173–232

Schneider L, Ferrara CR, Vogt RC, Schaffer C (2016) Subsistence-level chelonian exploitation on the Rio Negro and one viable alternative. Chelonian Conserv Biol 15:36–42 https://doi.org/10.2744/CCB-1188.1

Silva-Forsberg MC, Forsberg BR, Zeidemann VK (1999) Mercury contamination in humans linked to river chemistry in the Amazon Basin. Ambio 28:519–521

Thome-Souza M, Forsberg BR, Vogt RC, Sabbayrolles MGP, Peleja JRP (2004) The use of mercury and stable isotopes to investigate food chain structure in aquatic ecosystems in the Amazon. In: Presented at the VI International congress on the biology of fish. Fish Communities and Fisheries, Manaus, pp 235–239

Vogt RC (2001) Turtles of the Rio Negro. In: Chao NL, Petry P, Prang G, Sonneschien L, Tlusty M (eds) Conservation and Management of Ornamental Fish Resources of the Rio Negro Basin, Amazonia, Brazil. Editora Universidade do Amazonas, Manaus, AM. Brazil, pp 245–262

Vogt RC (2008) Amazon turtles. Grafica Biblos, Lima. In: Peru

Vogt RC (2014) Chattering turtles of the Rio Trombetas. Tortoise 1:118–127

Vogt RC, Sever DM, Moreira G (1998) Esophageal papillae in Pelomedusid turtles. J Herpetol 32:279–282

Part VI
Plants

Chapter 12
Do the Igapó Trees Species are Exclusive to this Phytophysiognomy? Or Geographic Patterns of Tree Taxa in the Igapó Forest – Negro River – Brazilian Amazon

Veridiana Vizoni Scudeller

12.1 Introduction

Amazonian forests are characterized by high alpha diversity, expressed in high species richness within uniform habitats (Tuomisto et al. 1995). Prance (1982) and Gentry (1982) estimated that the richness of woody species in the Amazon can reach between 30,000 and 50,000 species, a number based only on existing collections in herbariums or monograph groups, and recent data estimate about 16,000 tree species only (Ter Steege et al. 2013).

However, there is still controversy regarding the reason – explanation of this wealth – and two theories are put forward to explain this great diversity of Neotropical flora, a "museum model" which suggests that the climatic stability of the tropics allows the species to accumulate over due to low extinction rates in the absence of major environmental disturbances (Stebbins 1974). However, it is known that there was climatic stability in the Neotropics in the last 2 million years ago, during the Pleistocene (Whitmore and Prance 1987), and fossil records of the Tertiary in the Neotropics show a diversity of angiosperm families similar to the current floras (Burnham and Graham 1999). However, the records are not complete enough to access species richness (Richardson et al. 2001). This led to the second theory, which proposes that the current diversity of tropical forest species could be more recent, resulting from speciation through allopatric differentiation of populations into separate refuges (Haffer 1982). Other recent geological phenomena that resulted in speciation in the Neotropical region would be the uplift of the Andes Northwest (5 Ma) and the Panama Isthmus Bridge (3.5 Ma) (Gentry 1982).

If studies carried out in Amazonian mainland forests are not enough to state the richness and diversity of this environment, those of seasonally flooded forests are

V. V. Scudeller (✉)
Departamento de Biologia ICB, Universidade Federal do Amazonas – UFAM, Manaus, Brazil
e-mail: vscudeller@ufam.edu.br

R. W. Myster (ed.), *Igapó (Black-water flooded forests) of the Amazon Basin*,
https://doi.org/10.1007/978-3-319-90122-0_12

infinitely less conclusive, and little is known about this type of environment so characteristic of Amazonia. Alfred Russel Wallace (1853) was the first naturalist to divide the waters of the Amazon basin into three limnological systems: white water, clear water and black water. According to Goulding et al. (1988), for Wallace the colour ratio of the second system is easily deduced by the geology of the terrain that drains, however, the black waters, which drain Paleozoic and Pre-Cambrian soils of the Guiana Shields and Central Brazil, are in relatively close areas to clear water rivers; the causes of this peculiar colour appear to be produced by a combination of decaying fallen leaves, roots and other plant materials present in the water.

Periodically flooded forests are affected by a long and predictable flooding period (Junk 1989). This dynamic imposes a strong impact on the structure and composition of plant species, changing continuously along the river channel (Rosales et al. 2001; Wittmann et al. 2004; Albernaz et al. 2012). The high level of precipitation, its seasonality and the low slope in the lowlands of the Amazon lead to a great area that are flooded periodically along the great rivers. Trees established in these areas make use of several adaptations to survive the lack of oxygen in flooded soils and the almost complete submersion (Montero 2012). According to Sioli (1967) and Prance (1980), forests flooded by white or muddy water, such as the Solimões, are called várzea and the ones flooded by black and acid waters, such as the Rio Negro, and transparent waters, such as the Tapajós, are the igapós. The high sedimentation rate in white water rivers, the scarcity of nutrients in black and transparent water rivers and the anoxic conditions in the rhizosphere accentuate the complex phyto-ecological interaction that trees of flooded environments are exposed to. Consequently, the wealth of trees in flooded environments is lower than on "terra firme".

In terms of species richness, the Negro River igapó has an average of 63 species. ha^{-1} and is considered one of the poorest forest types in the Amazon (Montero 2012). However, recent studies have shown that flooded areas of the Amazon are richer in terms of species than any flooded environment in the world (Wittmann et al. 2006). The authors speculate that it can be explained by the high habitat richness, established by water chemistry, as well as geohydrological and biogeographic factors, and by the relatively stable condition in the Amazon basin over millions of years. Even assuming tectonic and climatic changes during Tertiary and Quaternary, it is likely that Amazonian wetlands exist over a long period of geological time (Junk and Piedade 2010).

In the várzea area, Wittmann et al. (2006) studied almost 63 ha and recognized a dominance of generalist species of wide distribution. Albernaz et al. (2012), also in a study in the floodplain, covering approximately 2800 km, recognized three regions, one from the source of the Solimões to the encounter with the Negro River, a second from the beginning of the Amazon to the junction with the Xingu River and the third between this stretch and the mouth in the sea, and commented that only 49 (9.04%) species (of the 542 taxa analysed) occurred in the three regions, that is, they presented a wide distribution. Montero (2012), analysing the distribution of species over 600 km of igapó, found that 62.3% of the individuals sampled belonged to 30 species, and over 60% of the values of relative abundance were composed by only 15 species. However, very little is known about the distribution of the tree species of igapó nor what factors are influencing this distribution.

Thus, this chapter aims to answer a simple question about the distribution of tropical tree species from periodic flooded areas and to understand that local or regional processes have influenced this pattern, focusing on the description and interpretation of the patterns of local, regional and global distribution. The guiding questions were: What would the geographic patterns of arboreal species be in the Brazilian igapó forests? Which plant taxa would be the most constant in the Brazilian igapó forests? Would the most constant taxa be exclusive to Brazilian igapó forests, or would they also occur in their neighbour forest and non-forest formations? Which taxa would these be? Could there be a highly constant and exclusive taxon or group of taxa characteristic of this forest formation, or would its tree flora be so heterogeneous in space that it could only be characterized by its physiognomy? We aim at answering these questions as well as contributing to the knowledge of the geographic distribution of the tree taxa in Brazilian igapó forest. We consider this the first step to help characterize this forest formation.

12.2 Material and Methods

In order to answer the questions presented, floristic and phytosociological works were carried out in periodically flooded forests of the Rio Negro basin, Brazilian Amazon. Nineteen floristic surveys were used, which clearly showed (1) the methods used during sampling and species listing, (2) indication of vegetation formation raised, (3) providing information about the growth habit of raised plants (shrub or arboreal) and (4) identification of at least 50% of the material collected up to the species level. We used 19 surveys in our analysis (Fig. 12.1). The spelling of the scientific names was duly corrected, with the help of the International Plant Names Index (www.ipni.org), Tropicos.org (www.tropicos.org) and Flora do Brasil list (http://floradobrasil.jbrj.gov.br), and updated according to the APGIV (APG 2016) and consult from the SpeciesLink site.

For purposes of comparison, taxa missing from the determination (cf) were included in the list of duly determined species, as well as indeterminate and taxonomic taxa (aff) were grouped in the different localities, so that we can work with the greater similarity between the studied sites and lower potential wealth. Secondly, ter Steege et al. (2013) widely distributed the taxonomic and identification problems in the floristic inventories carried out in the Amazon. However, they affect species with a broader distribution, because the chance of erroneous determination is less. Despite this, the authors recognized that some individuals were well-determined and assumed that this error occurs within acceptable limits, especially for common species. Therefore, it was decided to consider these species in our matrix.

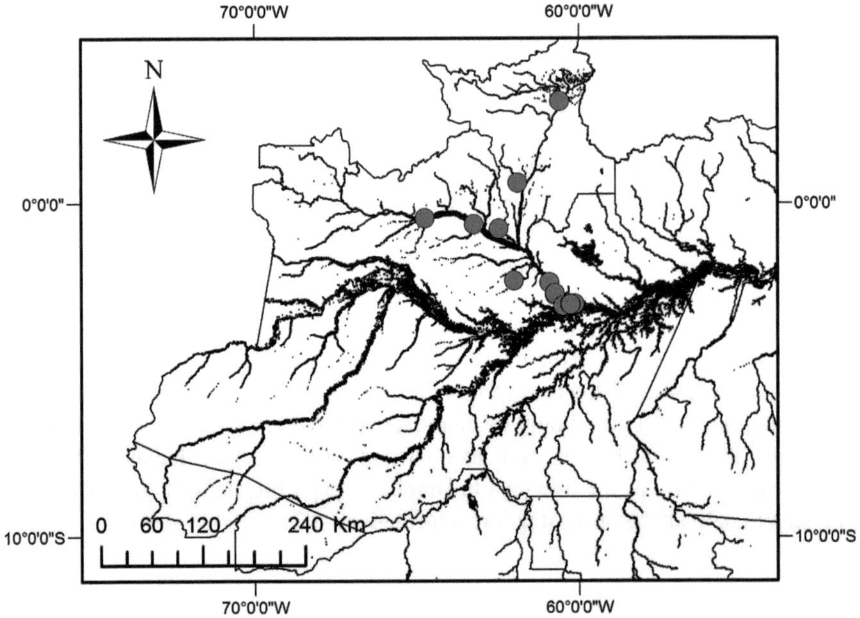

Fig. 12.1 Selected sites in this study where floristic surveys were carried out in the Negro River basin – Amazon

12.2.1 Study Area

The forests of igapó cover an area of 118,000 km^2 (Melack and Hess 2010) with a flood area of 35,944 km^2 in the low-water period and 152,721 km^2 in the high-water period (Frappart et al. 2005), forming the largest plain of flooding with world forests (Montero 2012). The Rio Negro receives more than 500 small tributaries, but the largest is the Branco River, which originates in the Roraima massif and transports sand and fine sediments to the Negro River (Goulding et al. 2003; Latrubesse and Franzinelli 2005), altering the physical-chemical qualities of the water of the Negro River (Santos et al. 1984).

The Negro River is the largest black water river in the world, covering 1700 km from the source of pre-Andean Colombia to the drainage in the Solimões (Goulding et al. 1988), and the headwaters in the Guiana Shield, via Rio Branco and Rio Uraricoera, two main tributaries of the Rio Negro, drain more than 1500 km. The Black River basin is connecting the rivers in the north, south, east and west. The most famous connection, the Rio Casiquiare, connects the upper Rio Negro as the Orinoco system. Geomorphologically it is a tributary of the Orinoco and a tributary of the Negro River, which in the recent geological history was captured by the changes in the elevation of the terrain in that region, facilitating Casiquiare to flow from the south to north.

The Negro River contributes approximately 15% of the waters that flow in the Amazon River in the Atlantic (Meade et al. 1979). Rio Negro is the largest tributary of the Amazon River, draining approximately 700,000 km^2 from Colombia (10%), Venezuela (6%), Guyana (2%) and Brazil (82%) (Fig. 12.1). According to Frappart et al. (2005) in the Black River basin, the floodplain covers an area of 35,944 km^2 in the low-water period and 152,721 km^2 in the high-water period. It incorporates the two largest river archipelagos in the world, Mariuá and Anavilhanas, located in the middle and lower reaches of the river, respectively (Melack and Hess 2010). The basin flows over three main geological formations, the former Guiana Shield, Upper Cretaceous and Lower Tertiary plains and Quaternary alluvial plains (Goulding et al. 1988).

The granulometry and sediment chemistry along the Negro River basin also vary from the upper portion of the river, which is composed basically of thicker materials of very low fertility, influenced by the western extension of the Guiana Shield and the lower portion with more sediments fertility. In the lower Negro River, the influence of the Branco River, which transports finer sediments to the main channel and the damming effect of the Solimões River, results in changes in the nutritional status of its waters near its mouth (Goulding et al. 1988; Junk et al. 2011).

Most of the Negro River is in the average annual precipitation isoeta of 2000–2200 mm; however this increases to 3500 mm in the northwest of Amazonas, near the equator, in the Uaupes or Vaupués river region, as it is called in Colombia (SUDAM 1984). The Negro River is chemically one of the poorest rivers in the world. Sioli (1967) refers to the Negro River as "slightly contaminated distilled water". In absolute concentration of the great elements, the waters of the Negro River are very similar to the waters of the rain (Sioli 1967). This black water river is extremely poor in nutrients because it drains geologically very impoverished soils (Goulding et al. 1988). The pH of the Rio Negro varies from 3.6 to 5.8 depending on the exact location and time of year (Furch 1997; Junk et al. 2011). The values increase in the vicinity of the confluence with some tributaries, especially the white river, which drain the shield of Guyana. The transparency of the Rio Negro, measured in Secchi disk, ranges from 0.9 to 1.5 m (Sioli 1967).

12.2.2 Analysis

We divided the occurrence area of the tree species on the Amazon biome using the phytogeographic patterns established by Prance (1977; Fig. 12.2a): (1) Costa Atlantica, (2) Jari/Trombetas, (3) Xingu/Madeira, (4) Roraima/Manaus, (5) Noroeste Alto Rio Negro, (6) Solimões/Amazonia Ocidental and (7) Sudeste (see Fig. 12.2). Considering the different occurrence possibilities of a taxon in the Amazon biomes, we defined three possible tree species geographic patterns: AA (ample amazon), taxon occurrence in all biome regions; FP (floodplain), taxon occurrence in floodplain area; and BW (black water), taxon occurrence in Negro River basin.

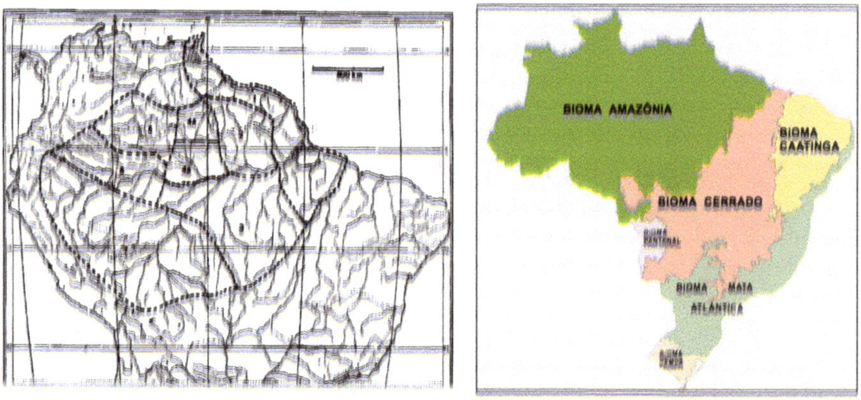

Fig. 12.2 Phytogeographic regions. (**a**) The phytogeographic Amazon biome using the patterns established by Prance (1977); (**b**) IBGE classification of the six biomes in Brazil

For each taxon we calculated the frequency of occurrence (FC or range of distribution, as used by Ter Steege et al. 2013) in the sample plots as follows: < 5.5% frequency, rare; 5.5–24%, occasional; 25–49%, common; 50–74%, moderately frequent; and >75%, frequent (Mueller-Dumbois and Ellenberg 1974). The highest values of frequency of occurrence were used to select taxa as examples in each phytogeographic pattern. For each phytogeographic pattern, we investigated the occurrence of each taxon in other formations out of the Amazon forest, according to the five biomes by the Brazilian Institute of Geography and Statistics (IBGE) classification: cerrado (CE), caatinga (CA), Pantanal (PA), Atlantic rain forest (Mata Atlântica, MA) and e Pampa (PP) – see Fig. 12.2b.

12.3 Results

In all, 19 surveys were selected (Table 12.1), and a specific richness of at least 636 woody taxa was recognized, since the indeterminate taxa were grouped, totalling 78 botanical families along with the almost 1200 km travelled. Of these, 380 species are duly determined and belong to 211 genera and 62 families.

The great majority of species (218) was restricted to only one occurrence site (FC < 5.5% = rare), and only 4 species (*Swartzia polyphylla* DC., *Pouteria elegans* (A. DC.) Baehni, *Licania apetala* (E.Mey) Fritsch and *Campsiandra comosa* Benth.) were considered moderately frequent (HR > 50%). No species was found in more than 12 (63.15%) sites sampled (Table 12.2). Only 50 species were considered common (over 25% of the occurrence), and 112 species presented occasional distribution. Of the species considered common, only 9 are restricted to black water (BW) and 7 to floodplain (FP), and this species not occur in other formations out of Amazon, except to *Himatanthus attenuatus* that was recorded to CE and MA (Table 12.2).

Table 12.1 Floristic surveys conducted along the Rio Negro channel and tributaries analysed in this study

Locality	F	Sp	Lat	Long	Fonte
Rio Negro – Santa Isabel do Rio Negro	34	107	−0.45	−64.77	Montero (2012)
Rio Jufari – Barcelos	33	83	−0.77	−62.48	Montero (2012)
Rio Negro – Barcelos	31	88	−0.63	−63.25	Montero (2012)
Parque Nacional do Jaú – Novo Airão	10	20	−2.33	−62.01	Ferreira (1997)
Parque Nacional do Jaú – Novo Airão	11	19	−2.33	−62.01	Ferreira (2000)
Rio Branco – Bonfim	28	56	3.05	−60.60	Carvalho (unpublished data)
Rio Branco – Caracaraí	37	100	0.60	−61.90	Carvalho (unpublished data)
Parque Nacional de Anavilhanas – Novo Airão	35	102	−2.77	−60.75	Montero (2012)
Parque Nacional de Anavilhanas – Novo Airão	27	52	−2.39	−60.92	Parolin et al. (2003)
Lago Miuá – Parque Nacional de Anavilhanas, Novo Airão	19	31	−2.39	−60.92	Piedade (1985)
Lago Prado – Parque Nacional de Anavilhanas, Novo Airão	23	41	−2.72	−60.75	Piedade (1985)
Praia Grande – Novo Airão	40	88	−3.08	−60.49	Revilla (1981)
Rio Tarumã – Manaus	11	20	−3.03	−60.13	Ferreira (2000)
Rio Tarumã – Manaus	18	54	−3.03	−60.13	Keel and Prance (1979)
Rio Tarumã – Manaus	15	35	−3.03	−60.13	Keel and Prance (1979)
Rio Tarumã-Mirim – Manaus	35	73	−3.03	−60.13	Ferreira and Almeida (2005)
Rio Tarumã-Mirim – Manaus	22	48	−3.03	−60.28	Parolin et al. (2004)
Reserva de Desenvolvimento Sustentável do Tupé – Manaus	21	53	−3.04	−60.26	Hamaguchi and Scudeller (2011)
Reserva de Desenvolvimento Sustentável do Tupé – Manaus	44	158	−3.04	−60.26	Scudeller and Souza (2009)

F, wealth of families found in each survey; *Lat*, latitude; *Long*, longitude; *Sp*, total wealth

Campsiandra comosa, *Swartzia polyphylla* and *Pouteria elegans* are not endemic in Brazil and are widely distributed in the Amazon (AA). *Licania apetala* occurs in the Amazon rainforest, in the cerrado (CE) and in the Atlantic Forest (MA), with registration in the states of Acre, Amazonas, Amapá, Pará, Rondônia, Roraima and Tocantins; Bahia, Ceará, Maranhão, Piauí and Sergipe; and Federal District, Goiás, Mato Grosso and Rio de Janeiro (Flora of Brazil).

The order of species richness and the families are practically the same; Fabaceae maintains the first position (79 species), followed by Sapotaceae (20); Myrtaceae (18); Annonaceae and Euphorbiaceae (17 each); Chrysobalanaceae (15); Lauraceae, Lecythidaceae and Rubiaceae (14 each); Apocynaceae (13); and Moraceae with 11 species. Twenty families (32.25%) were represented by only one species. *Eugenia* was the richest genus (10 species), followed by *Eschweilera*, *Pouteria*, *Swartzia* (with 9 each); *Ocotea* (8), *Guatteria* (7), *Buchenavia*, *Licania*, *Macrolobium* and *Tachigali* (6 each), *Zygia*. Six genera were represented by 4 species, 20 by 3 species, 41 by 2 spp and 133 by only 1 species.

Table 12.2 Species considered with frequency of common occurrence (over 25% of the occurrence in the 19 localities analysed) and their distribution in Amazon forest

Species	Amazon	CE	MA	CA	PA	PP	Frequency
Pouteria elegans (A. DC.) Baehni	AA						12
Campsiandra comosa Benth.	AA						11
Licania apetala (E. Mey.) Fritsch	AA	x	x				11
Swartzia polyphylla DC.	AA						11
Caraipa grandifolia Mart.	AA						9
Macrolobium acaciifolium (Benth.) Benth.	AA			x			9
Swartzia laevicarpa Amshoff	AA						9
Virola elongata (Benth.) Warb.	AA						9
Astrocaryum jauari Mart.	AA	x					8
Couepia paraensis Benth. ex Hook. f.	AA	x					8
Aldina latifolia Spruce ex Benth.	BW						8
Mollia speciosa Mart.	BW						8
Heterostemon mimosoides Desf.	FP						8
Hevea spruceana (Benth.) Müll.Arg.	FP						8
Acosmium nitens (Vogel) Yakovlev	AA	x	x				7
Calophyllum brasiliense Cambess.	AA	x	x	x	x		7
Licania heteromorpha Benth.	AA						7
Mabea nitida Spruce ex Benth.	AA	x					7
Maprounea guianensis Aubl.	AA	x	x	x			7
Parkia discolor Spruce ex Benth.	BW						7
Alchornea discolor Poepp.	AA	x	x		x		6
Caryocar glabrum (Aubl.) Pers.	AA						6
Licania micrantha Miq.	AA		x				6
Nectandra amazonum Nees	AA	x			x		6
Unonopsis guatterioides R.E.Fr.	AA	x			x		6
Zygia juruana (Harms) L. Rico	AA						6
Aldina heterophylla Spruce ex Benth.	BW						6
Elvasia calophyllea DC.	BW						6
Swartzia argentea Spruce ex Benth.	BW						6
Handroanthus barbatus (E. Mey.) Mattos	FP						6
Leopoldinia pulchra Mart.	FP						6
Ocotea brenesii Standl.							6
Aspidosperma nitidum Benth. ex Müll. Arg.	AA						5
Calyptranthes cuspidata DC.	AA						5
Crudia amazonica Spruce ex Benth.	AA						5
Garcinia madruno (Kunth) Hammel	AA	x	x	x			5
Gustavia augusta L.	AA	x		x			5
Hirtella racemosa Lam.	AA	x	x	x			5
Mabea caudata Pax and K. Hoffm.	AA						5
Macrolobium angustifolium (Benth.) R.S. Cowan	AA						5
Macrolobium multijugum (DC.) Benth.	AA						5
Peltogyne paniculata Benth.	AA						5

(continued)

Table 12.2 (continued)

Species	Amazon	CE	MA	CA	PA	PP	Frequency
Virola calophylla (Spruce) Warb.	AA						5
Virola surinamensis (Rol. ex Rottb.) Warb.	AA	x		x			5
Anacampta rupicola (Benth.) Markgr.	BW						5
Cynometra spruceana Benth.	BW						5
Himatanthus attenuatus (Benth.) Woodson	BW	x	x				5
Burdachia prismatocarpa A. Juss.	FP		x				5
Ormosia excelsa Benth.	FP						5
Pentaclethra macroloba (Willd.) Kuntze	FP						5

AA, ample distribution; *BW*, black water floodplain; *FP*, distribution restricted to periodic floodplain area in Amazon. Out of Amazon, *CA*, caatinga; *CE*, cerrado; *MA*, Atlantic rain forest; *PA*, Pantanal; *PP*, Pampas Do Sul

We checked 231 species with wide distribution on the Amazon (AA), 65 occur only on floodplain (FP), 61 occur only near black water river (BW), 7 are restricted to one formation on Amazon and 16 species are not registered on Species Link site. The majority of the species BW are considered rare (59%) and only 12 species (19.7%) occur in other Brazilian biome. We observed the same with floodplain species (FP), only 12.3% also occur in other Brazilian biome (Table 12.3).

Of the species considered AA, 41.5% (96 species) also occur on cerrado (CE), 9.2% of the FP and 11.5% of the BW also occur on CE – Fig. 12.3. Only five species were recorded in all Brazilian biomass. When we analysed together FP and BW, only 11% of the species occur on neighbour formation (CE) showing a strong specificity for the environment, probably due to its conditions of variation of the water level throughout the year.

In the generic level, we observed that the 43.1% are classified as rare, 31.7% as occasional, 18% as common, 6.2% as moderately frequent and only 9.5% as frequent (Table 12.4). Only *Aldina* presented distribution restricted to black water (BW) and both *Campsinandra* and *Guatteria* also occur only in Amazon forest. Of the 28 genera that occur in BW, only 5 (17.8%) occurred in another Brazilian biome and the 26 genera in FP, only 7 (26.9%) also occur out of Amazon (Table 12.5).

Only six families were classified as frequent: Fabaceae (100% of the surveys), followed by Apocynaceae, Chrysobalanaceae, Euphorbiaceae and Lecythidaceae (in 84.2% of the surveys each) and Sapotaceae (in 15 surveys) – see Tables 12.6 and 12.7. Moderately frequent we found 11 families, common 13, occasional and rare 16 each. Only 19 families had their representatives occur exclusively in Amazonia and of these, only 4 were restricted to FP and only Thymelaeaceae restricted to BW. Humiriaceae also was the family that only occurred in BW, but not exclusively because occurred in CE, MA and CA too (Table 12.8).

We called exclusive the taxa that did not occur in other Brazilian biome, but only in the surveys from the Amazon forest. This "exclusiveness" must be regarded carefully, because we verify only the 380 species that was recorded in the 19 surveys analysed in the species link site. About 85.5% of the families presented ample distribution on Amazon forest, less than expected since family is a taxonomic level of

Table 12.3 Species that presented distribution restricted to FP, BW or only one formation on Amazon forest, excluding the 231 species with AA distribution

Species	Amazon	CE	MA	CA	PA	PP	Frequency
Bombax aquaticum (Aubl.) K. Schum.	1						1
Coccoloba pichuana Huber	3						1
Duguetia sessilis (Vell.) Maas	3		x				1
Tibouchina grandifolia Cogn.	3		x	x			1
Franchetella crassifolia Pires and W.A. Rodrigues	7						2
Couepia macrophylla Spruce ex Hook. f.	7						1
Ficus anthelmintica Mart.	7		x				1
Aldina latifolia Spruce ex Benth.	BW						8
Mollia speciosa Mart.	BW						8
Parkia discolor Spruce ex Benth.	BW						7
Aldina heterophylla Spruce ex Benth.	BW						6
Elvasia calophyllea DC.	BW						6
Swartzia argentea Spruce ex Benth.	BW						6
Anacampta rupicola (Benth.) Markgr.	BW						5
Cynometra spruceana Benth.	BW						5
Himatanthus attenuatus (Benth.) Woodson	BW	x	x				5
Duroia velutina J. D. Hook. ex Schumann	BW						4
Macrosamanea discolor Britton and Killip	BW						4
Poecilanthe amazonica (Ducke) Ducke	BW						4
Swartzia reticulata Ducke	BW						4
Clathrotropis nitida (Benth.) Harms	BW						3
Ocotea cinerea van der Werff	BW						3
Peltogyne excelsa Ducke	BW						3
Schistostemon macrophyllum (Benth.) Cuatrec.	BW						3
Acmanthera latifolia (A. Juss.) Griseb.	BW						2
Buchenavia suaveolens Eichler	BW	x					2
Erisma calcaratum (Link) Warm.	BW						2
Neoxythece elegans (A. DC.) Aubrév.	BW						2
Remijia tenuiflora Benth.	BW						2
Tachigali hypoleuca (Benth.) Zarucchi and Herend.	BW	x					2
Triplaris surinamensis Cham.	BW		x		x		2
Zygia claviflora Barneby and J.W. Grimes	BW						2
Amanoa gracillima W.J.Hayden	BW						1
Aspidosperma sandwithianum Markgr.	BW						1
Blastemanthus grandiflorus Spruce ex Engl.	BW						1
Bombacopsis nervosa (Uittien) A. Robyns	BW						1
Buchenavia capitata (Vahl) Eichler	BW	x	x	x			1
Buchenavia fanshawei Exell and Maguire	BW						1
Centrolobium paraense Tul.	BW						1
Cordia nitida Vahl	BW						1

(continued)

Table 12.3 (continued)

Species	Amazon	CE	MA	CA	PA	PP	Frequency
Cybianthus quelchii (N.E. Br.) G. Agostini	BW						1
Cybianthus reticulatus (Benth. ex Miq.) G. Agostini	BW						1
Dipteryx oppositifolia (Aubl.) Willd.	BW						1
Elaeoluma glabrescens (Mart. and Eichler) Aubrév.	BW	x					1
Elizabetha speciosa Ducke	BW						1
Erythroxylum spruceanum Peyr.	BW						1
Eugenia cachoeirensis O. Berg	BW			x			1
Eugenia longiracemosa Kiaersk.	BW						1
Guatteria phanerocampta Diels	BW						1
Havetiopsis flavida (Benth.) Planch. and Triana	BW						1
Henriquezia nitida Spruce ex Benth.	BW						1
Lasiadenia rupestris Benth.	BW						1
Lecythis barnebyi S.A. Mori	BW						1
Licania discolor Pilg.	BW						1
Martiodendron excelsum (Benth.) Gleason	BW	x					1
Matayba opaca Radlk.	BW						1
Mezilaurus synandra (Mez) Kosterm.	BW						1
Micropholis humboldtiana T.D. Penn.	BW		x				1
Microplumeria anomala (Müll. Arg.) Markgr.	BW		x				1
Mouriri angulicosta Morley	BW						1
Odontadenia funigera Woodson	BW						1
Pera pulchrifolia Ducke	BW						1
Pouteria pimichinensis T.D. Penn.	BW						1
Roupala obtusata Klotzsch	BW						1
Simaba obovata Spruce ex Engl.	BW						1
Swartzia sericea Vogel	BW		x				1
Ternstroemia candolleana Wawra	BW	x					1
Toulicia acuminata Radlk.	BW						1
Heterostemon mimosoides Desf.	FP						8
Hevea spruceana (Benth.) Müll. Arg.	FP						8
Handroanthus barbatus (E. Mey.) Mattos	FP						6
Leopoldinia pulchra Mart.	FP						6
Burdachia prismatocarpa A. Juss.	FP		x				5
Ormosia excelsa Benth.	FP						5
Pentaclethra macroloba (Willd.) Kuntze	FP						5
Amanoa oblongifolia Müll. Arg.	FP	x					4
Eschweilera tenuifolia (O. Berg) Miers	FP						4
Tococa subciliata (DC.) Triana	FP						4
Tovomita macrophylla (Poepp.) Walp.	FP						4
Vatairea guianensis Aubl.	FP						4
Duguetia uniflora (DC.) Mart.	FP						3

(continued)

Table 12.3 (continued)

Species	Amazon	CE	MA	CA	PA	PP	Frequency
Eschweilera atropetiolata S.A. Mori	FP						3
Eugenia teffensis O. Berg	FP						3
Macrosamanea amplissima Barneby and J.W. Grimes	FP						3
Micrandra siphonioides Benth.	FP						3
Pouteria gomphiifolia (Mart. ex Miq.) Radlk.	FP						3
Salacia impressifolia (Miers) A.C. Sm.	FP						3
Swartzia macrocarpa Spruce ex Benth.	FP						3
Tachigali venusta Dwyer	FP						3
Calyptranthes multiflora Poepp. ex O. Berg	FP						2
Eschweilera bracteosa (Poepp. ex O. Berg) Miers	FP						2
Eugenia inundata DC.	FP	x					2
Exellodendron coriaceum (Benth.) Prance	FP						2
Malouetia furfuracea Spruce ex Müll. Arg.	FP						2
Mezilaurus itauba (Meisn.) Taub. ex Mez	FP	x	x				2
Ocotea fasciculata (Nees) Mez	FP		x				2
Parkia panurensis Benth. ex H.C. Hopkins	FP						2
Quiina rhytidopus Tul.	FP						2
Styrax guyanensis A. DC.	FP	x	x				2
Tovomita longifolia (Rich.) Hochr.	FP		x				2
Xylopia calophylla R.E. Fr.	FP						2
Coccoloba charitostachya Standl.	FP						1
Combretum laurifolium Mart.	FP						1
Dicorynia paraensis Benth.	FP						1
Diospyros bullata A.C. Sm.	FP						1
Diospyros glomerata Spruce	FP						1
Diospyros vestita Benoist	FP						1
Diplotropis martiusii Benth.	FP						1
Eschweilera odora (Poepp. ex O. Berg) Miers	FP						1
Etaballia dubia (Kunth) Rudd	FP						1
Guatteria amazonica R.E. Fr.	FP						1
Guatteria chrysopetala Miq.	FP						1
Ilex inundata Poepp. ex Reissek	FP						1
Laetia corymbulosa Spruce ex Benth.	FP						1
Licaria chrysophylla (Meisn.) Kosterm.	FP						1
Mabea occidentalis Benth.	FP		x				1
Macrolobium suaveolens Spruce ex Benth.	FP						1
Macrolobium unijugum Pellegr.	FP						1
Mora paraensis (Ducke) Ducke	FP						1
Mouriri subumbellata Triana	FP						1
Ocotea opifera Mart.	FP						1
Ormosia discolor Spruce ex Benth.	FP						1

(continued)

Table 12.3 (continued)

Species	Amazon	CE	MA	CA	PA	PP	Frequency
Ouratea spruceana Engl.	FP						1
Pagamea coriacea Spruce ex Benth.	FP						1
Piranhea trifoliata Baill.	FP	x					1
Platymiscium duckei Huber	FP						1
Protium hebetatum D.C. Daly	FP						1
Stachyarrhena spicata Hook. f.	FP						1
Strychnos jobertiana Baill.	FP						1
Swartzia corrugata Benth.	FP						1
Talisia cupularis Radlk.	FP						1
Tassadia trailiana (Benth.) Fontella	FP						1
Tococa caudata Markgr.	FP						1
Ocotea brenesii Standl.							6
Croton lanjouwiansis Jabl.							3
Eugenia chrysobalanoides DC.							2
Swartzia coriacea Desv.							2
Amorphospermum schomburgkianum (Miq.) Baehni							1
Andira amazonum Mart. ex Benth.							1
Annona duckei Diels							1
Eschweilera ovalifolia (DC.) Nied.							1
Ferolia guianensis Aubl.							1
Guatteria alata Maas and Setten							1
Lecythis jarana (Huber ex Ducke) A.C. Sm.							1
Myrcia jauensis							1
Ocotea megacarpa van der Werff							1
Pachira glabra Pasq.		x	x				1
Pouteria gamblei (C.B.Clarke) Baehni							1
Rollinia deliciosa Saff.							1

high inclusiveness, and only 3.2% was restricted to igapó forest. Few species (4) showed moderately frequency (more than 50% of the sites investigated), however, 60.8% showed ample distribution on Amazon forest. Generally, the most constant taxa showed ample distribution patterns, but the 4 species above related only *Licania apetala* also occur out of the Amazon forest (Table 12.2).

12.4 Discussion

According to Mueller-Dumbois and Ellenberg (1974), species with intermediate constancy may be differential species and characterized by different communities; highly constant species are considered indifferent or opportunist, and species with low constancy are considered accidental. In our analyses, most taxa showed rare

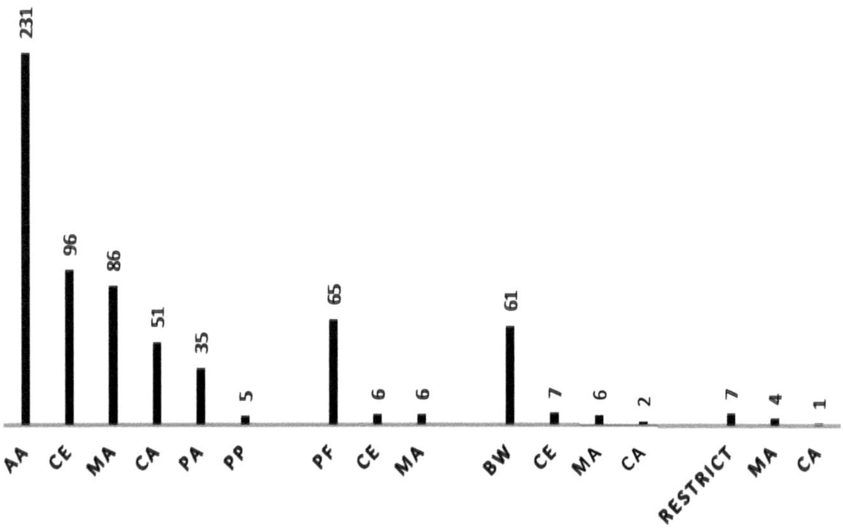

Fig. 12.3 Number of species recorded on igapó of the Negro River basin, classified on wide distributed on Amazon (AA), only on floodplain of the rivers (FP), only igapó (BW) and restricted to only one formation (restricts) that also occur in other Brazilian biomes (CE, cerrado; MA, Atlantic rain forest; CA, caatinga; PA, Pantanal; PP, pampas)

Table 12.4 Genus that occur in igapó forest of Negro River basin with frequency >50% in the 19 surveys analysed

Genus	Amazon	CE	MA	CA	PA	PP	Frequency
Macrolobium	AA			x			16
Swartzia	AA		x				15
Aldina	BW						14
Eschweilera	AA	x					13
Licania	AA	x	x				13
Virola	AA	x		x			13
Astrocaryum	AA	x					12
Campsiandra	AA						12
Mabea	AA	x					12
Pouteria	AA	x	x	x	x		12
Couepia	AA	x	x				10
Guatteria	AA						10
Hevea	AA	x	x	x			10
Mollia	AA	x					10
Ocotea	AA	x	x	x			10

Table 12.5 Genus that presented distribution restricted to FP, BW or only one formation on Amazon forest, excluding the 147 genus with AA distribution

Genus	Amazon	CE	MA	CA	PA	Frequency
Bombax	1					1
Tibouchina	3		x	x		1
Franchetella	7					2
Ficus	7		x			1
Aldina	BW					14
Elvasia	BW					6
Himatanthus	BW	x	x			6
Anacampta	BW					5
Duroia	BW					4
Poecilanthe	BW					4
Clathrotropis	BW					3
Schistostemon	BW					3
Acmanthera	BW					2
Lecythis	BW					2
Neoxythece	BW					2
Remijia	BW					2
Triplaris	BW		x		x	2
Blastemanthus	BW					1
Bombacopsis	BW					1
Centrolobium	BW					1
Dipteryx	BW					1
Elaeoluma	BW	x				1
Elizabetha	BW					1
Havetiopsis	BW					1
Henriquezia	BW					1
Lasiadenia	BW					1
Martiodendron	BW	x				1
Matayba	BW					1
Microplumeria	BW		x			1
Odontadenia	BW					1
Roupala	BW					1
Toulicia	BW					1
Heterostemon	FP					8
Parkia	FP					8
Handroanthus	FP					6
Leopoldinia	FP					6
Tovomita	FP					6
Amanoa	FP	x	x			5
Burdachia	FP		x			5
Pentaclethra	FP					5

(continued)

Table 12.5 (continued)

Genus	Amazon	CE	MA	CA	PA	Frequency
Diospyros	FP	x				4
Vatairea	FP					4
Mezilaurus	FP	x				3
Micrandra	FP		x			3
Exellodendron	FP					2
Quiina	FP					2
Styrax	FP	x	x			2
Dicorynia	FP					1
Diplotropis	FP					1
Etaballia	FP					1
Ilex	FP					1
Mora	FP					1
Pagamea	FP					1
Piranhea	FP	x				1
Platymiscium	FP					1
Stachyarrhena	FP					1
Strychnos	FP					1
Tassadia	FP					1

Table 12.6 Summary of the classification of species, genera and families in relation to frequency of occurrence in igapó forest, Negro River Brazilian basin

Taxon	Rare	Occasional	Common	Moderately frequent	Frequent	Total
Species	218	112	46	4	–	380
Genus	91	67	38	13	2	211
Families	16	16	13	11	6	62

frequency or low constancy, but we do not interpret this by indicating that most of the tree flora of igapó forest consisted of accidental species. Instead, the predominance of constant species may indicate that most of them are very finely tuned to small environmental variations between sites, that is, they would have small ecological niches and could be considered highly specialized species.

In general, tropical forests are composed of a small number of oligarchic species that are frequent in the region, abundant locally and dominate in the different surrounding habitats (Pitman et al. 2001; Macía and Svenning 2005). This hypothesis predicts that tropical forests, like temperate forests, may be characterized by a limited number of species that represent the majority of individuals. They also state that dominant species at the local level are also at broader, regional or even continental scales (Pitman et al. 2001, 2013). This hypothesis is closely related to the positive correlation between local abundance and species distribution amplitude, one of the best studied macroecological patterns (He and Gaston 2000). However, Scudeller and Vegas-Vilarrúbia (unpublished date) found that the same only four species (1.04%) were moderately frequent (occurrence above 50%) but did not

Table 12.7 Families that occur in igapó forest of Negro River basin with frequency >50% in the 19 surveys analysed

Families	Amazon	CE	MA	CA	PA	PP	Frequency
Fabaceae	AA	1	1	1	1	1	19
Apocynaceae	AA	1	1	1	1		16
Chrysobalanaceae	AA	1	1	1			16
Euphorbiaceae	AA	1	1	1	1		16
Lecythidaceae	AA	1		1			16
Sapotaceae	AA	1	1	1	1		15
Annonaceae	AA	1	1	1	1		14
Caryocaraceae	AA						13
Clusiaceae	AA	1	1	1	1		13
Myristicaceae	AA	1		1			13
Rubiaceae	AA	1	1	1	1	1	13
Malvaceae	AA	1	1	1			12
Lauraceae	AA	1	1	1	1		11
Melastomataceae	AA	1	1	1	1		11
Araceae	AA	1	1		1		10
Myrtaceae	AA	1	1	1	1	1	10
Salicaceae	AA	1	1	1	1		10

Table 12.8 Families that presented distribution restricted to FP and BW or only one formation on Amazon forest, excluding the 53 families with AA distribution

Families	Amazonia	CE	MA	CA	Frequency
Humiriaceae	BW	x	x	x	5
Thymelaeaceae	BW				1
Arecaceae	FP				6
Phyllanthaceae	FP	x			5
Quiinaceae	FP				2
Styracaceae	FP	x	x		2
Aquifoliaceae	FP				1
Loganiaceae	FP				1
Picrodendraceae	FP	x			1

present high abundance; the majority (57.55%) were restricted to only one survey of occurrence (rare); thus, there are no oligarchic or hyperdominant species in the igapó.

According to Wittmann et al. (2010), 112 of the most frequent central Amazonian várzea tree species are generalist species (or 60.21%) which also occur in other Neotropical ecosystems. Prance (1979) and Worbes (1997) stated that some of these generalist tree species (e.g. *Pseudobombax munguba*, *Leonia glycycarpa* Ruiz and Pav. and *Duroia duckei*) are widely distributed throughout the neotropics and occur particularly in semi-deciduous forests and savannas. The authors, however, showed that of the 186 most common tree species of the central Amazonian várzea, only 12

(6.45%; mostly *Malvaceae* and *Fabaceae*) occur in ecosystems with climatic- and/or edaphic-induced aridity, such as the cerrado, the caatinga or the Pacific (western) slope of the Andes. Twenty-seven (12.90%) várzea species also occur in Southeastern Brazilian Atlantic rainforests. The majority of Amazonian várzea tree species are found in adjacent moist terra firme forests (100 tree species, 53.76%). Only 21 (18.75%) are restricted to várzea and igapó, whereas 36 (32.14%) occur in all Amazonian ecosystems, i.e. várzea, igapó and moist terra firme forests.

Montero (2012) mentions that differences between igapó of the Negro River and the várzea in the Central Amazon can already be detected at the family level. Fabaceae, Euphorbiaceae, Sapotaceae and Lecythidaceae are important families of both forest ecosystems but with much greater importance in the igapó. In this study, the richest and most frequent families were Fabaceae, Euphorbiaceae, Rubiaceae, Annonaceae, Apocynaceae, Chrysobalanaceae, Lecythidaceae and Myrtaceae, different from those found by Montero (2012), which states that Malpighiaceae, Combretaceae and Ochnaceae which are among the ten most important families in igapó are of very little importance in lowland forests.

In a review of 44 floristic inventories (62.34 ha) of várzea forests scattered over the Amazon basin, Wittmann et al. (2006) stated that the Fabaceae were the most important várzea tree family, followed by the Malvaceae (including former Bombacaceae, Sterculiaceae and Tiliaceae), Euphorbiaceae, Moraceae, Palmae and Salicaceae (including the former Flacourtiaceae). The family importance, however, depends strongly on the location of the forest along the flood-level gradient, the successional stage and the geographic location of the inventories (Wittmann et al. 2010).

The predominance of low-constant taxa also implies the predominance of restricted distribution patterns. Restricted distribution is likely to be a general geographic pattern in Brazilian woody formations. Analysing tree species distribution among 106 localities, Oliveira-Filho and Ratter (1995) found 54.3% of species occurring only in one locality in Central Brazil. More than half of the 973 confidently identified species occurred in up to two localities of the 78 analysed by Castro et al. (1999) in their study of the Brazilian "cerrado" flora. The predominance of taxa with restricted geographic distribution patterns in Brazilian woody formations imposes special planning for conservation, in which few large reserves established in few geographic regions would not be enough to maintain diversity.

In general, the most constant/moderately frequent species in the igapó forest showed large geographic amplitude occurring in other Brazilian biomes. A great proportions of them also occurred in cerrado (CE), followed by Atlantic rain forest (MA). The fact that about 37.6% of all species occurred also in other formations, 43.1% of genera and 69.4% of families in igapó forest also occur out of Amazon forest. This may indicate that (a) all these formations remained in contact during the evolutionary time (b) exchange of floristic elements between them is likely to have occurred and (c) elements coming from a formation might have speciated in another formation. The evolution of South American flora was influenced by events near the end of Cenozoic, such as the impact of Andean orogeny and climatic oscillations between the glacial and interglacial times (van der Hammen 1988). It is admitted that savanna (cerrado CE) formations are very ancient and were already present in the early Cenozoic and that these formations maintained contact during evolutionary time (Sarmiento 1983). The riparian connectivity and the highly adapted dispersal mechanisms of floodplain trees would have reduced species losses at regional scales (Wittmann et al. 2006), with the floodplains themselves acting as linear refuges for sensitive terra firme species during periods with postulated dryer climatic conditions (Pires 1984).

According to Junk and Piedade (2010), the geologic and paleoclimatic history of Amazonia is closely related to water. The Amazon Basin is part of a very old depression that already existed in the Gondwana continent and then opened to the west. When South America separated from Africa, during the Early Cretaceous period, about 110 million years (Ma) before present (BP), the basin was already closed in the west by the Early Andes, except for an opening to the Pacific (Marañon Portal or Guayaquil Gap) that closed probably during the Late Cretaceous period (73 Ma). Rivers drained to the west into a depression along with the eastern border of the Early Andes that opened to the Caribbean Sea. With the uplift of the Andes, the pre-Andean depression was subjected to marine ingressions in the Late Cretaceous (83–67 Ma), the Early Tertiary (61–60 Ma) and the Late Tertiary (11.8–10 Ma) periods, as indicated by marine sediments. Following the interruption of marine transgressions, the depression became covered by rivers, lakes and extended wetlands. Large freshwater lakes were formed in the Tertiary period and were filled with sediments of riverine origin from the Andes and the shields of Central Brazil and the Guianas. In the Late Miocene (8 Ma), the connection to the Caribbean Sea and the Orinoco basin was closed by the Vaupes Arch. The Amazon River opened its way to the Atlantic Ocean by breaching the Purus Arch, and the modern Amazon drainage system incised large valleys and floodplains in the soft sediments (Lundberg et al. 1998).

According to Jaramillo et al. (2010), the record of plant diversity in the Amazons is still incomplete. Nevertheless, palynological and palaeobotanical data reveal that during the Neogene Amazonia, it already was covered by a highly diversified and multistratified forest that varied in composition and distribution over time under the influence of the major events (Hoorn 1993, 1994a,b, 2006). The potential effect on Amazonian forests of global cooling and possible associated changing precipitation patterns over the last 5 million years is unclear. Preliminary evidence suggests a major reduction in area from that formerly covered by rainforest. Areas in northern Venezuela, which were floristically similar to Amazonia during the Late Miocene, became isolated by the rise of the Andes and subsequently underwent a transformation to dry vegetation. There was also an extensive development of tropical savannas, which further encroached on the Amazonian rainforest. The overall effect of this reduction in forested area on Amazonian vegetation is unclear, but it might have caused a loss in diversity. However, it is now evident that the Quaternary glacial cycles did not significantly affect diversity in Amazonia (Bush 2004; Rull 2008). Furthermore, most of the species dated using molecular techniques indicate origination ages older than 2 million years ago (Rull 2008).

These climatic oscillations between the glacial and interglacial times would have caused variations in humidity, rainfall and temperature, provoking alternate expansions and contractions of the areas occupied sometimes by savannas and sometimes by forests, as represented by the model of ecological microrefuges (Rull 2008). The "displacements" of whole biomes along continental distances would have propitiated intense speciation in forests, mainly during the drier fluctuations, and in cerrado, mainly in the more humid fluctuations (Sarmiento 1983). Hence, it is plausible to think that the high richness occurring presently in these formations has at least partially resulted from dynamic evolutionary interactions among them. There could not be such a high richness in Brazilian cerrado (Castro et al. 1999) if there were not also a high richness in the forests (Sarmiento 1983) and vice versa.

Acknowledgements We would like to thank CAPES/CSF BEX 3936/13-9 for the postdoctoral scholarship and the Federal University of Amazonas for the institutional support. We also thank Tiago Carvalho, who provided us with his data from the Branco River.

References

Albernaz AL, Pressey RL, Costa LRF, Moreira MP, Ramos JF, Assunção PA, Franciscon CH (2012) Tree species compositional change and conservation implications in the white-water flooded forests of the Brazilian Amazon. J Biogeogr 39(5):869–883

APG (Angiosperm Phylogeny Group) (2016) An update of the Angiosperm Phylogeny Group classification for the orders and families of flowering plants: APG IV. Bot J Linn Soc 181(1):1–20

Burnham RJ, Graham A (1999) The history of neotropical vegetation: new developments and status. Ann Mo Bot Gard 86:546–589 http://www.jstor.org/stable/2666185

Bush M (2004) Amazonian speciation: a necessarily complex model. J Biogeogr 21:5–17

Castro AAJF, Martins FR, Tamashiro JY, Shepherd GJ (1999) How rich is the flora of Brazilian cerrados? Ann Mo Bot Gard 86:192–224

Ferreira LV (1997) Effects of the duration of flooding on species richness and floristic composition in three hectares in the Jaú National Park in floodplain forests in Central Amazonia. Biodivers Conserv 6:1353–1363

Ferreira LV (2000) Effect of flooding duration on species richness, floristic composition and forest structure in river margin habitats in Amazonian black water floodplain forests: implications for future design of protected areas. Biodivers Conserv 9:1–14

Ferreira LV, Almeida SS (2005) Relação entre a altura de inundação, riqueza específica de plantas e o tamanho de clareiras naturais em uma floresta inundável de igapó, na Amazônia Central. Rev Árvore 29:445–453

Frappart F, Seyler F, Martinez JM, León JG, Cazenave A (2005) Floodplain water storage in the Negro River basin estimated from microwave remote sensing of inundation area and water levels. Remote Sens Environ 99:387–399

Furch K (1997) Chemistry of várzea and igapó soils and nutrient inventory of their floodplain forests. In: Junk WJ (ed) The central amazon floodplain. Springer-Verlag, Berlin, Heidelberg, pp 47–67

Gentry AH (1982) Neotropical floristic diversity: phytogeographical connections between Central and South America, Pleistocene climatic fluctuations, or an accident of the Andean orogeny? Ann Mo Bot Gard 69:557–593

Gibbs RJ (1971) Mechanisms controlling world water chemistry: Evaporation-crystallization process [reply to J.H. Feth]. Science 172:871–872

Goulding M, Carvalho ML, Ferreira EJG (1988) Negro River, rich life in poor water: Amazonian diversity and food chain ecology as seen through fish communities. SPB Academic Publishing, The Hague
Goulding M, Barthem R, Ferreira EJG (2003) The Smithsonian atlas of the Amazon. Princeton Editorial Associates, London
Haffer J (1982) General aspects of the Refuge Theory. In G.T Prance: Biological Diversification in the Tropics. pp. 6–24. New York. Columbia University Press
Hamaguchi JO, Scudeller VV (2011) Estrutura arbórea de uma Floresta de Igapó no Lago Tupé, Manaus, AM. In: Santos-Silva EN, Scudeller VV, Cavalcanti MJ (org) BioTupé: Meio Físico, Diversidade Biológica e Sociocultural do Baixo Negro, Amazônia Central, vol 3. INPA, Manuas, pp 83–97. Rizoma Editorial
He F, Gaston KJ (2000) Estimating species abundance from occurrence. Am Nat 156(5):553–559
Hoorn C (1993) Marine incursions and the influence of Andean tectonics on the Miocene depositional history of Northwestern Amazonia: results of a palynostratigraphic study. Palaeogeogr Palaeoclimatol Palaeoecol 105:267–309
Hoorn C (1994a) Fluvial palaeonvironments in the intracratonic Amazonas Basin (Early Miocene-early Middle Miocene, Colombia). Palaeogeogr Palaeoclimatol Palaeoecol 109:1–54
Hoorn C (1994b) An environmental reconstruction of the palaeo-Amazon River system (Middle-Late Miocene, NW Amazonia). Palaeogeogr Palaeoclimatol Palaeoecol 112:187–238
Hoorn C (2006) Mangrove forests and marine incursions in Neogene Amazonia (Lower Apaporis River, Colombia). PALAIOS 21:197–209
Jaramillo C, Hoorn C, Silva SAF, Leite F, Herrera F, Quiroz L, Dino R, Antonioli L (2010) The origin of the modern Amazon rainforest: implications of the palynological and palaeobotanical record. In: Hoorn C, Wesselingh FP (eds) Amazonia, landscape and species evolution: a look into the past, 1st edn. Blackwell Publishing, Oxford
Junk WJ (1989) Flood tolerance and tree distribution in central Amazonian floodplains. In: Holm-Nielsen LB, Nielsen IC, Balslev H (eds) Tropical forests: botanical dynamics, speciation and diversity. Academic Press, London, pp 47–64
Junk WJ, Piedade MTF (2010) An introduction to South America wetland forest: distribution, definitions and general characterization. In: Junk WJ et al. (eds) Amazonian floodplain forests: ecophysiology, biodiversity and sustainable management. Springer, Dordrecht, the Netherlands, pp 3–25
Junk WJ, Piedade MTF, Schöngart J, Cohn-Haft M, Adeney JM, Wittmann F (2011) A classification of major naturally occurring Amazonian lowland wetlands. Wetlands 31(4):623–640
Keel SHK, Prance GT (1979) Studies of the vegetation of a white-sand black water igapó (Rio Negro, Brazil). Acta Amazon 9(4):645–655
Latrubesse EM, Franzinelli E (2005) The late Quaternary evolution of the Negro River, Amazon, Brazil: implications for island and floodplain formation in large anabranching tropical systems. Geomorphology 70:372–397
Lundberg JG, Marshall LG, Guerrero J, Horton B, Malabarba MCSL, Wesselingh F (1998) The stage for Neotropical fish diversification. In: Malabarba LR, Reis RE, Vari RP, Lucena ZMS, Lucena CAS (eds) Phylogeny and classification of Neotropical fishes. EDIPUCRS, Porto Alegre, pp 13–48
Macía MJ, Svenning JC (2005) Oligarchic dominance in western Amazonian plant communities. J Trop Ecol 21:613–626
Melack JM, Hess LL (2010) Remote sensing of the distribution and extent of wetlands in the Amazon basin. In: Junk WJ et al (eds) Amazon floodplain forests: ecophysiology, biodiversity and sustainable management. Ecological studies. Springer, Dordrecht, p 4360
Meade RH, Nordin CF, Curtis WF, Costa Rodrigues FM, do Vale CM, Edmond JM (1979) Sediment Loads in the Amazon River. Nature, 278:161–163
Montero JC (2012) Floristic variation of the Igapó Forests along the Negro River, Central Amazonia. Faculty of Forest and Environmental Sciences Albert-Ludwigs Universität, Freiburg, Germany

Mueller-Dumbois D, Ellenberg H (1974) Aims and methods of vegetation ecology. Wiley, New York 574pp

Oliveira-Filho AT, Ratter JA (1995) A study o the origin of central Brazilian forests by the analysis of plant species distribution patterns. Edinb J Bot 52(2):141–194

Parolin P, Adis J, Silva MF, Amaral I, Schimidt L, Piedade MTF (2003) Floristic composition of a floodplain forest in the Anavilhanas archipelago. Brazilian Amazonia Amazoniana 17(3/4):399–411

Parolin P, Adis J, Rodrigues WA, Amaral I, Piedade MTF (2004) Floristic study of an igapó floodplain forest in Central Amazonia, Brazil (Tarumã-Mirim, Negro River). Amazoniana 18(1/2):29–47

Piedade MTF (1985) Ecologia e biologia reprodutiva de *Astrocaryum Jaúari* Mart. (Palmae) como exemplo de população adaptada às áreas inundáveis do Rio Negro (igapós). MSc Thesis, Instituto nacional de Pesquisas da Amazônia, Manaus, Brazil

Pires JM (1984) The Amazonian forest. In: Sioli H (ed) The Amazon – limnology and landscape ecology of a mighty tropical river and its basin. Junk, Dordrecht, pp 581–602

Pitman NCA, Terborgh JW, Silman MR, Núñez P, Neill DA, Cerón CE, Palacios WA, Aulestia M (2001) Dominance and distribution of tree species in upper Amazonian terra firme forests. Ecology 82:2101–2117

Pitman NCA, Silman MR, Terborgh JW (2013) Oligarchies in Amazonian tree communities: a ten-years review. Ecography 36:114–123

Prance GT (1977) Floristic inventory of the tropics: where do we stand? Ann Mo Bot Gard 64:661–684

Prance GT (1979) Notes on the vegetation of Amazonia. 3. The terminology of Amazonian forest types subject to inundation. Brittonia 31:26–38

Prance GT (1980) A terminologia dos tipos de florestas amazônicas sujeitas a inundação. Acta Amazon 10(3):495–504

Prance GT (1982) Forest refuges: evidence from woody in Angiosperms. In: Prance GT (ed) Biological diversification in the tropics. Columbia University Press, New York, pp 137–158

Revilla JDC (1981) Aspectos florísticos e fitosociólogicos do igapó de Praia Grande, Negro River, Amazonas. MSc Thesis, Instituto Nacional de Pesquisas da Amazônia, Manaus, Brazil

Richardson JER, Pennington TD, Pennington PM (2001) Forest trees rapid diversification of a species-rich genus of Neotropical rain. Science 293:2242–2245

Rosales J, Petts G, Knab-Vispo C (2001) Ecological gradients in riparian forests of the lower Caura River, Venezuela. Plant Ecol 152(1):101–118

Rull V (2008) Speciation timing and neotropical biodiversity: the Tertiary-Quaternary debate in the light of molecular phylogenetic evidence. Mol Ecol 17(11):2722–2729. https://doi.org/10.1111/j.1365-294X.2007.03789.x

Santos UM et al (1984) Rios da bacia Amazonica I. Afluentes do rio Negro. Acta Amazon 14(1–2):222–237

Sarmiento G (1983) The savannas of tropical America. In: Burlière F (ed) Ecosystems of the world. Tropical savannas, vol 13. Elsevier, Amsterdam, pp 245–288

Scudeller VV, Souza AM (2009) Composição florística de uma floresta de igapó na Amazônia Central. In: Santos-Silva EN, Scudeller VV (org) Biotupé -meio físico, diversidade biológica e sociocultural do Baixo Negro River, Amazônia Central, vol 2. UEA, Manaus, pp 97–108

Scudeller VV, Vegas-Vilarrúbia T. Distribution and beta diversity of tree species in igapó forests (Negro River basin – Brazilian Amazon) submitted to Journal of Vegetation Science

Sioli H (1967) Hydrochemistry and geology in the Brazilian Amazon region. Amazoniana 1:267–277

Stebbins GL (1974) Flowering plants: evolution above the species level. Harvard University Press, Cambridge, MA

SUDAM (1984) Projeto de Hidrologia e Climatologia da Amazônia Brasileira. In: Atlas climatológico da Amazônia brasileira, vol 39. SUDAM Publica, Belém 125p

Ter Steege H, Pitman NCA, Sabatier D, Baraloto C, Salomão RP, Guevara JE, Phillips OL, Castilho CV, Silman MR (2013) Hyperdominance in the Amazonian tree flora. Science 342:325–335

Tuomisto H, Ruokolainen K, Kalliola R, Linna A, Danjoy W, Rodrigues Z (1995) Dissecting Amazonian biodiversity. Science 269:63–66

Van der Hammen T (1988) The tropical flora in historical perspective. Taxon 37:515–518

Wallace AR (1853) A narrative of travels on the Amazon and Rio Negro, with an account of the native tribes and observation on the climate, geology, and natural history of Amazon Valley. Reeve & Co., London

Whitmore TC, Prance GT (1987) Biogeography and Quaternary history in tropical America. Clarendon Press (Oxford Monographs on Biogeography), Oxford 214p

Wittmann F, Junk WJ, Piedade MTF (2004) The várzea forests in Amazonia: flooding and the highly dynamic geomorphology interact with natural forest succession. For Ecol Manag 196:199–212

Wittmann F, Schöngart J, Montero JC, Motzer T, Junk WJ, Piedade MTF, Queiroz H, Worbes M (2006) Tree species composition and diversity gradients in white-water forests across the Amazon basin. J Biogeogr 33:1334–1347

Wittmann F, Junk WJ, Schöngart J (2010) Phytogeography, species diversity, community structure and dynamics of central Amazonian floodplain forests. In: Junk WJ, Piedade MTF, Parolin P, Wittmann F, Schöngart J (eds) Central Amazonian floodplain forests: ecophysiology, biodiversity and sustainable management. Ecological studies. Springer Verlag, Heidelberg

Worbes M (1997) The forest ecosystem of the floodplains. In: Junk WJ (ed) The central Amazon floodplain: ecology of a pulsating system. Ecological studies, vol 126. Springer, Berlin/Heidelberg/New York, pp 223–265

Chapter 13
The Fishes and the *Igapó* Forest 30 Years After Goulding

Mauricio Camargo Zorro

13.1 Introduction

The natural flow regime of the rivers is crucial for maintenance of processes as water dynamics, flood pulse, nutrient cycling, passive transport of carbon sources, and top-down control (Poff and Zimmerman 2010; Winemiller et al. 2014). About 14% of the Amazon Basin is covered by floodplains, which are key to biological production, biogeochemical cycling, and trophic flows (Luize et al. 2015).

In the three kinds of Amazon waters, the regular and prolonged lateral flooding is the main determinant factor of phenological patterns in *várzea* (flooded forests by white waters) (Junk et al. 2010) and *igapó* (flooded forest by black or clear waters) (Haugaasen and Peres 2005; Camargo et al. 2015a). In the first months of flooding season, the riparian forest has peaks of fructification (Parolin et al. 2004; Haugaasen and Peres 2005; Camargo et al. 2015a), providing food resources and temporally habitats for fish. These hydrological dynamics maintain resident aquatic biodiversity.

Floods stimulate nutrient remineralization as well as primary and secondary production in floodplain habitats (Welcomme 1985; Junk et al. 1989). The new habitats created by floods and the movement of fishes into flooded forests cause a dynamic seasonal rearrangement of fish communities, and many fish species are adapted to take advantage by reproducing at the beginning of the wet season, which allows early life stages to feed and grow within inundated floodplain habitats (Lowe-McConnell 1987).

In contradiction, oligotrophic waters of the Negro River support a higher fish diversity (750 described species; Petry and Hales 2013) and sustain important fisheries (Inomata and Freitas 2015); yet what energy sources maintain the blackwater

M. C. Zorro (✉)
Instituto Federal de Educação, Ciência e Tecnologia da Paraíba, Belem, Brazil
e-mail: camargo.zorro@gmail.com

R. W. Myster (ed.), *Igapó (Black-water flooded forests) of the Amazon Basin*,
https://doi.org/10.1007/978-3-319-90122-0_13

fish production? During the lateral inundation, the *igapó* of clear and black waters provides important links for food chains, transferring energy from the riparian forest and other terrestrial low trophic levels like arthropods for the fishes (Camargo et al. 2014; Correa and Winemiller 2018). As a response to periodical allochthonous terrestrial sources, omnivorous and opportunistic feeding habits may have evolved.

13.2 Food Web Structure in Black Waters

Food webs are diagrammatic descriptions of trophic connections among species in communities (Fig. 13.1). Understanding the main sources of primary production that support the trophic levels in the Amazon waters is crucial to know the structure of food webs, perceive ecosystem flows, and promote sustainability of the natural processes. For the Negro River basin, detritus and other allochthonous sources are the main pathways for energy and nutrients that are provided by the forest to transfer into higher consumer levels (Goulding et al. 1988).

Isotopic studies performed in the middle Negro River concluded that the main autotrophic sources are the flooded forest, aquatic herbaceous vegetation, and algae (periphytic and phytoplanktonic), which utilize a C3 photosynthetic pathway (Marshall et al. 2008). Thus, the importance of phytoplankton in blackwater environments in trophic food webs could be underestimated. Isotope studies have demonstrated that phytoplankton and periphytic algae are more important than their abundance suggests (Hamilton et al. 1992; Forsberg et al. 1993).

Both algae and aquatic macrophytes enter in aquatic food webs mostly in form of detritus (fine and coarse particulate organic matter) or being transported by water flow and settling onto substrates (Winemiller 2004). Particulate organic matter in the stream of rapids and waterfalls is mostly associated with biofilm and epilithic diatoms that grow on rocks, submerged wood, and herbaceous plants and compose the main energy sources for macroinvertebrates and other trophic links (Camargo 2009a).

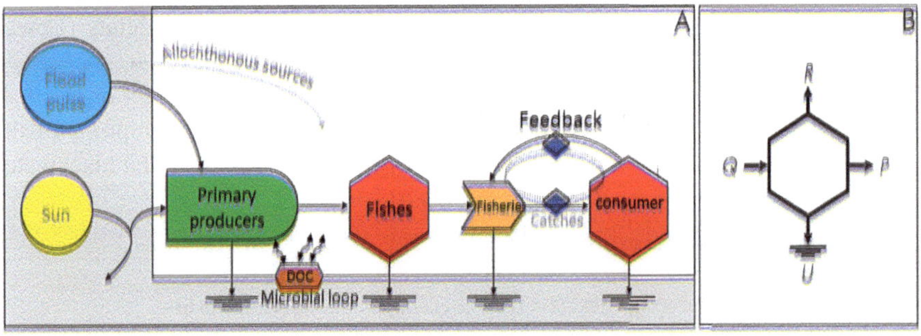

Fig. 13.1 (**a**) Simplified model flow of energy in Amazon River during the flood period. (**b**) Flows to each box: Q=Consumption; P=Production; R=Respiration; U=Unassimilated

In tropical blackwater rivers, the aquatic insect abundance is low with most species and biomass concentrated in leaf litter and wood debris providing important habitats (Benke et al. 1984) but supporting large populations of atyid and palaemonid shrimp (Winemiller 2004). Isotope mixing models in a blackwater system concluded that the first link of secondary production during the annual flood is composed by terrestrial arthropods that bond with forests and fishes and made a greater proportional contribution to fish biomass (Correa and Winemiller 2018).

The blackwater systems have patterns with high richness of fish species, low primary production, and a large piscivore abundance (Winemiller and Jepsen 2004). For the Negro basin, even an amount in fish exceeding 750 described species (Petry and Hales 2013), partitions of abundance in the communities are usually skewed having very few highly abundant species and outnumbered rare (uncommon) species (Magurran and Henderson 2003). Those abundant species have an important role transferring most material in food webs and short food chains (Winemiller 2004).

The average of isotopic ^{13}C for in situ blackwater fish assemblage revealed that benthivorous fishes feed primarily on burrowing midge larvae that are fed mostly with organic matter derived from terrestrial plants and midwater characids that consume fruits and seeds of terrestrial plants plus smaller fractions of terrestrial insects (Winemiller and Jepsen 2004). Through values of ^{13}C for *Semaprochilodus insignis*, the view of Fernandez (1993) who concluded that the route of carbon assimilation is consistent with the fish migration from white water to black water for Negro and Amazonas rivers was confirmed. With fish migration, ecosystems can be linked by transfer production from rich nutrient whitewater to poor nutrient blackwater systems (Winemiller and Jepsen 2004).

Large piscivorous like *Cichla temensis* (Aguiar-Santos et al. 2018), *Hidrolycus* and *Serrasalmus* in a wider range of habitats allowing consumption of a great range of prey sizes. It should have greater potential to exert top-down effects on the food web. Moreover, reduction of body size (miniaturization) (Weitzman and Vari 1988) seems to constitute a very common trait among fishes that inhabit environments with food limitation like the black waters of the Negro River. That evolutionary strategy is a response to minimize the energetic demand among fishes.

13.3 The *Igapó* of the Middle Xingu River

The stretch of the Xingu River between Xingu-Irirí confluence and Belo Monte waterfalls is a homogenous geomorphologic region named middle Xingu (Camargo et al. 2004), comprising a unique geographic feature of lower Amazon Basin, with its relative steep terrain and sinuous configuration (Fig. 13.2). This physiography is resulted by ancient orogenic processes, dating two billion years ago. The numerous rapids carved by the river into the faults of giant granite and gneisses blocks created interfacial environments that suffer considerable seasonal modifications resulting in annual variations in volume of water (Camargo et al. 2013).

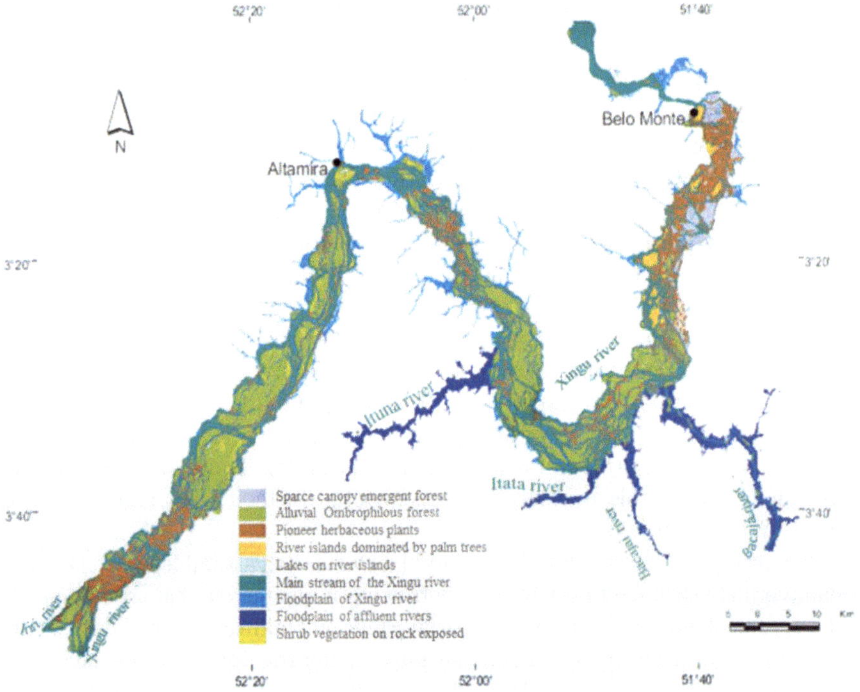

Fig. 13.2 The stretch of the Xingu River between Xingu-Irirí confluence and Belo Monte waterfalls

The middle Xingu River differs in a geomorphological way from the lowland Amazon floodplain. By its wavy geomorphology and rocky bottom that expands or contracts in function of the river flow regime, there is a definition of a flooding limited area. By the way, the inundate areas correspond to their own waterway that, by a high dynamics of alluvial transport and sedimentation, generates flooded terraces with many succession of vegetable stages in pioneer ways till the alluvial dense ombrophilous forest (Estupiñan and Camargo 2009).

The largest flooding area is composed by outnumbered fluvial islands representing 21.89 Km^2 of the studied area. Only 8.3% of the total area is covered by a strait riparian ombrophilous forest and 1.0% by pioneering regularly flooded shrub vegetation (Estupiñan and Camargo 2009). The increase of the river's flow rate and the rise of waters in rainy season favor the progressive flooding margin areas that are topographically lower turning into waterways, locally denominated *sangradouros*. Those, while flooding, turn into important access channels and connect alluvial forest areas with the river's main stream (Giarrizzo and Camargo 2009).

13.3.1 *The* Igapó *and the* Piracemas *in the Middle Xingu River*

The use of the word *piracema* is still misunderstood. For some, the term is related with the rise of fishes through the rivers till its spring for spawning. The other way associates the *piracema* with the season when fishes go into reproduction. Etymologically, the word comes from Tupi language, with the assembling of the prefix *pirá* (fish) and *sema* (grouping), which means "meeting of fishes for spawning."

In the vision of local Xingu fishermen, a *piracema* is understood as an aquatic habitat located on the margins of fluvial islands and even in the river, where shoals of many species like the piaus (Anostomidae), curimatã and ariduia (Prochilodontidae), branquinas (Curimatidae), sardinhas (*Triportheus*) and even trairas (Erithrynidae) and other fish species meet for spawning. Then, the term applies for fishes, in general, that migrate small distances for reproduction.

In the Amazon River system, the hydrological regime strongly influences all the ecological processes (Poff and Allan 1995); by the way, the periodical flood pulse constitutes an inductive process for reproduction of many fishes (Junk et al. 1989). So, a big group of fishes migrate in search of appropriate sites for spawning in synchrony with the start of regional floods in a more or less restrictive season (Welcomme 1979; Araujo-Lima et al. 1995). During this migration, fishes use temporally the flooded forest, which warranties a habitat rich in food and refuge from predators (Goulding et al. 1988; Waldhoff et al. 1996).

Piracema **waterways or** *sangradouros*

13.3.2 The Source of Piracema Fishes

The association of the phenomenon denominated by fishermen as *repiquete*, which consists in a sudden increase on water level in the main waterway still in the end of dry season (November), and the flooding of the *sangradouros* in the pluvial islands of Xingu stimulate a process of short migration and spawning in small periods of time (hours), in superficial depths (30 cm). This process can be explained by the possible changes on some parameters of the aquatic environment, which works as a start signal for fishes to get ready physiologically and behaviorally for spawning.

In that way, the association of factors like: I. The currency of migrant adult fishes concentrated in the waterfall areas of Xingu in dry season; II. Environmental signals indicating changes on the physicochemical attributes of aquatic environment, and III. the flooding of *sangradouros* by the sudden increase of the water level with the *repiquete*, causing short displacements in countercurrent to reach those *sangradouros*, for spawning.

Even with the existence of seasonal insular lakes, forming a surface of approximately 5.0 km^2, representing 0.31% of the studied area in Xingu, these and the *sangradouros* possibly don't support in an exclusive way the fishery production of the middle Xingu. A hypothesis proposal in relation to the source of migrant fishes is associated with the displacement of metapopulations in Xingu's tributary rivers, like Bacajá, Bacajaí, Itatá, and Ituna that drain into its margin. All those rivers have big traces of marginal floodable forests covering an area of approximately 152 km^2 (an estimated extension of 20 km) representing 10% of the researched area.

Another peculiarity of the middle Xingu River is its altitude variation. Its dip is 85 m in an extension of 160 km, and the pluvial islands topographical heterogeneity demonstrate that the migration process doesn't occur in a synchronized way in this stretch of the river. While in the main channel, some kilometers upstream, the *sangradouros* is being flooded, in the downstream way, some waterways even have achieved its minimal level to induce fishes that are spawning. In shallow flooded areas, the fertilized eggs form big floating clouds in the surface. There is a belief that with this mechanism, species can minimize mortality by predators and maximize juveniles' chance of survival with water level decrease, occupying marginal areas.

Afterward, some species stay in flooded areas throughout the flood period or just turn back to the main waterway. This cross-border migration cycle (river > *sangradouros* > flooded areas > *sangradouros* > river) is known by the local population. Some riverside communities take advantage of those concentration of fishes and build "fences" or wood barriers allowing the entrance of fish that are ready to spawn but limit its exit, making an easy target to be captured, and impairing the reproduction process of migrant fishes.

"Fences" or wood barrier allowing the entrance of fish that are ready to spawn

13.3.3 Is Expected a Reproduction Pattern to Piracema Fishes?

The available info about the behavior of reproductive migration of the Amazon fishes have been described for white, clear, and darkwater environments (e.g., Fernandes 1997; Brito Ribeiro 1983; Carvalho and Merona 1986). Although some study relate reproductive behavior and kinds of migrant fishes in different Amazonian environments (Welcomme 1979; Junk et al. 1983; Tejerina-Garro et al. 1998), those were made in environments where the flood surface is significantly big and the flood of the river lasts for months.

With the help of local fishermen, a cartographical survey about *piracemas* was made, along the middle Xingu (Giarrizzo and Camargo 2009). Totalling 198 places were recorded where *piracema* occurs. Nearly 58% of identified places (114) are registered in fluvial islands, and the others at the river margins. Most part of *piracemas* (73%) happen in places with ombrophilous forest, those acting as an incident solar radiation filter over water surface, providing food and used as refuge and protection from predators.

These findings show that the same fish species with migratory habits for reproduction have a broad plasticity in its response to perform long and short displacements according to the regional hydrological dynamics. In this way, for some Amazon environments, migratory fishes like curimatã (*Prochilodus nigricans*) need a long displacement through the main stream for reproduction. In the middle Xingu River, the curimatã develops its gonads in rapids and after doing small displacements entering in the available narrow waterways with the water level increasement at the *repiquete* time, in the flooded forest (Fig. 13.3).

It's possible that in the middle Xingu River the stocks of migratory species have a homing ability to return in spawning season, for its area of nursery and growth in the same *piracema* channel. This assumption is supported by fishermen that know different morphotypes inside the same species and associate with a particular habitat of *piracema*. Other fishes, like the pacus (Serrasalmidae), even spawning in synchrony with the flooding season, can show an asynchrony spawn peak, during dry season in the Xingu main stream. The occurrence of juveniles during the regional drought indicates this asynchrony in some fishes that are spawning (Camargo and Lima Jr. 2007).

Mature fishes come into the *piracema* **habitat**

Migrations made by fishes during annual flooding warrant, even reproduction success, the access to primary allochthonous sources. High rates of primary annual production of plant fractions that make up the alluvial forest and the tight sync in fruit production with the flood indicate a mutualism where plants provide food to fishes and release seeds in other areas by ichthyochory (Gottsberger 1978).

From 26 abundant plant species recorded in the alluvial forest of the middle Xingu River, 22 feature fruiting in synchronic with the flood, when fishes inhabit flooded areas (Fig. 13.4). Araça (*Myrcia* sp.), caferana (*Quiina florida*), piranheira (*Piranha trifoliata*), and seringueira (*Hevea brasiliensis*) are included as a preference of frugivorous fish (Giarrizzo and Camargo 2009), while other pioneer plant fruits like embauba (*Cecropia* sp.), even if digested, do not constitute the main food base of many fishes, probably because of its low nutrition value, when compared with primary forest plants (Waldoff et al. 1996).

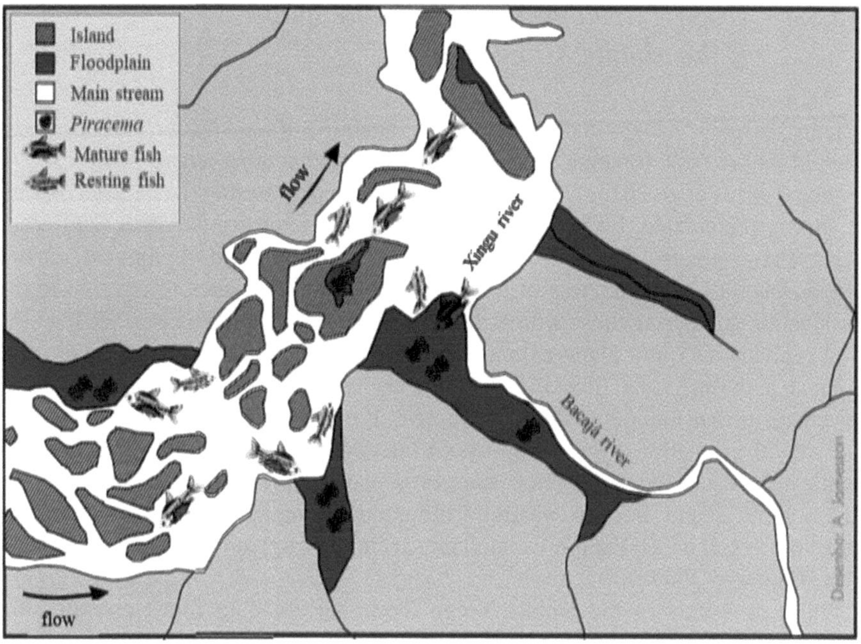

Fig. 13.3 Model of lateral migration – fish reproduction in the middle Xingu River

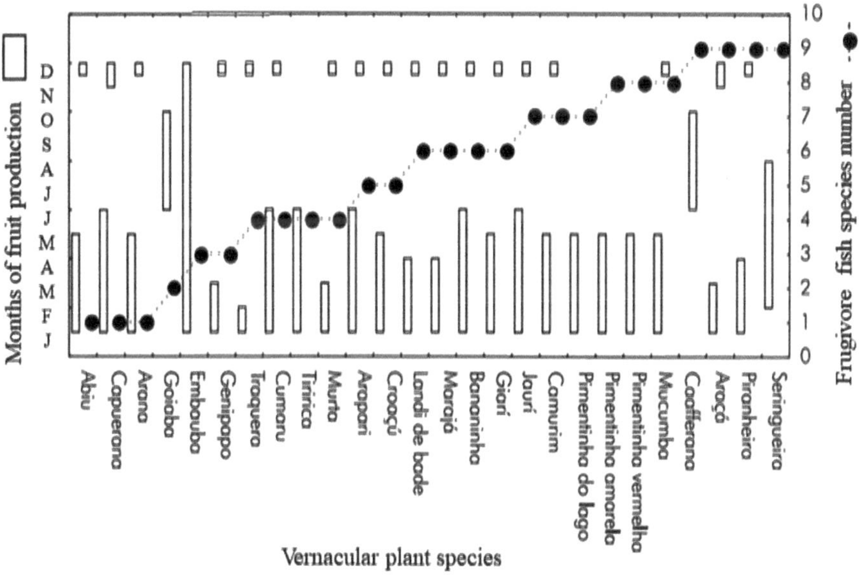

Fig. 13.4 Synchrony of fruit production, seasonal flooding, and use by frugivore fishes in the middle Xingu River

13.3.4 *Trophic Ecology of Fishes in the* Igapó *of the Middle Xingu River*

Facing seasonal changes in the quality and availability of food resources, consumers should adjust their foraging behavior in a way to maximize energy and nutrient intake (Correa et al. 2015a). Floodplains of the Amazon Basin are characterized by a strong, predictable, hydrological seasonality providing fruits, seeds, and other terrestrial-originated food resources. Therefore, rivers with rhythmical annual floods have a higher richness of fish species and elevated rates of riparian forest production comparing those with arrhythmic flood pulses (Jardine et al. 2015).

Frugivory is a key plant–animal interaction that contributes to maintain biological and functional diversity. Ichthyocory is a seed long-distance dispersal mechanism that maintains the reproductive dynamics of plant communities and possibly the radiation of early angiosperms in flooded forests (Correa et al. 2015a), while fishes gain access to rich sources of carbohydrates, lipids, and proteins (Horn et al. 2011). In clear waters of the middle Xingu, Serrasalmidae were the most diverse and also the more abundant species as frugivore fish with 0.63 g.m^{-2} d.w (Camargo 2009b).

In clear waters of the middle Xingu River, the flooding forest consumption includes other plant parts such as leaves and flowers, as well as terrestrial and aquatic invertebrates. Juveniles of pacus (*Myleus*, Serrasalmidae) were observed jumping some centimeters over the water to consume seeds and flowers of marginal grasses (Camargo 2004). Litterfall production per month, during the seasonal flooding, achieved 33 g.m^{-2} of dry weight (d.w) for fruits ($r = 0.48 \pm 0.32$ s.d), 137 g.m^{-2} d.w for leaves ($r = 0.42 \pm 0.09$ s.d), and 10 g.m^{-2} d.w for flowers ($r = -0.61 \pm 0.02$ s.d) (Camargo et al. 2015b).

Feeding studies for 126 fish species of main stream backwater, riffles, fluvial lakes, and riparian forest environments in the middle Xingu stretch revealed 8 trophic guilds (Table 13.1). A PCA based on the relative abundances (CPUEg) of guilds exposes four spatial patterns: I the insular lakes, similar composition of iliophages (preference of epilithon-benthic algae), carnivores, piscivores, and insectivores; II. Main stream of Xingu and Iriri river environments are very similar by the abundance of piscivores, frugivores, and omnivores; III backwaters near waterfalls, the higher affinity was defined, mainly, by piscivore fishes; IV riffles with iliophage habits (preference on biofilm and epilithic organism associated with rocks, for macroinvertebrates and another trophic link as important energy sources) (Camargo 2009a).

A wide spectrum of trophic guilds recorded in the middle Xingu River indicated a development system with various base sources of energy and with rapid transference along the different links, involving physical and ecological processes. The physical processes are related with the water flow, rains, and transport–deposition of particulate material along the stream. Ecological processes include decomposition and recycling of biogenic material via microbial-loop and their availability to primary and secondary production. The abundance of piscivores in the spatial-temporal

Table 13.1 CPUEg (g.m^{-2}.h^{-1}) – trophic guild and habitat in seasonal variation – middle Xingu River

Guild/habitat	Main stream	Fluvial lakes	Tributary stream	Waterfalls
Detritivores	0.015–0.020	0.013–0.027	0.012–0.035	0.013–0.014
Iliophagous fishes	0.014–0.017	0.041–0.021	0.020–0.017	0.013–0.011
Plankton feeders	0.006–0.013	–	–	0.082–0.00
Frugivores	0.034–0.043	0.007–0.002	0.048–0.00	0.014–0.03
Insectivores	0.005–0.009	0.011–0.005	0.002–0.027	0.002–0.013
Omnivores	0.039–0.022	0.012–0.012	0.064–0.027	0.005–0.003
Carnivores	0.027–0.014	0.035–0.021	0.013–0.010	0.015–0.004
Piscivores	0.034–0.036	0.027–0.038	0.027–0.052	0.032–0.032
Total	0.173–0.176	0.126–0.145	0.187–0.162	0.176–0.107

variation of the Xingu River confirms the important role of the top-down control in the system production.

Comparison of the middle Xingu riffles and the blackwater rapids of the Negro River in relation with rheophilic fish assemblages evidenced that the former is more complex. The Xingu River revealed a higher diversity of feeding tactics, different foraging areas, and periods of activity (Zuanon 1999; Camargo 2004). The main light incidence on submerged rocks in clear waters, associated with deposition of fine particulate material, promotes a major microbial-loop via detritus and the development of the biofilm on rocks. Most of this production is derived from algae and, during the wet season, from vascular plants. Limited light penetration caused by high levels of humic acids that stain in the black waters reduces its transparency and restricts growth of epilithon.

13.3.5 Food Webs in the Middle Xingu

Studies in the fluvial compartments of the ecosystem in the middle Xingu River identified its main functional groups tracing and quantifying the flow of biomass through the system (Camargo et al. 2004; Brito et al. 2009; Camargo et al. 2009; Costa et al. 2009; Estupiñan and Camargo 2009; Jesus et al. 2009; Giarrizzo and Camargo 2009; Camargo 2009a, b; Bastos et al. 2011; Camargo et al. 2015a, b) of 24 components which were analyzed using Ecopath with Ecosim software (Camargo et al. 2014).

The model was developed for wet seasons (December–February), when the riparian forest is flooded. The results indicate that the main flow of biomass into the food web is made by alluvial forest. Also in the backwater, the floating submerged leaves of marginal vegetation are colonized by dense aggregations of epiphytes. In contrast with the riparian forest, on long stretches of rapids and waterfalls, rocks of the riffles support other sources of primary production, such as the mats of epiliths, which in turn provide substrate for the complex of organisms composing the "auf-

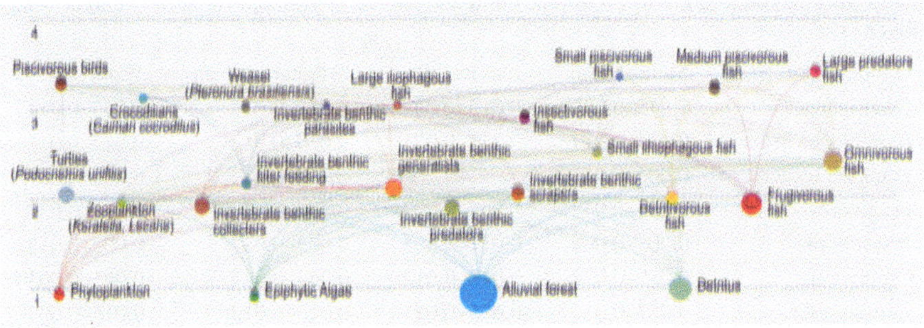

Fig. 13.5 Trophic model of the aquatic ecosystem of the middle Xingu River including the area affected by the Belo Monte hydroelectric project. The size of each circle corresponds to its contribution to biomass of each compartment. The numbers correspond to the trophic level of each compartment

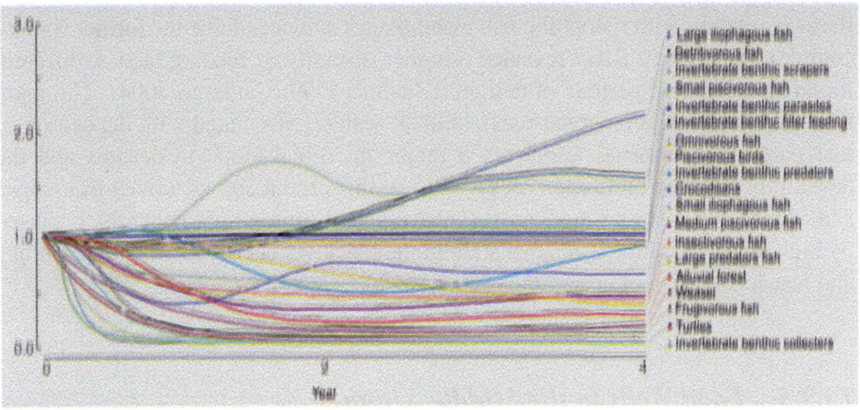

Fig. 13.6 Simulation of reduction on biomass of the alluvial forest and its effects on selected compartments of the ecosystem

wuchs" that generates important carbon sources supporting a large rheophilic fish biomass. The model had four distinct trophic levels (Fig. 13.5).

A simulation developed to predict the impacts of the Belo Monte dam on the middle Xingu River revealed alterations to the natural hydrological regime of the river and indicates major impacts on the vegetation associated with the rocky outcrops and the areas of fluvial forest. The effects of gradual processes of suppression of the alluvial forest, following the filling of the Belo Monte reservoir (Fig. 13.6), indicated a rise of epilithon, in the case of primary producers, with an increase of benthic scrapers and iliophagous fish, while frugivores fish, which constitutes an important source of protein for local communities, declined.

In addition to a general loss of biodiversity, the model indicates clearly the potential loss of fishery productivity on the stretch of the river that will be affected by the inundation of the Belo Monte reservoir. The reduction in these habitats, which are dependent on the annual flood pulse, would likely result in a drastic reduction in abundance of commercially important frugivores, such as pacus (Serrasalmidae), and large-body detritivores like the curimatã and ariduia (Prochilodontidae), leading to major changes in the composition of local catches (Camargo 2009b).

13.4 Amazon Wetlands: Implications for Conservation and Management

Overfishing could eliminate the most effective seed dispersers from floodplain ecosystems, with negative consequences for plant regeneration dynamics, species diversity, and genetic flow among plants in wetland habitats (Correa et al. 2007; Harrison et al. 2013; Kurten 2013; Correa et al. 2015b).

A review about the consequences of suppressing riparian forest in drainage areas of freshwater ecosystems, made by Pinheiro et al. (2015), exposed their important role on maintaining the local biodiversity acting as a corridor for animal migration (Metzger 2010), promoting deposition of particulate and dissolved organic matter (POM and DOM) and nutrients via mineralized accumulated litterfall (Moss 2010), removing pesticides and fertilizers inserted at adjacent agricultural areas (Wantzen et al. 2008; Gücker et al. 2009), and decreasing their harmful effects on biotic components (Ramírez et al. 2008; Winemiller et al. 2008).

In addition, Pinheiro et al. (2015) alert that watershed deforestation may diminish the DOM and nutrient inflow into aquatic systems and change water chemical composition, unbalancing the food chains and decreasing primary productivity (Lampert and Sommer 2007). Likewise, the authors highlight that large-scale removal of riparian forest may cause negative impacts on the resident biota and also on the communities of aquatic bodies. For example, intense deforestation of marginal areas along the Negro River may interfere with DOM input (Moss 2010), reducing the concentration of humic compounds and, therefore, decreasing the water darkness and allochthonous biomass.

Brazil's original Forest Code of 1965 established a proportion of rural land that should be maintained permanently as forest (legal reserves) and also prohibited the clearing of vegetation in fragile areas – such as along the margins of rivers and streams (APP, areas of permanent protection). In turn, the new 2012 forest law, promoted by rural and agribusiness interests, allows opening wide new forest areas for agriculture. Areas formerly held for being too steep or vital on protecting watersheds and watercourses are now open to destruction. A study made by IPEA, an organization linked with the federal government, concluded that the area that could be deforested due to the changes in the Brazilian Forest Law could be sized as much as 79 million hectares with a release amount of 28 billion tons of carbon release into the atmosphere (WWF 2018).

For the restoration and conservation of riparian forest, Pinheiro et al. (2015) proposed to implement ecological corridors in the deforested regions, allowing natural restoration and thus ensuring that the seed bank or allochthonous propagules can help in the recovery of suppressed vegetation, adding them to the APP strips and renouncing some of the taxes paid by landowners.

References

Aguiar-Santos J, de Hart PAP, Pouilly M, Freitas CEC, Siqueira-Souza F (2018) FK trophic ecology of speckled peacock bass *Cichla temensis* Humboldt 1821 in the middle Negro River, Amazon, Brazil. Eco Fresh Water Fish 00:1–11

Araújo-Lima CARM, Agostinho AA, Fabré NN (1995) Trophic aspects of fish communities in Brazilian rivers and reservoirs. In: Tundisi JG, Bicudo CEM, Matsumura-Tundisi T (Eds.). Limnology in Brazil. ABC/SBL, Rio de Janeiro, pp 105–136

Bastos ASM, Costa VB, da Costa S, Martins-da-Silva RC, Paiva RS, Camargo M (2011) Abundância e frequência de ocorrência de algas epilíticas das localidades Boa Esperança e Arroz Cru do setor do médio rio Xingu/PA-Brasil. Boletim Técnico-Científico do CEPNOR 11:59–70

Benke AC, van Arsdall TC, Gillespie DH, Parrish FK (1984) Invertebrate productivity in a subtropical Blackwater river: the importance of habitat and life history. Ecol Monogr 54:25–63

Brito Ribeiro MCL (1983) As migrações dos jaraquis (Pisces: Prochilodontidae) no rio Negro, Amazonas, Brasil. MSc. Dissertation, INPA. Manaus, p 192

Brito SAC, Melo NF, Camargo M (2009) Consumidores invertebrados: O Zooplâncton In: Camargo M, Ghilardi R Jr (Eds.). Entre a Terra, as Águas e os Pescadores do Médio Rio Xingu – Uma Abordagem Ecológica. Belém, pp 119–156

Camargo M (2004) A comunidade ictica e suas interrelações troficas como indicadores de integridade biológica na área de influencia do projeto hidrelétrico Belo Monte, rio Xingu. DSc. UFPA, Belem, p 183

Camargo M (2009a) Teias alimentares e fluxos de energia. In: Camargo M, Ghilardi R Jr (Eds.). Entre a Terra, as Águas e os Pescadores do Médio Rio Xingu – Uma Abordagem Ecológica. Belém pp 297–329

Camargo, M (2009b) Os Consumidores: Peixes: Ecologia trófica In: Camargo M, Ghilardi R Jr (Eds). Entre a Terra, as Águas e os Pescadores do Médio Rio Xingu – Uma Abordagem Ecológica. Belém pp 193–214

Camargo M, Giarrizzo T, Isaac VJ (2004) Review of the geographic distribution of fish fauna of the Xingu river basin. Brazil Ecotropica 10:123–147

Camargo M, Giarrizzo T, Jesus AJS (2015a) Effect of seasonal flooding cycle on litterfall production in alluvial rainforest on the middle Xingu River (Amazon basin, Brazil). Braz J Biol 175:250–256

Camargo M, Giarrizzo T, Isaac VJ (2015b) Population and biological parameters of selected fish species from the middle Xingu River, Amazon Basin. Braz J Biol 175:112–124

Camargo M, Giarrizzo T, Jamesson A (2014) Assessing the impact of hydrolectric dams on Amazonian rivers using Ecopath with Ecosim: a case study of the Belo Monte dam. Fisheries Centre Research Reports 22:168–169

Camargo M, Gimênes H Jr, de Sousa LM, Rapp Py-Daniel L (2013) Loricariids of the middle Rio Xingu – Loricariiden des mittleren Rio Xingu. Panta Rhei, Hannover, p 304

Camargo M, Gonçalves AP, Carneiro CC, Castro GTN (2009) Pesca de consumo In: Camargo M, Ghilardi R Jr (Eds.). Entre a Terra, as Águas e os Pescadores do Médio Rio Xingu – Uma Abordagem Ecológica. Belém, pp 265–282

Camargo M, Lima WMA Jr (2007) Aspectos da biologia reprodutiva de seis espécies de peixes de importância comercial do médio Rio Xingu – bases para seu manejo. Revista Uakari 3:64–77

Carvalho JL, Merona B (1986) Estudos sobre dois peixes migratorios do baixo Tocantins, antes do fechamento da barragem de Tucuruí. Amazoniana 9:595–607

Correa SB, Winemiller K, López-Fernández H, Galetti M (2007) Evolutionary perspectives on seed consumption and dispersal by fishes. Bioscience 57:748–756

Correa SB, Winemiller KO (2018) Terrestrial-aquatic trophic linkages support fish production in a tropical oligotrophic river. Oecologia 186:1069–1078

Correa SB, Costa-Pereira R, Fleming T, Goulding M, Anderson JT (2015a) Neotropical fish-fruit interactions: eco-evolutionary dynamics and conservation. Biol Rev 90:1263–1278

Correa SB, Araujo JKM, Penha JMF, Nunes da Cunha C, Stevenson PR, Anderson JT (2015b) Overfishing disrupts an ancient mutualism between frugivorous fishes and plants in Neotropical wetlands. Biol Conserv 191:159–167

Costa V, Costa S, Camargo M (2009) Os produtores primários: O fitoplâncton e o epilíton In: Camargo M, Ghilardi R Jr (Eds.). Entre a Terra, as Águas e os Pescadores do Médio Rio Xingu – Uma Abordagem Ecológica. Belém, pp 73–115

Estupiñán RA, Camargo M (2009) Ecologia da paisagem natural. In: Camargo M, Ghilardi R Jr (Eds.). Entre a Terra, as águas e os Pescadores do Médio Rio Xingu – Uma Abordagem Ecológica. Belém, pp 33–53

Fernandes CC (1997) Lateral migration of fishes in Amazon floodplains. Ecol Freshw Fish 6:36–44

Fernández JM (1993) Fontes autotroficas de energia em juvenis de Jaraqui, *Semaprochilodus insignis* (Schomburgk, 1841) e curimata, *Prochilodus nigricans* Agassiz, 1829 (Pisces: Prochilodontidae) de Amazonia central. Masters Thesis, INPA/Fed Univ de Amazonia, Manaus, Brazil, p 56

Forsberg BR, Araújo-Lima CARM, Martinelli LA, Victoria RL, Bonassi JA (1993) Autotrophic carbon sources for fish of the Central Amazon. Ecology 74:643–652

Giarrizzo T, Camargo, M (2009) As Píracemas In: Camargo M, Ghilardi R Jr (Eds.). Entre a Terra, as Águas e os Pescadores do Médio Rio Xingu – Uma Abordagem Ecológica. Belém, pp 283–295

Goulding M, Carvalho ML, Ferreira EJG (1988) Rio Negro, rich life in poor water: Amazonian diversity and foodchain ecology as seen through fish communities. SPB Academic Publishing, The Hague, p 200

Gottsberger G (1978) Seed dispersal by fish in the inundated region of Humaita, Amazonia. Biotropica 10:170–183
Gücker B, Boëchat IG, Giani A (2009) Impacts of agricultural land use on ecosystem structure and whole-stream metabolism of tropical cerrado streams. Freshw Biol 54:2069–2085
Hamilton SK, Lewis WM Jr, Sippel SJ (1992) Energy sources for aquatic animals in the Orinoco river floodplain: evidence from stable isotopes. Oecologia 89:324–330
Harrison RD, Tan S, Plotkin JB, Slik F, Detto M, Brenes T, Itoh A, Davies SJ (2013) Consequences of defaunation for a tropical tree community. Ecol Lett 16:687–694
Haugaasen T, Peres C (2005) Tree phenology in adjacent Amazonian flooded and unflooded forests. Biotropica 37:620–630
Horn MH, Correa SB, Parolin P, Pollux BJA, Anderson JT, Lucas C, Widmann P, Tjiu A, Galetti M, Goulding M (2011) Seed dispersal by fishes in tropical and temperate fresh waters: the growing evidence. Acta Oecologica 37:561–577
Inomata SO, de Freitas CE C (2015) A pesca comercial no médio rio Negro: aspectos econômicos e estrutura operacional. Bol Inst Pesca 41:79–87
Jardine TD, Bond NR, Burford MA, Kennard MJ, Ward DP, Bayliss P, Davies PM, Douglas MM, Hamilton SK, Melack JM, Naiman RJ, Pettit NE, Pusey BJ, Warfe DM, Bunn SE (2015) Does flood rhythm drive ecosystem responses in tropical riverscapes? Ecology 96:684–692
Jesus AJS, Camargo M, Aquino TCH, Barros E (2009) Consumidores invertebrados: Macroinvertebrados aquáticos In: Camargo M, Ghilardi R Jr (Eds.). Entre a Terra, as Águas e os Pescadores do Médio Rio Xingu – Uma Abordagem Ecológica. Belém, pp 157–192
Junk WJ, Bayley PB, Sparks RE (1989) The flood-pulse concept in river-floodplain systems. In: Dodge DP (Ed.) Proceedings of the international Large River Symposium (LARS), Canadian Journal of Fisheries and Aquatic Sciences Special Publication, vol 106. NRC research press, Ottawa, pp 110–127
Junk WJ, Piedade MTF, Parolin P, Wittmann F, Schöngart J (2010) Ecophysiology, biodiversity and sustainable management of Central Amazonian floodplain forest: a synthesis. In: Junk WJ, MTF P, Wittmann F, Schöngart J, Parolin P (Eds.) Amazonian floodplain forests: Ecophysiology, Biodiversity and Sustainable Management, Ecological Studies 210. Springer, Berlin, pp 511–540
Junk WJ, Soares MGM, Carvalho FM (1983) Distribution of fish species in a Lake Amazon river floodplain near Manaus lago Camaleão with special reference to extreme oxygen conditions. Amazoniana 7:397–431
Kurten EL (2013) Cascading effects of contemporaneous defaunation on tropical forest communities. Biol Conserv 163:22–32
Lampert W, Sommer U (2007) Limnoecology: the ecology of lakes and streams. Oxford University Press, New York, p 324
Lowe-McConnell RH (1987) Ecological studies in tropical fish communities. Cambridge University Press, Cambridge, p 382
Luize BG, Venticinque EM, Silva TSF, Novo EMLM (2015) A floristic survey of angiosperm species occurring at three landscapes of the Central Amazon *várzea*, Brazil. Check List 11:1789
Magurran AE, Henderson PA (2003) Explaining the excess of rare species in natural species abundance distributions. Nature 422:714–716
Marshall BG, Forsberg BR, Thomé-Souza MJF (2008) Autotrophic energy sources for *Paracheirodon axelrodi* (Osteichthyes, Characidae) in the middle Negro River, Central Amazon, Brazil. Hydrobiologia 596:95–103
Metzger JP (2010) O código florestal tem base científica? Natureza & Conservação 8:92–99
Moss B (2010) Ecology of freshwaters: a view for twenty-first century Hong Kong. Wiley-Blackwell, Hoboken, p 480
Parolin P, Adis J, Rodrigues WA, Amaral I, Piedade MTF (2004) Floristic study of an igapó floodplain forest in Central Amazonia, Brazil (Tarumã-Mirim, Rio Negro). Amazoniana 18:29–47
Petry P, Hales J (2013) Freshwater ecoregions of the world. Ecoregion 314:Rio Negro. http://www.feow.org/ecoregions/details/314

Pinheiro MHO, Carvalho LN, Arruda R, Guilherme FAG (2015) Consequences of suppressing natural vegetation in drainage areas for freshwater ecosystem conservation: considerations on the new "Brazilian forest code". Acta Bot Bras 29:262–269

Poff NL, Allan JD (1995) Functional organization of stream fish assemblages in relation to hydrologic variability. Ecology 76:606–627

Poff NL, Zimmerman JKH (2010) Ecological responses to altered flow regimes: a literature review to inform environmental flows science and management. Freshw Biol 55:194–120

Ramírez A, Pringle CM, Wantzen KM (2008) Tropical stream conservation. In: Dudgeon D (Ed.) Tropical stream ecology, London, pp 107–146

Tejerina-Garro FL, Fortini R, Rodriguez MA (1998) Fish community structure in relation to environmental variation in floodplain lakes of the Araguaia river, Amazon Basin. Environ Biol Fish 51:399–410

Waldhoff D, Saint-Paul U, Furch B (1996) Value of fruits and seeds from the floodplain forest of Central Amazonia as food resource for fish. Ecotropica 2:143–156

Wantzen KM, Yule CM, Tockner K, Junk WJ (2008) Riparian wetlands of tropical streams. In: Dudgeon D (Ed.). Tropical stream ecology, London, pp 199–217

Weitzman SH, Vari RP (1988) Miniaturization in South American freshwater fishes; an overview and discussion. Proc Biol Soc Wash 101:444–465

Welcomme RL (1979) Fisheries ecology of floodplain rivers, New York, p 317

Welcomme RL (1985) River fisheries. Food and Agriculture Organization of the United Nations, FAO Fisheries Technical Paper 262. Rome, p 330

Winemiller KO (2004) Floodplain river food webs: generalizations and implications for fisheries management. In: Welcomme RL, Petr T (Eds.). Proceedings of the second international symposium on the Management of Large Rivers for Fisheries Volume II Regional Office for Asia and the Pacific. Bangkok, pp 285–309

Winemiller KO, Roelke DL, Cotner JB, Montoya JV, Sanchez L, Castillo MM, Montaña CG, Layman CA (2014) Top-down control of basal resources in a cyclically pulsing ecosystem. Ecol Monogr 84:621–635

Winemiller KO, Agostinho AA, Caramaschi EP (2008) Fish ecology in tropical streams. In: Dudgeon D (Ed.) Tropical stream ecology, London, pp 107–146

Winemiller KO, Jepsen DB (2004) Migratory neotropical fish subsidize food webs of oligotrophic blackwater rivers. In: Polis GA, Power ME, Huxel GR (Eds.) Food webs at the landscape level. University of Chicago Press, Chicago, pp 115–132

World Wide Fund For Nature (2018) Brazilian Forest Law – What is happening? http://wwf.panda.org/wwf_news/brazil_forest_code_law.cfm. Consulted in: 10/07/2018

Zuanon JAS (1999) Historia natural da icitiofauna de corredeiras do rio Xingu, na região de Altamira, Pará. Ph.Sc. – Thesis, UNICAMP, Campinas, p 198

Chapter 14
Structure, Composition, Growth, and Potential of the Forest in Temporary or Periodically Flooding Forests by Sewage, Near Iquitos

Juan Celidonio Ruiz Macedo, Dario Dávila Paredes, and Rodil Tello Espinosa

14.1 Fundamentals

Deforestation of primary forests in developing countries is unquestionable, particularly, in tropical regions like Peru. Even extractors already use flooded forests. Consequently, populations of plants of commercial value have declined, putting their genetic diversity at risk (FAO 1993: 10–30; Namkoong et al. 1996). In the Peruvian Amazon, this problem is alarming, because the flooded alluvial plains represent more than 12% of the territory, with more than 60,000 km^2, with the aggravating circumstance that the exuberant forest that it presents, develops in very poor soils (Higuchi et al. 2005), as occurs in the forests of the Nanay River. The water in this river is black and very poor in silt, so the additional supply of nutrients is almost zero during the flood. To these areas, Burga (2007) mentions that it is a meandering plain characterized by the presence of ridges and semilunar bars, encompassing flatlands with slopes of less than 2%, where restinga and lowland are observed.

The flooded plain of the Nanay River, of Amazonian origin, is very narrow. The fluvial dynamics is less intense and presents greater stability; according to Rasanen et al. (1998), there are complexes of banks with alternating restinga, low elongation, and bajiales. The black color of the Nanay River waters is due to the high content of fulvic and humic acids and low sediment load in suspension. For these characteris-

J. C. R. Macedo (✉)
Herbarium Amazonense (AMAZ), Centro de Investigacion de Recursos Naturales (CIRNA) de la Universidad Nacional de la Amazonia Peruana (UNAP), Iquitos, Peru

D. D. Paredes
Facultad de Ingenievia Forestal, Universidad Nacional de la Amazonia Peruana (UANP), Iquitos, Peru

R. T. Espinosa
Biology Department, Facultad de Ingenieria Forstal, Universidad Nacional de la Amazonia Peruana (UANP), Iquitos, Peru

R. W. Myster (ed.), *Igapó (Black-water flooded forests) of the Amazon Basin*,
https://doi.org/10.1007/978-3-319-90122-0_14

tics, it is assumed that diversity patterns are also different from those reported in other sub-ecoregions. They are soils with great decomposition of leaves and trunks; the vegetation presents a structure that induces to define it as gallery forest, conformed by species with xeromorphic adaptations, fuccreas roots, suberous or exfoliating bark, and seeds with floating accessories (Encarnacion and Salazar 2004).

The alluvial plain of the Nanay River forms a complex ecosystem, completely interrelated with each other, with an increasing anthropic activity, which influences the structure, composition, potential, and growth of the forest and the environmental, phytogeographic, and altitude factors, combined from different landscapes or forestry strata of importance for industry, forest management, land management, and environmental or scientific zoning, among others. In this scenario the question arises: What are the characteristics of the structure, growth, floristic composition, and forest potential of the alluvial forest of the Nanay River, Iquitos, Peru, for sustainable management purposes?

Responding is complex, because biotic and abiotic systems and man are divided into subsystems, vertically structured (plant strata) and horizontally (De Lima 2003). The environment is considered as a set of microsystems (contexts) that are interconnected, which the individual participates sequentially or simultaneously (Carrizosa 2001). The tree is seen as an active and adaptable unit, and a forest is made of a vast number of such units, interacting with each other with soil and climate factors (Higuchi et al. 2005). The ecosystem approach is a strategy for the integral management of land, water, and biodiversity, which promotes the conservation and sustainable use of resources, as well as the fair and equitable distribution of derived benefits (GTZ/FUNDECO/IE 2001). Sustained management is based on four pillars of sustainability: technical, economic, ecological, and social (Higuchi et al. 2005). Specifically, for the evaluation of the structure and composition of the forests, as well as the estimation of their potential, forest sampling is required by inventing trees with diameter at breast height (DBH) $\geq$ 10 cm. And for the study of the dynamics of vegetation, permanent sampling plots (PPM) are used (Ferreira 1995), subdividing them into subplots of 100 m^2, in which the trees are evaluated. The information is useful to make the proposal of sustainable use of the forest based on fast-growing species, also to maintain the ecosystem with species of interest to the population and to contribute to environmental cleanup (decrease atmospheric CO_2), convenient for the health of the inhabitants of the city of Iquitos and for the whole world (FAO 2001).

Water plays a preponderant role in the Amazonian landscape, forming ecosystems with a narrow floodplain, called várzea in Brazil. In Peru, the várzeas and their flood forests are defined by the rivers and other bodies of water in the lower jungle of Peru, as well as the portions of land covered by water, whether permanent or temporary and stagnant or running through the overflow nature of the aquatic bodies. Adjacent areas of poor drainage, such as aguajales and marshes, are included.

14.2 The Forests of the Flooded Alluvial Plain

Flooded alluvial plain forests may retain minimal species richness, bearing in mind that many of these are not established or survive in flooded places. The forests of the alluvial floodplain different to the Nanay River basins show a high productive potential and biodiversity conservation, with basal areas between 2.6 and 10.0 m^2/ha and volumes between 59 and 240 m^3/ha (Nebel et al. 2000a,b,c). Families Leguminosae, Euphorbiaceae, Annonaceae, and Lauraceae and abundance of *Maquira coriacea*, *Guarea macrophylla*, *Terminalia oblonga*, *Spondias mombin*, *Ceiba pentandra*, *Hura crepitans*, *Eschweilera* spp., *Campsiandra angustifolia*, *Pouteria* sp., *Licania micrantha*, *Parinari excelsa*, and *Calycophyllum spruceanum* are abundant (Nebel et al. 2000b). However, because of the easy access to the most remote areas, the potential of valuable species slowly decreases to become rare, as is the case with *Swietenia macrophylla*.

In the Nanay River, the flood influences its own structure and floristic composition. Ecologically it is a biotope occupied mainly by arboreal mass (Vicen and Vicen 1996), or an extensive group of trees in thickets. By definition, a tree is a perennial, erect, and strongly lignified, with stems less than 7 m, usually with clear differentiation of the stem or shaft and crown.

For the purposes of forest management, it is important to know the abundance, dominance, and frequency of the species; biologically they indicate the horizontal occupation of the soil (Matteucci and Colma 1982); weighting of these parameters with the method of Curtis and Mcintosh (1950) is the rate of importance value of species (IVI), whose values reveal essential aspects of the floral composition (Lamprecht 1964; Delgado et al. 1997). The vertical structure informs about the floristic composition of the strata that allow to recognize the signification of the species and the laws that regulate the relations of the organisms with the way of life of the species. A table containing the names of the species and/or the family would be sufficient to show the floristic composition (Lamprecht 1964).

The importance value index (IVI) is calculated by taking the average of the abundance of the species as a percentage of the total number of stems within a geographical unit (N), the basal area of species as a percentage of the total within the geographical unit (G), and the frequency of the species (proportion of plots in the sample in which each is present) as a percentage of all frequencies (F): $(N + G + F)/3$ (Curtis and Mcintosh 1950).

In the Lamprecht document (1964), abundance indicates the number of individuals of each species within the plant association by a unit of area, either in its absolute or relative values, referred to the total of recorded trees; in dominance, these values are calculated based on the basal area, while the frequency indicates in how many plots of the survey area there is a species. It is absolute when expressed as a percentage of the parcels in which it occurs with respect to the total number of parcels (100%) and relative when the percentage is calculated based on the total of the absolute frequencies.

Depending on whether the distribution of the number of trees per diameter class is a different ecological guild sciophilous, long-lived pioneer, ephemeral heliophilous and understory (Louman et al. 2001; Fredericksen et al. 2001) and what Palacios (2004) calls forestry unions. According to Louman et al. (2001), the guild reflects the behavior of the species before the most important environmental gradients within forest ecosystems such as light and soil; but, in tropical forests, the light factor is the one that most limits successful regeneration.

On the other hand, Peters (1994) classifies the species by their density of individuals per hectare (*N*) in high density ($N \geq 10$), average density ($N \geq 5$ and <10), and low density ($N < 5$). Taking into account the topography of the soil, which is an important gradient that influences the structure and composition of the species, Nebel et al. (2000b) report three types of forest: high restinga, low restinga, and tahuampa, characterized in part by an annual flood of 1, 2, and 4 months/year, respectively.

The concept of diversity refers to the variety of species that occur in a defined space-time dimension, resulting from the interaction between species that are integrated in a process of selection, mutual adaptation, and evolution, within a historical framework of environmental variations local. In this framework, these species constitute a complex structure, in which each element expresses an abundance dependent on the remaining elements (Ramirez 1999).

Regarding the forest inventory in Peru, it is recommended to use systematic inventories, with plots of 0.5 ha (INRENA 2004), which are sufficient and much more efficient for forest inventory (Hughell 1997) and suitable for the forest inventory of the flooded alluvial plain (Rios and Burga 2005). The diameter at breast height (DAP) is the length of the line that joins two points of the circumference passing through its center, and is easy to use, control in the data collection and reading. The height is the total length of the tree measured from the ground level to its apex (Ferreira 1995). The basal area and the volume are expressions of the growth of the tree, depending on the diameter; the growth occurs in the secondary meristems located under the bark of the tree. Depending on the volume, growth is the gradual increase in the size of an organism (tree) and population (forest) in a given period. Based on the basal area, growth is a simple input and output system (Ferreira 1995; Lewis et al. 2004; Higuchi et al. 2005).

The basal area (*G*) is defined as the area of the cross section of the tree or as the projection of the DAP to the ground, and the volume (*V*) of wood is the result of multiplying the individual basal area by height and by stem factor of 0.65 for Peru (INRENA 2004). With this data the biomass is calculated after knowledge of the basic density of the wood; in Manaus, Brazil, the mean density was 0.704 ± 0.117 ($\mu \pm s$); this density is 15% higher in forests of the central and eastern Amazon, compared to that of the northeastern Amazon (Nogueira et al. 2005; Baker et al. 2004b). Even so, the density of each wood species is not known, but this does not significantly affect the estimation of biomass (Baker et al. 2004a,b).

The usual techniques for studying long-term forestry dynamics are periodic inventories, repetitive measurements in permanent plots, studies of dendrochronology, and development of simulation models (Brenes 1990). A permanent parcel is one that is established in order to remain indefinitely in the forest and whose proper demarcation allows the exact location of its boundaries and points of reference over time and of each of the individuals that make it up (Brenes 1990; Leano 1998).

Plots of 1 ha are widely used in many tropical countries. The parcel totals represent values per hectare so there are no conversion problems and can be compared approximately with the 2.5 acre parcels previously established in many countries. Additionally, the plots of 1 ha can be subdivided into 100 subplots of 0.01 ha, so that the totals of these subplots can be compared with the information of the 1/40 acre (average, square chain) used in many linear surveys of regeneration sampling and the 100 squares are numbered from 00 to 99 from the coordinates of the plot (Synnott 1991).

According to Synnott (1991), a network of permanent sampling plots (PPM) in a continuous forest inventory will provide reliable information to estimate (a) changes in the number, size, and species of the forest over time; (b) variation in the composition and production with respect to the site (soil, appearance, initial vegetation, and volume of the mass) and treatments (different degrees of reduction in the volume of the mass and types of intervention); (c) the relationships between the variables of individual trees (diameter, height, and position of the crown), stand (local basal area and volume), and increments (diameter, basal area, and volume per tree or per plot), which can be used to predict future brand and production volumes; and (d) long-term changes (improvement, degradation) in the site and its productive capacity.

In this process, the tree diameter measurement represents the most important and reliable tree variable for the field information survey and development of allometric equations, because there are difficulties to accurately measure the height of the tree. Consequently, the increase in diameter is an essential indicator to monitor and examine the dynamics of change of natural forests, which allows us to understand the current and future trends of the global carbon cycle (Baker et al. 2004a).

Another important aspect is the growth of the forest; understood as the progressive development of a living being, it is defined from a variable or parameter and the period elapsed in its estimation. Thus, depending on the diameter, the growth occurs in the secondary meristems located under the bark of the tree. Depending on the volume, growth is the gradual increase in the size of an organism (tree) and population (forest) in a given period. Based on the basal area, growth is a simple input and output system (Lewis et al. 2004; Higuchi et al. 2005).

The growth is produced by the physiological activity of the plant product of the processes of anabolism or synthesis and catabolism. The growth rate is determined by genetic factors, site, and time (Louman et al. 2001; Ferreira 1995). There are three types of growth (Ferreira 1995): (a) annual current increase, which corresponds to the average of the increase produced in a year; (b) average annual increase, which corresponds to the average of increases up to the present time; and (c) periodic increase, which corresponds to the increase produced in a period of time greater than 1 year.

The current and periodic increases obey the following laws of growth (Ferreira 1995):

(a) As long as the current increase is increasing, it will remain higher than the average increase, and this will also be increasing.
(b) The average increase has a maximum value when it reaches equal to the current increase, and since then it will remain higher than this.
(c) The average increase reaches its maximum value by cutting the current increase curve at this point; i.e., at the maximum, the average increase is equal to the current increase. This law is very important, because it will determine the age of biological rotation, when the increases refer to the volume.

Growth rates vary widely in tropical moist forests, among tree species and among individuals of an arboreal species; they also vary in terms of their relation to age and microclimatic conditions (Lewis et al. 2004).

Mortality is a consequence of the competition law, which can be detected through the repeated measurement of permanent plots (Louman et al. 2001, Ferreira 1995). In this study, mortality in the basal area is the sum of all trees with DBH ≥ 10 cm that died in the plot within the census interval (Lewis et al. 2004, Louman et al. 2001).

14.3 Population of Diversity

The population was constituted by the forest trees with a DAP ≥ 10 cm that occupied the forest of the floodplain of the Nanay River; the area was 400 ha of forests on soils whose greater use capacity is forestry. For the study of floristic composition, structure, forest potential, and potential, the sample size was 23 ha representing 5.7% sampling intensity to achieve an error of less than 15% of the total volume, which is acceptable in the elaboration of inventories for the elaboration of management plans. The sampling unit corresponded to a rectangular strip of 10 m width of variable length. The sampling was systematic in a single stage (Fig. 3.6). These plots were installed in the alluvial plain forest in 2001 (PPM 01 and 02) and 2004 (PPM 07 and 08), and each PPM (permanent sampling plot) was divided into 100 sub-units of 100 m^2.

14.4 Analysis Unit

The unit of analysis for the determination of the characteristics of the structure, floristic composition, forest potential, and potential was the hectare. The unit of analysis for the study of growth was the species within the permanent sampling plot.

14.5 Description of the Study Area

The study was conducted in the alluvial plain forest of the Nanay River, in the Province of Maynas, Loreto Region, in the intervention space of the Forestry Research and Teaching Center (CIEFOR) of the National University of the Peruvian Amazon (Fig. 14.1). The forest subject of study is located at coordinates 3° 49′ 48″ S and 73° 25′ 12″ W, at an altitude of 112 m. The place is reached through the

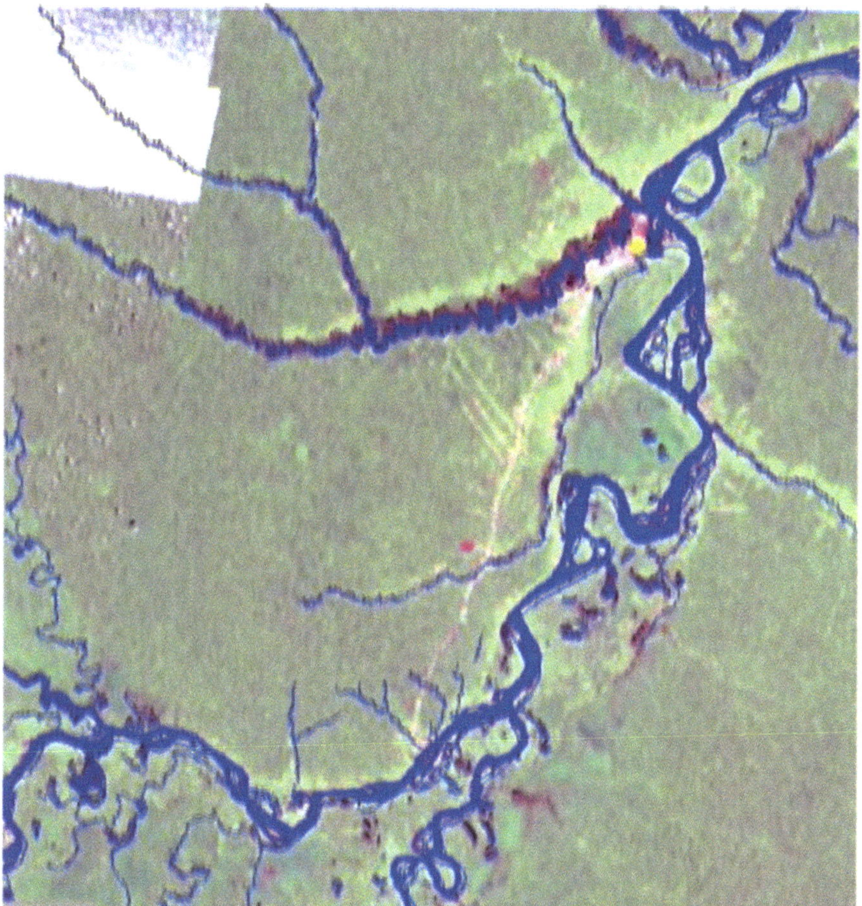

Fig. 14.1 Zoning of the alluvial forest of the Nanay River

Iquitos-Nauta highway (Fig. 14.2), taking the 8 km branch from Quistococha to Puerto Almendra, in the district of San Juan Bautista. The climate is from a tropical humid forest (BhT) (ONERN 1976).

The forest is located in the alluvial floodplain temporarily flooded (Malleux 1982); the flood with black waters occurs in the first quarter of each year, covering the lower parts of the riverbank, forming an ecosystem unlike any other tropical ecosystem in the Amazon in terms of species, reproduction, and function (Figs. 14.3 and 14.4). The soil is clayey, dark gray, with a protective cover of organic matter whose thickness varies between 2 and 5 cm; dry and fresh leaves on the ground, twigs, and others protrude (Fig. 14.5).

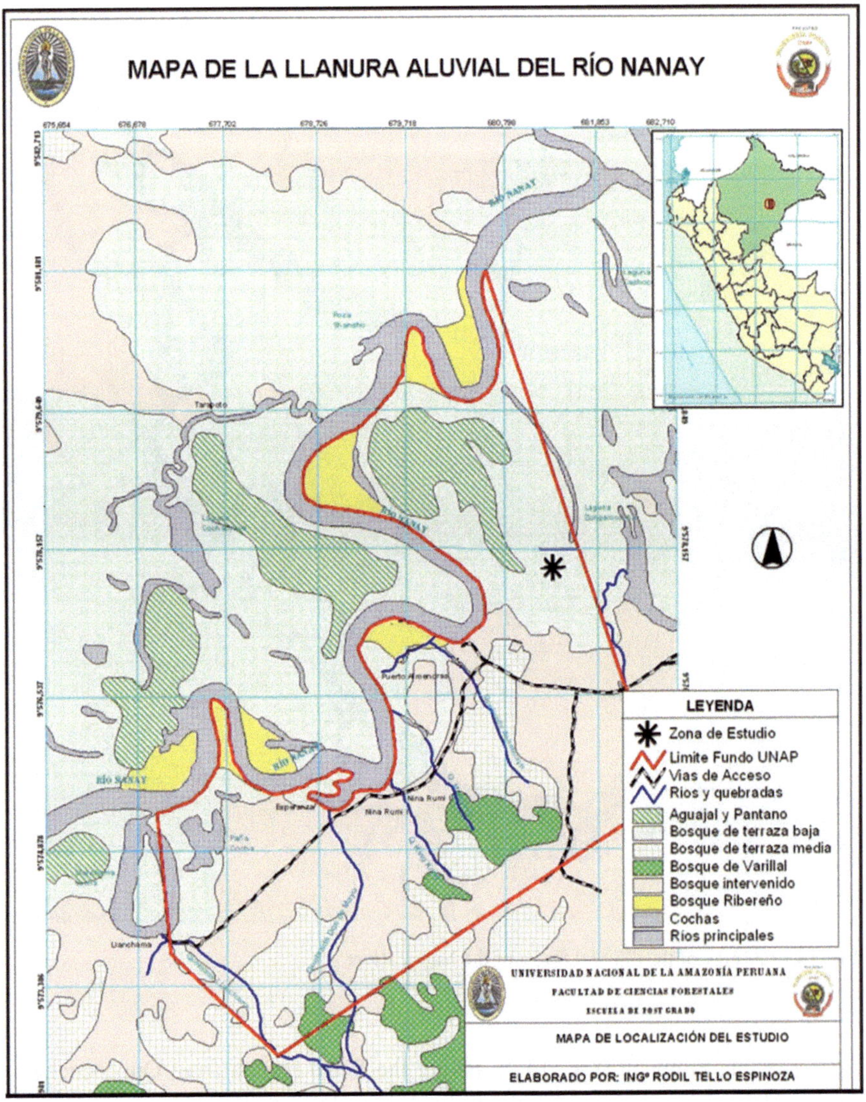

Fig. 14.2 Location of the study area in the area of influence of the Iquitos-Nauta highway. (Source: Imagen IIAP – BIODAMAZ)

Fig. 14.3 Alluvial forest of the Nanay River during the flood season. (Photo February 2007)

Fig. 14.4 Alluvial forest of the Nanay River during the emptying period

Fig. 14.5 Dry biomass in the soil of the alluvial forest of the Nanay River. (Photo May 2007)

14.6 Data Collection Instruments

In the systematic forest inventory, a field format was designed with the following fields: distance (m), common name, DAP, commercial height (hc), total height (ht), and cup lighting. Before this heading, the name of the components of the brigade, the azimuth of the belt, code of the sampling strip, number of the sampling unit, and inventory date were noted.

The field format to evaluate the recruitment, growth, and mortality of species had the following structure: number of the permanent plot, number of the subplot, number of the tree within the subplot, common name, family, species, DAP (cm), total height (m), and cup lighting, stem identification code (CIF). Before this heading, the name of the brigade components and date of the inventory were noted.

14.7 Procedure and Statistical Data Analysis

For the study of the structure and composition of the forest, the attributes of the species, horizontal structure, vertical structure, and value index of importance (at the family and species level) were described. For the forest potential, the attributes of the number of trees, basal area, volume of wood grouped by species, and diameter class were described; the biomass was considered for the carbon potential. The attributes of tree recruitment, growth, and mortality were also described to explain the dynamics of the forest; the DAP was measured with the forest calibrator (Forcípula) with an accuracy of 0.5 mm.

14.8 Study of the Structure, Composition, and Potential of the Forest

During the inventory, rectangular sampling strips of 10 m in width and variable length were used. For this, straight gates of 1.5 m wide were opened to facilitate the displacement of the personnel of the inventory brigade (Figs. 14.6 and 14.7).

For each tree (DAP ≥ 10 cm), the plant identified it by its common name; the species was identified in situ by a taxonomist of the Herbarium Amazonense (AMAZ) and collected botanical material for later identification (Fig. 14.8). An important variable was the diameter of the tree; it was measured generally at breast height with the forest calibrator (Fig. 14.9) at 1.30 m above the ground (WTP). The DAP reflects the length of the line joining two points of the circumference passing through the center of the tree; it is easy to measure and control (Ferreira 1995). The total height of the tree corresponded to the length between the level of the ground and the apex of the tree.

Due to the access to sunlight, each tree was classified as full when the glasses fully access to sunlight (total illumination); intermediate, when accessing partially or laterally; and inferior, when it indirectly accesses sunlight (Fig. 14.10).

Fig. 14.6 Displacement of personnel on the trail in the alluvial forest of the Nanay River. (Photo January 2007)

Fig. 14.7 Base trail of the inventory and access to the permanent plots during the growing season in the alluvial forest of the Nanay River. (Photo January 2007)

Fig. 14.8 Training of the inventory team within the alluvial forest of the Nanay River. (Photo December 2006)

Fig. 14.9 Measurement with the forest calibrator

Fig. 14.10 Illumination of the tree canopy in the alluvial forest of the Nanay River. (Photo 2007)

The data collected after reviewing its consistency was entered in an electronic spreadsheet of the Microsoft Excel, and then a thorough evaluation of the data was performed, to detect DAP and height values that were outside the theoretical range for this forest, for lighting of glass was checked if it is related to the height of the tree; this evaluation was done electronically.

The height class was generated electronically with the algorithm = integer (total height/2) × 2, where the value 2 is the width of the class; for the diameter class, the algorithm used was = integer (DAP/10) × 10, where 10 is the width of the class. The formulas of the basal area (G) and volume (V) are the following:

$$G = \frac{\pi \times (\text{DAP})^2}{40{,}000} \tag{14.1}$$

$$V = G \times \text{altura.total} \times 0.65 \tag{14.2}$$

To determine the strata of the forest, a frequency distribution was generated by height classes; each class had an interval of 2 m. Then the accumulated frequency was obtained; this accumulation started in the last class and ended in the first diametric class. Next, the ogive of the accumulated frequency with the diameter class was created. Within this warhead, the arboreal strata were identified, for which a sudden change in the trend of the ogive determined a stratum.

With these data, a dynamic table was generated by grouping the species by their diametric class, and a frequency distribution was generated by diametric class for each species. Because of the shape of the frequency distribution, the species were classified into ecological guilds: scryophytes (E) and understory scyophytes (E-SB), durable heliophytes (HD), and ephemeral heliophytes (HE) (Louman et al. 2001). Taking advantage of the total trees/ha of this table, it was classified as high tree density when the species had ≥10 trees/ha, average density trees when the total trees were ≥5 and <10 per ha, and low tree density when the species had <5 trees/ha (Peters 1994). The species with less than 1/3 trees/ha was considered as scarce.

The index of importance value (IVI) was calculated taking the average of the abundance of species as a percentage of the total number of stems within a geographical unit (N), the basal area of species as a percentage of the total within a geographical unit (G), and the frequency of the species as a percentage of the sum of all frequencies (F) (Curtis and Mcintosh 1950); the IVI formula is:

$$\text{IVI} = \frac{N + G + F}{3} \tag{14.3}$$

As an indicator of species richness per unit area, the species-area curve was graphed. The diversity of species according to the dominance of the species was calculated with the following Simpson index:

$$\text{Índice de Simpson} = 1 - \lambda = 1 - \sum p_i^2 \tag{14.4}$$

This index indicates the probability that two individuals taken at random are of the same species (Moreno 2002).

The diversity of the species in terms of abundance was calculated with the Shannon-Wiever index (H') (Moreno 2002) which has the following formula:

$$\text{Índice de Shannon-Wiever} = H' = -\sum p_i \ln p_i \tag{14.5}$$

Donde:

$$p_i = \frac{n_i}{N} \tag{14.6}$$

The Shannon-Wiever index measures the average degree of uncertainty to predict to which species an individual chosen at random from a collection will belong and acquires values between zeros, when there is only one species and the logarithm of S and when all the species are represented for the same number of individuals.

Dendrograms of the Morisita similarity index were constructed using the multivariate or conglomerate technique using the Estimate SWin 8.00 Program, after dividing the sampling area according to the flood gradient as transitional zone when the water column exceeds 2 m high and alluvial zone when the water column does not exceed 2 m in height.

Based on the hypothesis that some resource is scarce for the species, the expected abundance of trees was calculated under the resource allocation model (Franco et al. 2001), part of the initial assumption that the community has a total amount finite of a limiting resource and that the abundance of each species is proportional to the fraction of the resource that it can use or occupy. An order of priority was assigned to use the resource among the species according to their relative abundances, the most abundant species had priority 1, and each species used a fraction (*k*) of the available resources.

For the forest potential, species were classified by the use of wood as sawn, round, poles, firewood, coal, and other uses. The forest potential category was established based on the categories of Table 14.1 used by GEMA (2007).

For the estimation of forest biomass and carbon, the basic density of the species was collected from the UNAP wood technology laboratory and from various scientific articles, and for the species without data, the average value of 0.62 g/cm^3 was used, based on the criteria used by different authors.

For the calculation of the total aerial biomass, the mathematical method used expansion factors when working with forest tree inventory data (WTP $\geq$ 10 cm), which is summarized in the formula:

$$\text{Biomasa área total} = V \times \text{DB} \times \text{FE} \tag{14.7}$$

Table 14.1 Categories of forest potential

Categories	Power	Volume (m^3/ha)
I and II	Excellent	>150
III	Very good	120–150
IV	Good	90–120
V	Regular	60–90
VI	Poor	<60

Source: GEMA (2007: 50)

V = tree volume for DAP $\geq$ 10 cm DB = Basic wood density FE = Expansion factor (2.25).

Finally, according to the Intergovernmental Panel on Climate Change (IPCC 1996), the carbon content (CC) corresponds to half of the total aerial biomass. The tables were generated with the dynamic tables' option in Microsoft Excel 2003.

14.9 Growth, Recruitment, and Mortality Study

Four permanent sampling plots of 1 ha each were installed at random, the plots installed in 2001 had codes (PPM 01 and PPM 02), and those installed in 2004 had codes (PPM 07 and PPM 08); each PPM was divided into 100 sub-units of 100 m^2 (Fig. 14.11).

Within the permanent plots, the trees (DAP $\geq$ 10 cm) were marked with yellow paint, usually at chest height, so that the diameter was measured in only one place. In each subplot, the trees were numbered correlatively with yellow anticorrosive paint, and an aluminum plate was placed to identify the tree within the subplot and in the plot (Figs. 14.12 and 14.13). The number and the plate are only visible from the center of each subplot.

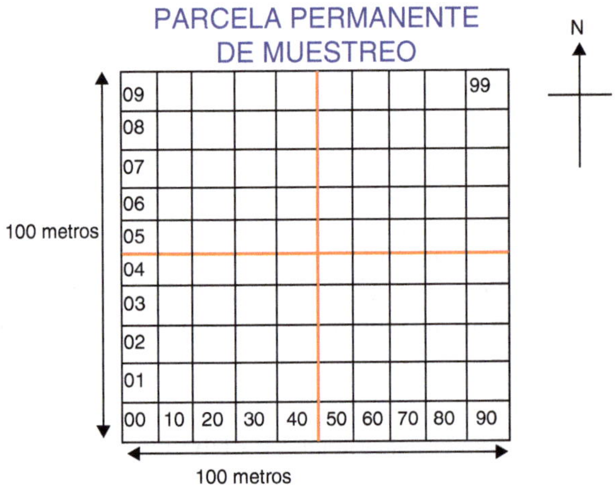

Fig. 14.11 Scheme of the plot

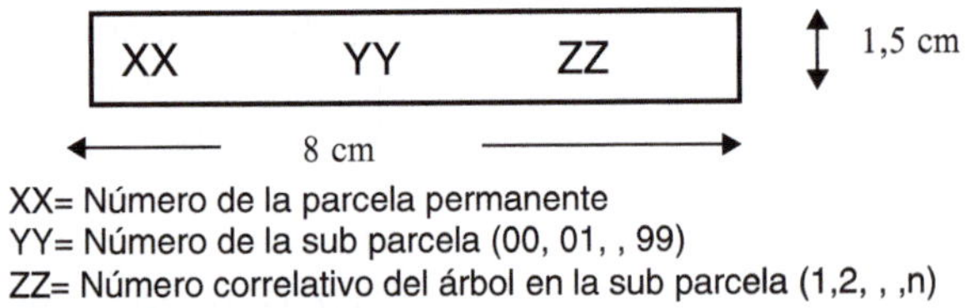

Fig. 14.12 Diagram of the aluminum plate

Fig. 14.13 Aluminum plate and tree number within the permanent sampling plot. (Photo 2007)

Fig. 14.14 Dead tree due to natural causes in the permanent sample plot. (Photo 2007)

Mortality was evaluated at the species level, within each of the permanent plots. To determine its impact on the basal area, it was assumed that mortality is the sum of all trees with DBH ≥ 10 cm, which died (Fig. 14.14) in the plot within the census interval (Lewis et al. 2004).

The one that today has a DAP ≥ 10 cm was considered a recruit tree and that in the first measurement, it had not reached the minimum DAP of 10 cm (Fig. 14.15).

Fig. 14.15 Recruit tree in the permanent sampling plot

The average annual increase was evaluated, which corresponds to the average of increases up to the present moment divided by the time elapsed or age. To examine and describe the dynamics of the population, the response variables were reported by the total number of dead trees (discharges) and recruits (income). The description of dynamic patterns at the level of populations of common species over time was expressed in three population categories: declining populations (recruits < mortality), static populations (recruits = mortality), and increasing populations (recruits > mortality).

For the analysis of the dimeric growth, we considered the individuals with DBH ≥ 10 cm that had an initial diameter and a diameter at the end per period (i.e., they had not died). Diametric increases that were beyond the established ranges between −2 mm and ≥40 mm were excluded, as these measurements were considered as measurement errors or not reliable. The criterion of selection of the common species was by the abundance of these. In this way, the one whose abundance was greater than or equal to 7 individuals/ha was taken as a common species.

The regression analysis was performed using the SPSS 13 software, and the tables were generated with the dynamic tables' option of the Microsoft Excel 2003.

14.10 Results

14.10.1 Composition and Structure of the Forest

The characterization of the floristic composition, richness, diversity, and structure of the Nanay River forest is based on the premise that the community of forest trees is the sum of the populations with their properties, plus the interactions between them; the community is a set of populations of different species that occur together in time and space (Louman et al. 2001).

20% of the forest species of the floodplain of the Nanay River had high density of trees per hectare; 13% average density/ha and 67% had low density per hectare (Fig. 14.16). The number of species per area (Fig. 14.17) shows that the richness (>70%) of species per unit area can be recorded in a sampling unit of 0.5 ha. On the other hand, the species *Nectandra* sp., *Hevea brasiliensis*, and *Ocotea argyrophylla*, with 0.2 trees/ha are considered scarce in the area.

The degree of similarity of the species between sampling strips was very high, and the degree of similarity, calculated with the Morisita index, showed 89% similarity between strips (Fig. 14.18).

The floristic composition is reflected in the 23 families and 81 inventoried species (Tables 14.2 and 14.3). The families Fabaceae, Lecythidaceae, Euphorbiaceae, Sterculiaceae, Chrysobalanaceae, and Sapotaceae are the most important of this

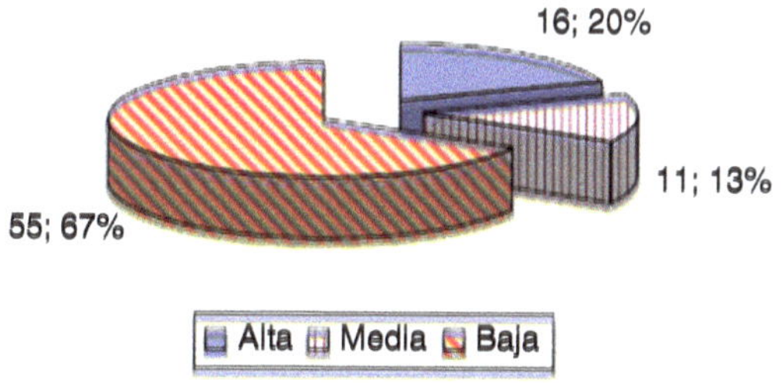

Fig. 14.16 Density of trees in the alluvial forest of the Nanay River

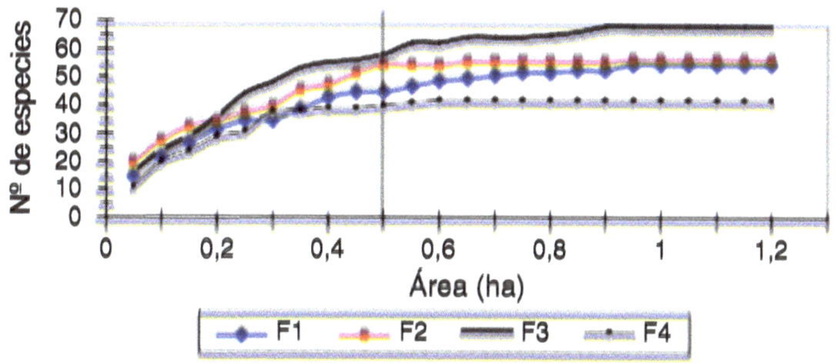

Fig. 14.17 Number of species per area of the alluvial forest of the Nanay River

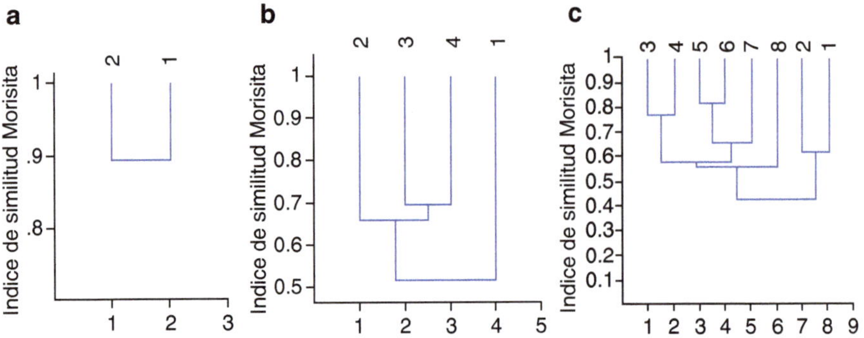

Fig. 14.18 Dendrograms of the morisite similarity index. Where: (**a**) Dendogram between high and low zone. (**b**) Dendogram between sampling belts. (**c**) Dendrogram between sub-bands dividing each strip in high zone (odd number) and low zone (no. Pair), strip 1 (no. 1 and 2), strip 2 (no. 3 and 4), strip 3 (no. 5 and 6), and band 4 (no. 7 and 8)

Table 14.2 Recruitment and mortality rate in the alluvial forest of the Nanay River

	Árboles ha^{-1}.año^{-1}	
Árbol	Reclutas	Muertos
En pie fuste completo	14.74	12.06
En pie fuste quebrado	0.13	0.75
Caído fuste completo	0.04	0.63
Caído fuste quebrado		0.72
Subtotal	14.91	14.16
Tocones		18.31
Total	14.91	32.47

forest, as they contributed half the value of the IVI. The Fabaceae family presented a greater abundance of trees with 89.45 per ha and greater dominance (*G*) (Table 14.2).

The probability that two individuals taken at random are of the same species is very low (0.029), indicating that in the alluvial plain forest, diversity was very high (Simpson Index = 0.97). This diversity, calculated with the Shannon-Wiever index, is also high, with a value of 3.88 close to the theoretical limit of 4.39, which corresponds to the Napierian logarithm of all species.

Of the 81 species recorded in the alluvial plain of the Nanay River, 16 species had greater ecological importance; together they contributed half of the value of IVI, 255.4 trees/ha and 13.7 m^2/ha of a basal area. By the IVI, the forest is characterized as an association of *Eschweilera coriacea*, *Theobroma glaucum*, *Campsiandra angustifolia*, *Licania harlingii*, *Caraipa densifolia*, *Pouteria cuspidata*, and *Cariniana decandra*. At the level of ecological guilds, it was observed that 12 shade-tolerant species (sciophytes) and 4 shade-intolerant species (durable heliophytes) were the most important within the IVI.

Table 14.3 Average annual increment (IMA) of the common species of the alluvial forest of the Nanay River

Especie	IMA cm	Densidad por ha	Gremio Ecològico	Uso
Aspaiosperma excelsum	0.58	Media	Heliófita durable	Redonda
Guatteria inundata	0.53	Media	Esciófita	Redonda
Guatteria citriodora	0.35	Media	Esciófita	Redonda
Pouteria glomerata	0.40	Alta	Heliófita durable	Postes
Brosimum guianense	0.54	Media	Heliófita durable	Leña y carbón
Inga sp.	0.52	Alta	Esciófita	Leña y carbón
Couepia ulei	0.51	Media	Esciófita	Leña y carbón
Mabea elata	0.48	Alta	Esciófita	Leña y carbón
Eschweilera coriacea	0.45	Alta	Esciófita	Leña y carbón
Campsiandra angustifolia	0.44	Alta	Esciófita	Leña y carbón
Calyptranthes pulchella	0.43	Alta	Heliófita efímera	Leña y carbón
Couratari oligantha	0.42	Media	Esciófita	Leña y carbón
Pouteria cuspidata	0.41	Baja	Heliófita durable	Leña y carbón
Licania harlingii	0.34	Alta	Esciófita	Leña y carbón
Zygia glomerata	0.33	Media	Heliófita durable	Leña y carbón
Hevea nitida	0.32	Media	Esciófita	Leña y carbón
Eschweilera albiflora	0.31	Media	Esciófita	Leña y carbón
Pouteria guianensis	0.30	Media	Esciófita	Leña y carbón
Terminaba amazónica	0.26	Baja	Heliófita durable	Leña y carbón
Parkia igneiflora	0.88	Baja	Heliófita efímera	Aserrada
Hymenaea courbaril	0.56	Alta	Heliófita durable	Aserrada
Cariniana decandra	0.42	Alta	Esciófita	Aserrada
Sapium glandulosum	0.41	Media	Esciófita	Aserrada
Caraipa densifolia	0.30	Alta	Esciófita	Aserrada
Vatairea erythrocarpa	0.23	Baja	Heliófita durable	Aserrada
IMA promedio (cm/año)	0.42			

Of the total inventoried species, in 15 species, the distribution of the number of trees by diametric class showed one (inverted J), characteristic of the scryophyte species, and 37 were understory sciophytes (Figs. 14.19, 14.20, and 14.21); together they represented 64.2% of the total. These species have the ability to tolerate low-light environments and to react to changes in light. Species with high light requirements, such as durable heliophytes and ephemeral heliophytes, accounted for 24.7% and 18.5%, respectively; three species were scarce in the area (Fig. 14.19).

Under the idea that the diversity of species is based on the assumption that co-occurring populations interact with each other and with the environment, which are manifested in the number of species present in the community and in the fraction with which it contributes said species (Fig. 14.22). In this figure, it is observed that 16 species with high density/ha apparently do not agree to use all the resources that would correspond to it and are being used by species with low density/ha as *Pouteria torta*, *Mabea subsessilis*, *Couepia bernardii*, *Carapa guianensis*,

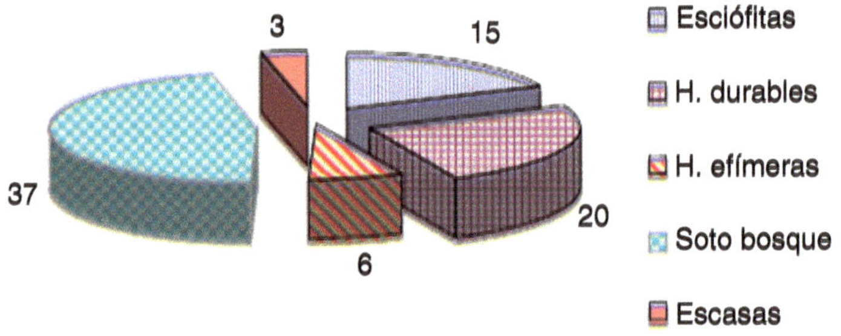

Fig. 14.19 Number of species per ecological guild in the alluvial forest of the Nanay River

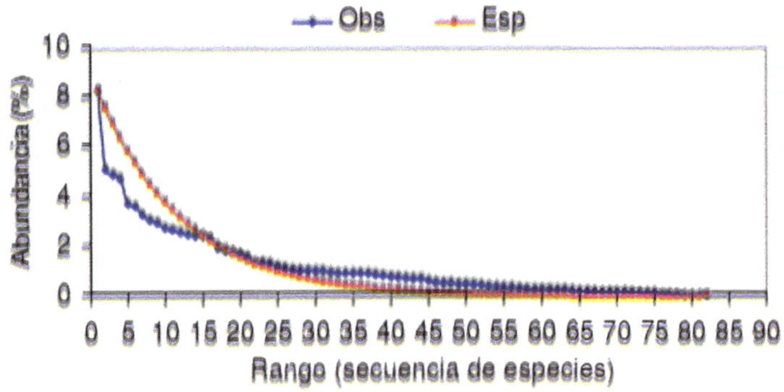

Fig. 14.20 Distribution of the number of trees by diametric class of four species in the alluvial forest of the Nanay River

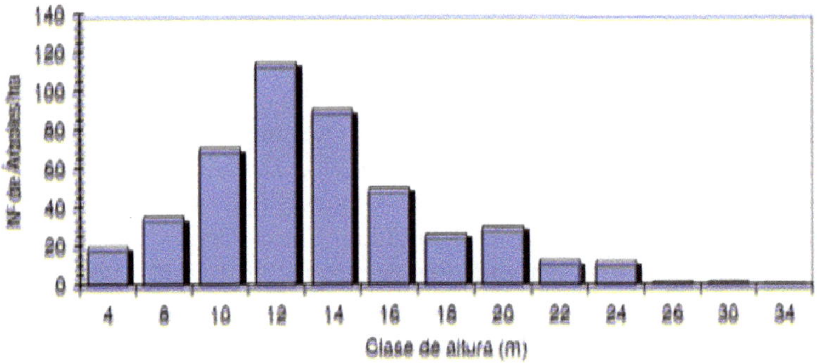

Fig. 14.21 Distribution of the scryophyte species (E) in the alluvial forest of the Nanay River

Symphonia globulifera, *Guatteria megalophylla*, *Pouteria glomerata*, *Brosimum guianense*, and *Hymenaea courbaril* mainly.

The distribution of the number of trees by height classes showed a normal distribution (Fig. 14.23), while the distribution of the basal area by diametric class

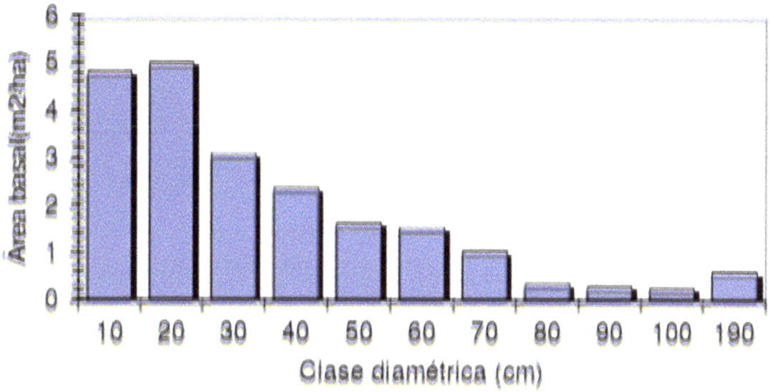

Fig. 14.22 Distribution of abundance by range of alluvial forest species of the Nanay River

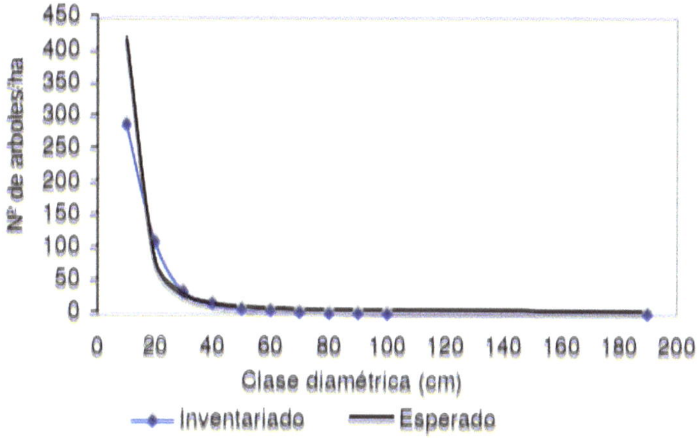

Fig. 14.23 Distribution of the number of trees by height class (m) in the alluvial forest of the Nanay River

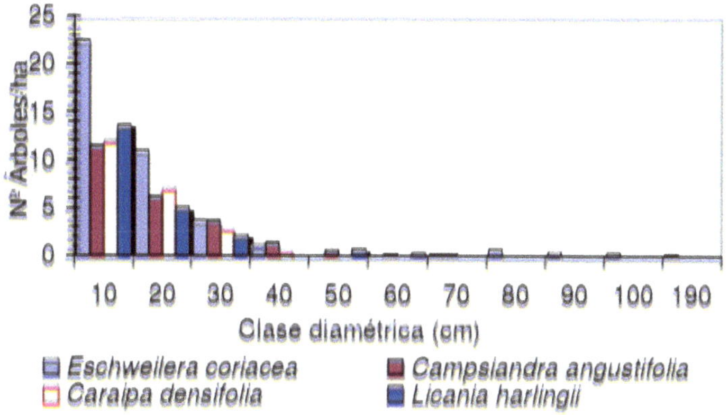

Fig. 14.24 Distribution of the basal area (m^2/ha) by diametric class (cm) in the alluvial forest of the Nanay River

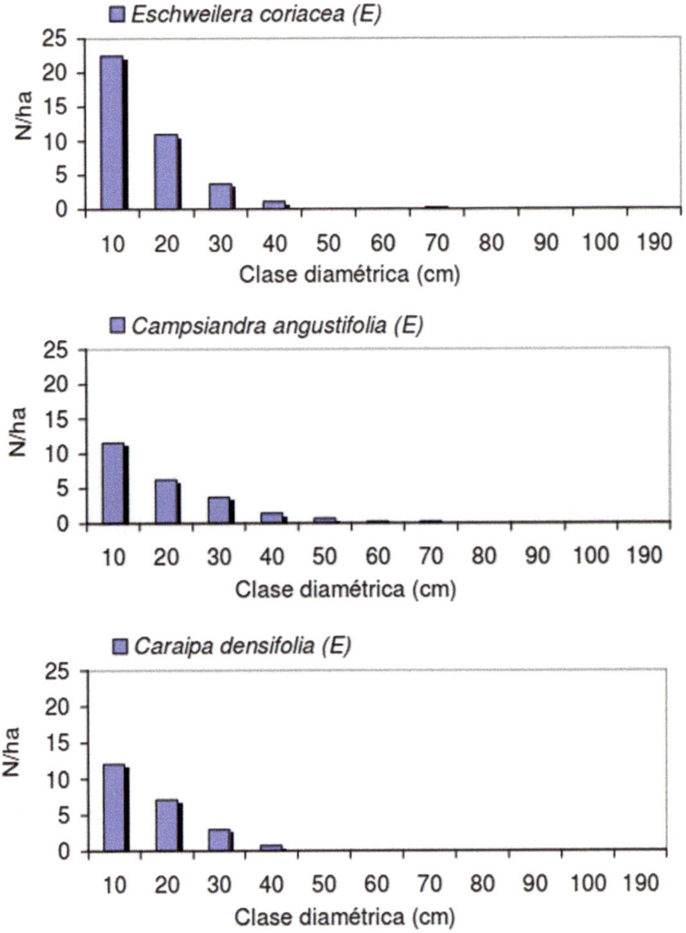

Fig. 14.25 Distribution of the number of trees by diametric class in the alluvial forest of the Nanay River

showed a negative exponential distribution with a slight increase in the last classes (Fig. 14.24).

In the distribution of the number of trees per DAP class (Fig. 14.25), the thin line corresponds to the trees and has been inventoried, and the thick line corresponds to the number of trees expected with the model $Y = 3{,}214{,}998{,}781 \times \text{DAP-}3.08$ with $R2 = 0.946$ $SX = 0.596$, typical of the discetaneus forest (distribution in the "inverted J" form of all species). Similar distribution patterns showed the four most important species of this forest such as *Eschweilera coriacea*, *Campsiandra angustifolia*, *Caraipa densifolia*, and *Licania harlingii* (Fig. 14.20), which were species with high density of trees per hectare.

The trees of the alluvial plain forest of the Nanay River were grouped into three strata, and the lower stratum corresponded to trees whose height does not exceed 11 m.

14.10.2 Forest Potential

The forest potential of the floodplain forests of the Nanay River is classified as category I (excellent); the 241.3 m^3/ha of wood (Table 14.4) is greater than 150 m^3/ha of Table 14.1. The largest volume of wood is suitable for firewood and charcoal with 138.1 m^3/ha. In second place were sawmills with 53.3 m^3/ha, 124.92 trees/ha, and 23 species (Table 14.4). Above the minimum cut diameter (DBH = 40 cm), there was 16.35 m^3/ha of wood; the species *Vochysia lomatophylla*, *Hymenaea courbaril*, *Sapium glandulosum*, *Cariniana decandra*, *Vatairea erythrocarpa*, and *Pterocarpus amazoniens* had 4.17, 2.78, 2.15, 1.62, 1.55, and 1.44 m^3/ha of wood, respectively (Table 14.5).

In roundwood 37.5 trees/ha were found, with 14.7 m^3/ha in ten species, and the most important were *Anaxagorea brevipes*, *Guatteria citriodora*, and *Aspidosperma rigidum* with 9.0, 8.2, and 6.3 trees/ha, respectively (Table 14.6).

Table 14.4 Tree stratum and number of trees (*N*/ha) per species in the alluvial forest of the Nanay River

	Estrato arbóreo			
Species	Inferior <11 m	Intermediate 11 a 19 m	Superior >19 m	*N*/ha
Eschweilera coriacea	5.64	23.61	9.27	38.52
Theobroma glaucum	5.84	7.12	0.84	13.80
Campsiandra angustifolia	4.59	12.88	6.34	23.81
Licania harlingii	2.89	17.09	2.00	21.98
Caraipa densifolia	4.24	14.99	3.57	22.81
Pouteria cuspidata	3.71	11.97	1.15	16.83
Cariniana decandra	3.46	8.79	5.09	17.34
Hevea nitida	2.30	9.60	2.39	14.29
Sapium glandulosum	1.06	8.51	5.67	15.24
Macrolobium angustifolium	1.88	5.65	3.29	10.82
Inga sp.	0.84	7.07	4.06	11.97
Couepia ulei	2.07	7.29	3.38	12.74
Mabea elata	1.61	7.27	3.61	12.50
Vochysia lomatophylla	2.59	4.23	1.05	7.87
Ocotea cernua	1.00	8.14	2.12	11.26
Pouteria glomerata	2.65	0.50	0.44	3.58
Subtotal	46.37	154.72	54.28	255.37
Otras especies	18.26	129.13	58.95	206.33
Total	64.63	283.85	113.23	461.70

Table 14.5 Ecological guild (GE), tree density (*D*/ha), volume (*V*), basal area (*G*), number of trees (*N*), and IVI per species of sawn wood in the alluvial forest of the Nanay River

			V	*G*	*N*	*F*	IVI
Especie	GE	*D*/ha	m^3/ha	m^2/ha	Árb/ha	%	%
Caraipa densifolia	E	High	10.0	0.9	22.8	100	3.7
Vochysia lomatophylla	HD	Medium	7.6	0.6	7.9	75	2.0
Cariniana decandra	E	High	7.4	0.7	17.3	75	2.8
Sapium glandulosum	E	High	5.8	0.6	15.2	100	2.7
Vatairea erythrocarpa	HD	Medium	4.4	0.4	6.3	100	1.7
Hymenaea courbaril	HD	Low	4.4	0.3	3.2	50	1.0
Ocotea cernua	E	High	3.2	0.3	11.3	100	2.0
Iryanthera tricornis	HE	Alto	2.2	0.3	11.6	100	1.9
Pterocarpus amazoniens	R	Low	2.0	0.2	0.7	75	0.8
Parkia igneiflora	HD	Low	1.6	0.2	4.4	100	1.2
Virola multinervia	HE	Medium	1.1	0.1	6.4	100	1.3
Total $V \geq 1$ m^3/ha			49.6	4.4	107.2		21.0
Total $V < 1$ m^3/ha			3.7	0.4	17.7		5.1
Total			53.3	4.8	124.9		26.1

E, escifix; *HD*, durable heliphy; *HE*, ephemeral H., *R*, absolute

Table 14.6 Ecological guild (GE), tree density (*D*/ha), volume (*V*), basal area (*G*), number of trees (*N*), and IVI per species of round wood in the alluvial forest of the Nanay River

			V	*G*	*N*	*F*	IVI
Especie	GE	*D*/ha	m^3/ha	m^2/ha	Árb/ha	%	%
Guatteria citriodora	E	Medium	2.24	0.25	8.20	75	1.47
Zygia basijugum	HD	Low	3.52	0.24	1.50	25	0.65
Aspidosperma rigidum	HD	Medium	3.44	0.31	6.30	100	1.58
Aspidosperma excelsum	HD	Low	1.26	0.14	2.89	75	0.91
Zygia glomerata	HD	Low	0.63	0.06	2.15	50	0.58
Anaxagorea brevipes	SB	Medium	2.04	0.22	9.00	25	1.17
Guatteria megalophylla	SB	Low	0.72	0.08	3.72	75	0.87
Mouriri sp.	SB	Low	0.46	0.05	2.05	25	0.39
Guatteria elata	SB	Low	0.23	0.03	1.19	75	0.61
Zygia sp.	SB	Low	0.16	0.01	0.50	25	0.21
Total madera redonda			14.70	1.39	37.49		8.43

E, escifix; *HD*, durable heliphy; *SB*, Soto forest

As for the wood for poles, it was found 28.2 trees/ha with 20.8 m^3/ha in five species; *Pouteria glomerata* (3.6 trees/ha) and *P. cuspidata* (16.8 trees/ha) contributed 9.6 and 9.4 m^3/ha of wood; the presence of hardwoods, such as *Diplotropis purpurea* and *Minquartia guianensis*, was very low in the area; it only represented 1.3 m^3/ha (Table 14.7).

The wood for firewood and coal was estimated at 243.5 trees/ha, with 138.1 m^3/ha in 39 species, and abundant *Eschweilera coriacea* (38.5/ha), *Campsiandra*

Table 14.7 Ecological guild (GE), tree density (*D*/ha), volume (*V*), basal area (*G*), number of trees (*N*), and IVI by wood species for poles in the alluvial forest of the Nanay River

			V	*G*	*N*	*F*	IVI
Especie	GE	*D*/ha	m^3/ha	m^2/ha	Árb/ha	%	%
Pouteria glomerata	HD	Baja	9.64	0.65	3.58	100	1.93
Pouteria cuspidata	HD	Alta	9.52	0.81	16.83	100	3.16
Diplotropis purpurea	HD	Baja	0.74	0.07	1.91	75	0.72
Minquartia guianensis	SB	Baja	0.54	0.06	1.75	25	0.38
Pouteria torta	SB	Baja	0.40	0.05	4.36	75	0.88
Total postes			20.67	1.63	28.19		7.07

HD, durable helihym; *SB*, forest grove

Table 14.8 Ecological guild (GE), tree density (*D*/ha), volume (*V*), basal area (*G*), number of trees (*N*), and IVI per species of wood for firewood and coal in the alluvial forest of the Nanay River

			V	*G*	*N*	*F*	IVI
Especie	GE	*D*/ha	m^3/ha	m^2/ha	Árb/ha	%	%
Theobroma glaucum	HD	High	36.7	2.5	13.8	100	5.7
Eschweilera coriacea	E	High	16.5	1.5	38.5	100	5.8
Campsiandra angustifolia	E	High	15.7	1.4	23.8	100	4.6
Licania harlingii	E	High	10.6	0.9	22.0	100	3.7
Inga sp.	E	High	7.2	0.6	12.0	100	2.4
Hevea nitida	E	High	6.7	0.6	14.3	100	2.7
Couepia ulei	E	High	5.7	0.5	12.7	100	2.4
Terminalia amazonica	HD	Medium	4.7	0.3	5.3	100	1.6
Brosimun guianense	HD	Low	4.2	0.3	3.3	50	1.1
Mabea elata	E	High	4.1	0.4	12.5	75	2.0
Perebea glabrifolia	HD	Low	3.7	0.3	4.4	50	1.2
Perebea guianensis	HD	Low	3.0	0.3	3.2	75	1.1
Licania longistyla	HD	Low	2.4	0.2	4.8	75	1.2
Sterculia sp.	E	High	2.1	0.3	11.3	100	1.9
Pouteria procera	SB	Medium	1.7	0.2	7.3	50	1.1
Mabea subsessilis	HD	Low	1.6	0.1	4.3	25	0.7
Inga punctata	HD	Low	1.6	0.2	5.0	100	1.3
Symphonia globulifera	HE	Low	1.5	0.1	3.9	75	1.0
Naucleopsis coccinea	HD	Low	1.4	0.1	2.5	25	0.5
Total $V \geq 1$ m^3/ha			130.9	10.9	205.0		41.8
Total $V < 1$ m^3/ha			7.0	0.8	38.1		9.9
Total			137.9	11.7	243.0		51.7

E, escifix; *HD*, durable heliphy; *HE*, ephemeral H.; *SB*, Soto forest

angustifolia (23.8/ha), *Licania harlingii* (22/ha), *Hevea nitida* (14.3/ha), *Theobroma glaucum* (13.8/ha), *Mabea elata* (12.5/ha), *Inga* sp. (12/ha), and *Sterculia* sp. (11.3/ha), with 16.5, 5.7, 10.6, 6.7, 36.7, 4.1, 7.2, and 2.1 m^3/ha, respectively (Table 14.8).

14.10.3 Growth, Mortality, and Tree Recruitment

The average annual increase, accumulated by diametric class of the most abundant species of the four permanent sampling plots, showed a sigmoid distribution in the segment between 10 and 90 cm of DBH (Fig. 14.26).

Based on the growth functions of application in the forest area, described by Kiviste et al. (2002), it was determined that the mathematical expression of the growth function for the alluvial plain forest (Fig. 14.26) was $y = -0.3163417925 + 0.0413202396 \times \text{DAP} + 0.0004563431 \times \text{DAP2} - 4.57815\text{E-006DAP3}$ ($R2 = 1.00$ Typical error = 0.012, significant for $\alpha = 0.01$).

It was also observed that before 33.3 cm of DAP trees grew without many restrictions of their environment, based on the intrinsic capacity of the species. After 40 cm of DAP, the environmental influences are greater, and the growth of the species decreased (Fig. 14.26). The distribution of the average annual increase with respect to the diameter of the trees biologically translates into the mathematical

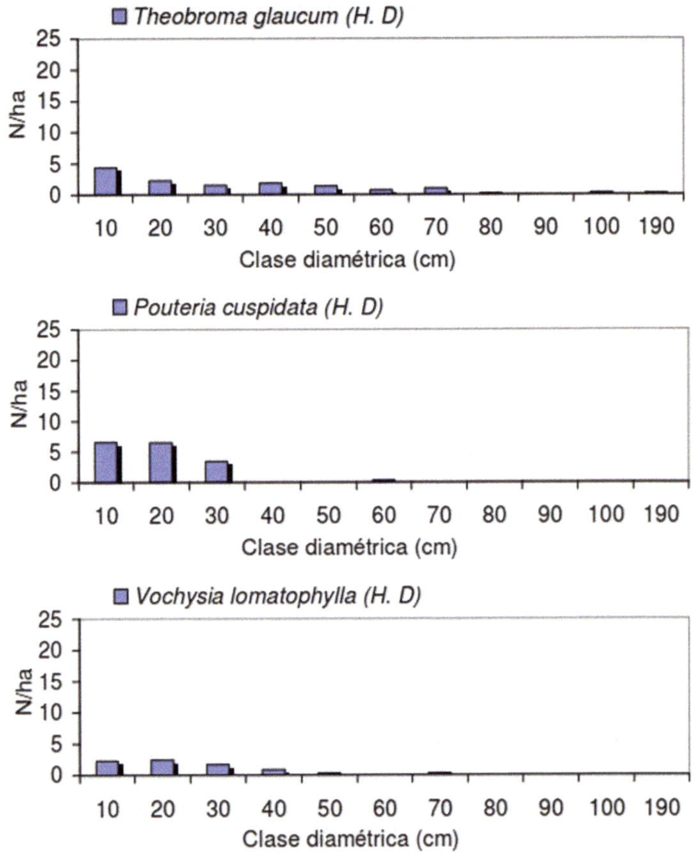

Fig. 14.26 Average annual increment (IMA) by diametric class and the projected IMA (IMA py) in the alluvial forest of the Nanay River

model: y = 0.05393755 + 0.03022102DAP-0.00052193DAP2 + 2.3308081E-006DAP3 (R2 = 0.883, typical error = 0.05 significant for $\alpha = 0.01$) (Fig. 14.27).

The financial maturity of the forest is achieved at 39.2 cm of DBH, after which the growth rate rapidly decreased to 75 cm of DBH, after which the growth of the trees is very slow; under no circumstances should the tree be cut if it does not reach the minimum diameter of 40 cm. The growth of trees in the alluvial plain of the Nanay River, expressed in terms of diameter, shows that the magnitude of the average annual increase (AMI) is 0.43 cm/year (Fig. 14.26). This increase was directly proportional to the increase in diameter of the tree, up to 40 cm in diameter, with a peak growth value of 0.62 cm/year; later, the increases gradually decreased due to the maturity of the trees.

On the other hand, the recruitment rate was 14.91 trees.ha^{-1}.year^{-1} and exceeds the rate at the mortality rate of 14.16 trees.ha^{-1}.year^{-1} (Table 14.2). The heliophyte species showed similar diameter growth, varied between 0.54 and 0.56 cm/year; the lowest growth was registered in the forest species with 0.39 cm/year (Fig. 14.28).

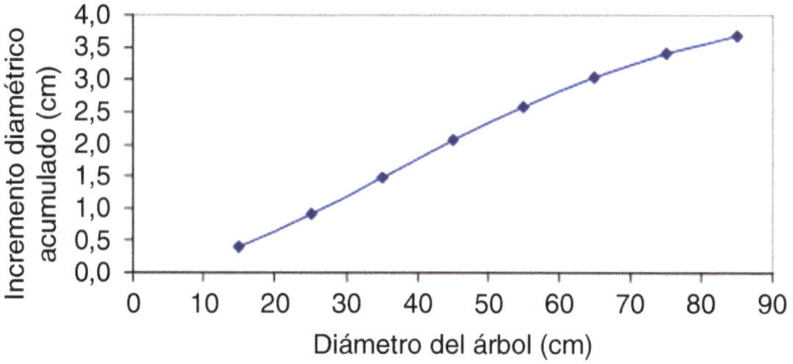

Fig. 14.27 Annual average increase (cm) per ecological guild in the alluvial forest of the Nanay River

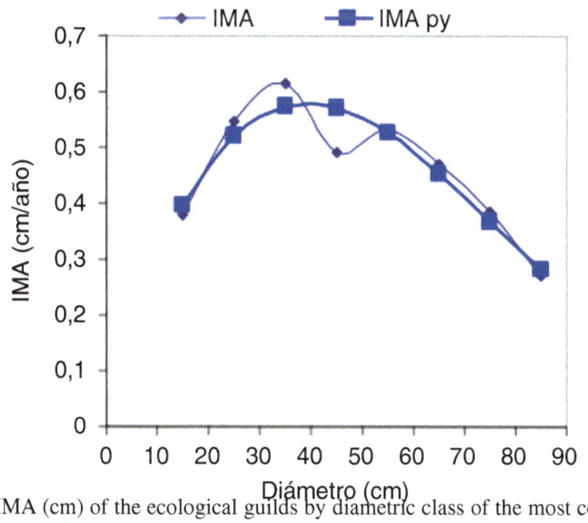

Fig. 14.28 IMA (cm) of the ecological guilds by diametric class of the most common species in the alluvial forest of the Nanay River

The growth of the species is not uniform; fluctuations of the annual average increase (AMI) per ecological guild were observed (Fig. 14.28); when the ephemeral heliophyte species reached a DBH between 30 and 39.9 cm, its growth was 1 cm/year; and durable heliophytes and scryophytes grew about 0.6 cm/year.

The average annual increase of the most common species of the alluvial plain forest is 0.42 cm/year. At the species level, this growth varies between 0.23 and 0.88 cm/year (Table 14.3). The highest diametric growth value was recorded in *Parkia igneiflora* (0.42) and the lowest in *Vatairea erythrocarpa* (0.88 cm/year).

14.11 Discussions

14.11.1 Composition and Structure of the Forest

In the forest of the alluvial plain, which has a heterogeneous topography, trees with a diameter at breast height were inventoried (DBH = 10 cm); this topography constitutes a gradient that influences the presence of a certain species, since in the highest zone, the water column and time of flooding were lower than the lowest zone. The dendrogram based on the Morisita index showed 89% similarity between zones (Fig. 14.18), and the similarity between strips 2, 3, and 4 (vertically crossing said zones) ranged between 65% and 70%; and the similarity with strip 1 is lower (51%). Subdividing each strip in the high and low zones, we observed focal differences of diversity; Guatteria citriodora, *G. elata*, and *G. megalophylla* were not found in the first strip (highest zone), similar to *Theobroma glaucum* and *Cariniana decandra. Perebea glabrifolia* and *P. guianensis* are concentrated in strips 1 and 2 (88%). This shows that the alluvial forest is the integration of different individuals where the characteristic curve of the number of trees per height class is similar to the normal distribution (Fig. 14.23), typical of discrete forests described by Malleux (1982: 52); but, when they are grouped in frequencies of sizes of diameters, they determine different curves to the normal by the conditioning of the environment (Fig. 14.24).

The forest had an average richness of species (81 species), which can be sampled using units of 0.5 ha, as shown by the curve the number of species per area (Fig. 14.17). This sample unit size was able to record the most representative richness of the alluvial forest. Similar results were found by Hughell (1997); Rios and Burga (2005) also report a similar size of area, confirming that 0.5 ha are adequate and sufficient for the forest inventory of the floodplain and compatible with the recommendations of INRENA (2004). By Simpson's equity index, it indicates that the probability of drawing two individuals of the same species is 0.029 (2.9%) and to get the other species is 0.97; this index is superior to the adjacent varillal forest, since Panduro del (1992) reported that Simpson's diversity index varied between 0.8 and 0.96. This diversity calculated with measures of abundance through the Shannon-Wiever index is 3.88 close to 4.39 (ln81), showing that there is equity among the species of this type of forest; the value of this index is within the range of 3.18 and 5.36 reported by Panduro del (1992) for the varillal forest near the area. These two areas showed a different floristic composition, so it is necessary to take into account the suggestions of Ruokolainen and Tuomisto (1998) that, for the management and sustainable use of biological resources, the recognition and mapping of such resources are important and urgent differences.

When compared with the density of trees per hectare of the species, it was observed that 67% have a low density per hectare (Fig. 14.16), and in some cases, such as *Nectandra* sp., *Hevea brasiliensis*, and *Ocotea argyrophylla*, with 0, 2 trees/ha was considered scarce in the area, confirming the results of diversity indices. For the conditioning of the environment, the floristic composition varies in space and time; the 23 families (Table 14.9) and 81 species (Tables 14.10, 14.11, 14.12, 14.13, and 14.14) recorded in the inventory reflect only the composition of a part of the area. The results showed a richness constituted by species of trees that interact with each other and with other organisms, whose presence and mixture are due to flooding alluvial soil with black waters with a different climate and a fluctuating environment, to which the species are adapted. In this regard, Nebel et al. (2000b) indicated

Table 14.9 Density (trees/ha), basal area (m^2/ha), frequency (%), and importance value index (%) (IVI) per family in the alluvial forest of the Nanay River

Familia	Árboles/ha	Área basal (m^2/ha)	Frecuencia (%)	IVI (%)
Fabaceae	89.45	4.78	100	**15.85**
Lecythidaceae	57.11	2.20	100	**9.40**
Euphorbiaceae	49.98	1.84	100	**8.29**
Sterculiaceae	25.12	2.77	100	**7.99**
Chrysobalanaceae	38.01	1.49	100	**6.89**
Sapotaceae	32.08	1.70	100	**6.78**
Sub total	291.74	14.78	600	55.18
Clusiaceae	36.47	1.24	100	6.37
Annonaceae	22.11	0.58	100	4.27
Myristicaceae	22.75	0.44	100	4.10
Moraceae	13.46	1.00	75	3.88
Lauraceae	17.36	0.45	100	3.72
Apocynaceae	9.19	0.44	100	3.13
Vochysiaceae	7.87	0.61	75	2.85
Combretaceae	5.31	0.33	100	2.67
Bombacaceae	6.73	0.17	100	2.51
Olacaceae	7.54	0.37	75	2.45
Melastomataceae	7.36	0.13	75	2.05
Myrtaceae	4.36	0.10	75	1.80
Burseraceae	2.69	0.07	75	1.63
Cecrcpiaceae	1.71	0.05	50	1.08
Elaeocarpaceae	0.73	0.01	50	0.95
Rubiaceae	3.41	0.09	25	0.84
Meliaceae	0.91	0.02	25	0.54
Total	461.70	20.88		100.00

Table 14.10 Floristic composition of the alluvial forest of the Nanay River: scarce species

Scientific name	Common name	Family
Hevea brasiliensis Muell.	Shiringa Moena	Euphorbiaceae
Nectandra sp. No Identificado	Mullaquillo	Lauraceae
Ocotea argyrophylla Ducke.	Moena	Lauraceae

Table 14.11 Floristic composition of the alluvial forest of the Nanay River: species with high density of trees per hectare

Scientific name	Common name vulgar	Family
Campsiandra angustifolia Spruce ex Benth.	Huacapurana	Fabaceae
Caraipa densifolia Mart.	Brea caspi	Clusiaceae
Cariniana decandra Ducke.	Cinta caspi	Lecythidaceae
Couepia ulei Pilger.	Parinari	Chrysobalanaceae
Eschweilera coriacea (A.DC.) S. Mori	Machimango	Lecythidaceae
Hevea nitida Mart. Ex Müll. Arg.	Chiringa	Euphorbiaceae
Inga sp.	Shimbillo	Fabaceae
Iryanthera tricornis Ducke.	Cumala blanca	Myristicaceae
Licania harlingii Prance.	Parinari	Chrysobalanaceae
Mabea elata Steyerm.	Polvora caspi	Euphorbiaceae
Macrolobium angustifolium (Benth.) R. S. Cowan.	Boa caspi	Fabaceae
Ocotea cernua (Nees) Mez.	Moena	Lauraceae
Pouteria cuspidata (A.DC.) Baehni.	Quinilla	Sapotaceae
Sapium glandulosum (L.) Moroni	Shiringarana	Euphorbiaceae
Sterculia sp.	Raton caspi	Sterculiaceae
Theobroma glaucum Karts.	Cacahuillo	Sterculiaceae

that the Amazonian alluvial forests are less rich in species per unit area than the forests of the mainland. The possible cause is the water stress attributed to the floods; in time it is subjected to annual floods (January to May) with a water column of 3–6 m in height (Fig. 14.7), around which the forest species they adapted respond to the fluvial dynamics produced during thousands of years. In that sense, Richards (1969) stated as a general rule that the locations with unfavorable increase in their conditions tend to make species less rich than those with optimal conditions, as shown by Nebel et al. (2000a,b,c) in the floodplain (low restinga) of the lower Ucayali that has white waters, richer than those of black waters. For this reason, soils are richer and with high primary production (Parolin 2002).

The influence of the environment was reflected in the number of trees per diametric class that decreases to an exponential proportion described in Malthus's law (Fig. 14.25), showing that the predefined state is movement (the exponential decline); in forestry it is known as distribution (inverted J) (Louman et al. 2001; Malleux 1982). Because of its highly significant relationship (ANVA for 05.0 = a), the model of Loetsch and Haller (1964) Y = 3,214,998,781 × DAP-0.308 with R × 781.3214998-308.0 = 0.956 gave a good estimate of the number of trees/ha. The

Table 14.12 Floristic composition of the alluvial forest of the Nanay River: species with low density of trees per hectare

Scientific name	Common name vulgar	Family
Alchornea latifolia Sw. *Aniba guianensis* Aublet. *Aniba* sp. *Aspidosperma excelsum* Benth. *Brosimum guianense* (Aubl.) Huber *Calophyllum brasiliensis* Camb. *Calyptranthes pulchella* DC. *Caraipa guianensis* Aublet.	Palometa huayo Moena Moena Remo caspi Tamamuri Lagarto caspi Guayabilla Brea caspi	Euphorbiaceae Lauraceae Lauraceae Apocynaceae Moraceae Euphorbiaceae Myrtaceae Clusiaceae
Cecropia ficifolia Snethl. *Clusia* sp. *Couepia bernardii* Prance. *Diplotropis purpurea* (Rich.) Amshoff. *Eriotheca globosa* (Aublet) Robyns *Eschweilera tessmannii* R.Knuth. *Garcinia macrophylla* Mart. *Garcinia* sp. *Guarea macrophylla* Vahl.	Cetico Chullachaqui caspi Parinari Chontaquiro Punga Machimango Charichuelo Charichuelo Requia	Cecropiaceae Clusiaceae Chrysobalanaceae Fabaceae Bombacaceae Lecythidaceae Clusiaceae Clusiaceae Meliaceae
Guatteria elata R. E. Fries.	Espintana	Annonaceae
Guatteria megalophylla Diles. *Hymenaea courbaril* L. *Licania longistyla* (Hook. F) Fritsch *Mabea maynensis* Spruce. *Mabea subsessilis* Pax & Hoffm. *Matisia dolichopoda* (A. Robyns) Cuatrec. *Minquartia guianensis* Aublet. *Mouriri* sp. *Naucleopsis coccinea* Aublet. *Nectandra pearcei* Mez. *Ocotea aciphylla* Mez.	Carahuasca Azucar huayo Apacharana Polvora caspi Shiringuilla Sacha zapote Huacapú Lanza huayo Chimicua Moena Moena	Annonaceae Fabaceae Chrysobalanaceae Euphorbiaceae Euphorbiaceae Bombacaceae Olacaceae Melastomataceae Moraceae Lauraceae Lauraceae
Ocotea myriantha Meisn. *Pachira aquatica* (Aubl) Schum. *Parkia igneiflora* Ducke. *Peltogyne paniculata* Benth.	Moena amarilla Punga Pashaco blanco Violeta	Lauraceae Bombacaceae Fabaceae Fabaceae

four most important species by the IVI (Fig. 14.20) showed a similar pattern of inverted J; this is important for forest management; according to Uslar (2003: 5), they have good regeneration.

This negative exponential growth (Fig. 14.25) showed that tree density was affected by limiting resources, the environment imposed a selective pressure, and the fittest survived, whose phenotypic variability is inherited to future generations (Darwin 1859), which explains the differences found in the composition and structure of the forest reporting for the arboretum (El Huayo), 36 families, 80 genera, and 162 species. For the alluvial forest of Nanay Burga (1994), it reported different species such as *Guatteria elata*, *Sapium marmierii*, *Manilkara* sp., *Virola* sp., and *Pithecellobium* sp.; Melendez (2000) for plot XII Arboretum reported that the most important families are Lecythidaceae (31.5%), Fabaceae (8.3%), Euphorbiaceae (10.7%), and Myristicaceae (10.6%), at the of species predominated *Eschweilera*

Table 14.13 Floristic composition of the alluvial forest of the Nanay River: species with low density of trees per hectare

Nombre científico	Nombre vulgar	Familia
Perebea glabrifolia (Ducke) Berg.	Chimicua	Moraceae
Perebea guianensis Aublet.	Chimicua	Moraceae
Pouteria glomerata (Miq.) Radlk.	Quinilla	Sapotaceae
Pouteria torta (Mart.) Radlk.	Quinilla	Sapotaceae
Protium nodulosum Swart.	Copal	Burseraceae
Protium subserratum Engls.	Copal	Burseraceae
Pterocarpus amazoniens (C. Martius ex Bentham) Amshoff.	Maria buena	Fabaceae
Remigia sp.	Cascarilla	Rubiaceae
Sloanea floribunda Spruce ex Benth.	Achiotillo	Elaeocarpaceae
Symphonia globulifera L.f.	Chullachaqui caspi	Clusiaceae
Tachigali paniculada Aublet,	Tangarana	Fabaceae
Tocoyena williamsii Standl.	Sacha huito	Rubiaceae
Virola calophylla Barck.	Cumala	Myristicaceae
Vismia amazonica Ewan (Stem).	Pichirina	Clusiaceae
Zygia sp.	Bushilla	Fabaceae

Table 14.14 Floristic composition of the alluvial forest of the Nanay River: species with average density of trees per hectare

Scientific name	Common name vulgar	Family
Anaxagorea brevipes Benth.	Carahuasca	Annonaceae
Aspidosperma rigidum Rugby.	Remo caspi	Apocynaceae
Guatteria citriodora Ducke.	Carahuasca	Annonaceae
Inga punctata Willd.	Shimbillo	Fabaceae
Macrolobium sp.	Boa caspi	Fabaceae
Miconia amazonica Triana.	Rifari Caimitillo	Melastomataceae
Pouteria procera (Mart.) T.D. Penn. *Tachigali bracteosa* (Harms) Zarucchi & Pipoly.	Tangarana	Sapotaceae Fabaceae
Terminalia amazonica (J.F. Gmel.) Exell.	Yacushapana	Combretaceae
Vatairea erythrocarpa (Ducke) Ducke	Mari mari	Fabaceae
Virola multinervia Ducke.	Cumala blanca	Myristicaceae
Vochysia lomatophylla Standl.	Quillosisa	Vochysiaceae

grandiflora (20.3%), *Eschweilera coriacea* (9.16%), *Eschweilera bracteosa* (4.3%), and *Pourouma tomentosa* (3.15%).

On the other hand, examining the temperament of forest species, it has been observed that 15 scryophyte species show a structure (inverted J) (Figs. 14.19 and 14.21), which indicates the intrinsic capacity of children and young people to develop under the shade of trees of greater size and age and that can survive under less illuminated conditions.

Table 14.15 Guild, cup lighting (%), and number of trees (*N*/ha) species in the alluvial forest of the Nanay River

	GE	*D*/ha	*V* m³/ha	*G* m³/ha	*N* Árb/ha	*F* %	IVI %
Pouteria glomerata	HD	Baja	9.64	0.65	3.58	100	1.93
Pouteria cuspidata	HD	Alta	9.52	0.81	16.83	100	3.16
Dipiotropis purpurea	HD	Baja	0.74	0.07	1.91	75	0.72
Minquartia guianensis	SB	Baja	0.54	0.06	1.75	25	0.38
Pouteria torta	SB	Baja	0.40	0.05	4.36	75	0.88
Total postes			20.67	1.63	28.19		7.07

In the forest of the Nanay River, it is natural to observe that the trees indistinctly of their ecological guild occupy different strata (Table 14.4); for this reason, in this study, it has been observed that between 15% and 34% of the trees of the scryophyte species were under shade; between 43.1 and 64.3 of the trees, their canopy received partial light; but between 15.3% and 33.8% of the trees, it showed full canopy illumination (Table 14.15). According to Louman et al. (2001), it is a response to the microenvironmental conditions present in the different profile heights, where temperature and humidity vary in each stratum, allowing species with different energy requirements to be located at the levels that best meet their needs; thus, Tello et al. (1992) verified the existence of silvicultural interaction of the species with respect to the degree of exposure of plants to sunlight. In the discrete forest, species that can present bell-shaped curves or bimodal distributions (with two or more peaks) usually correspond to light-demanding species that need larger clearings (Louman et al. 2001). In ephemeral heliophytes (Fig. 14.19), photosynthetic systems are very inefficient under shadow environments, so they do not have the capacity to regenerate under these conditions (DFF 2006). Durable heliophyte species are fast growing (Figs. 14.19 and 14.29), with intermediate photosynthetic capacity, and produce moderately light to moderate heavy woods (DFF 2006). The alluvial plain forest of the Nanay River is in a successional process characterized by high species richness, high tree density, and poor stratification of the canopy (Worbes et al. 1992; Wittmann et al. 2002).

The basal area estimated at 20.88 m²/ha (Table 14.9) is less than 23.39 and 31.93 m²/ha, reported by Perea (1995) for the wet varillal forest, and lower than that of the Mendrica plain, the basin of the lower Amazon reported by Rios and Burga (2005). Its distribution by diametric class reflected anthropogenic disturbances (extraction of wood DAP = 40 cm) and, according to Louman et al. (2001), constitutes a regression in the succession. This confirms that each species has a different response within the system, which translates into different rates of transformation of light energy into biomass, and the efficiency in the accumulation of energy in the trees depends on the level of adaptation to this fluctuating environment, as well as the place occupied by individuals in the light threshold that vary according to the species and the regeneration niche. For this reason, 16 species

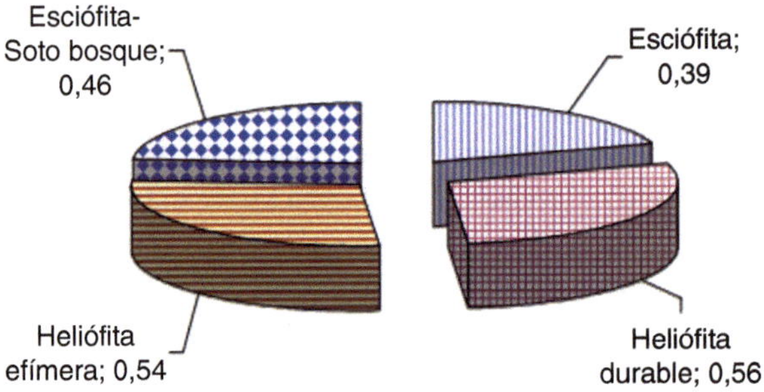

Fig. 14.29 Distribution of durable heliophyte species (HD) in the alluvial forest of the Nanay River

Table 14.16 Number of trees (*N*), basal area (*G*), and frequency (*F*); IVI (%), ecological guild (GE), and tree density (*D*) by alluvial forest species of the Nanay River

	N	*G*	*F*	IVI		
Especie	Árb./ha	m^2/ha	%	%	GE	*D*
Eschweilera coriacea	38.5	1.5	100	5.8	E	Alta
Theobroma glaucum	13.8	2.5	100	5.7	HD	Alta
Campsiandra angustifolia	23.8	1.4	100	4.6	E	Alta
Licania harlingii	22.0	0.9	100	3.7	E	Alta
Caraipa densifolia	22.8	0.9	100	3.7	E	Alta
Pouteria cuspidata	16.8	0.8	100	3.2	HD	Alta
Cariniana decandra	17.3	0.7	75	2.8	E	Alta
Hevea nitida	14.3	0.6	100	2.7	E	Alta
Sapium glandulosum	15.2	0.6	100	2.7	E	Alta
Macrolobium angustifolium	10.8	0.7	100	2.5	E	Alta
Inga sp.	12.0	0.6	100	2.4	E	Alta
Couepia ulei	12.7	0.5	100	2.4	E	Alta
Mabea elata	12.5	0.4	75	2.0	E	Alta
Ocotea cernua	11.3	0.3	100	2.0	E	Alta
Vochysia lomatophylla	7.9	0.6	75	2.0	HD	Media
Pouteria glomerata	3.6	0.6	100	1.9	HD	Baja
Subtotal	255.4	13.7		50.1		
Otras(66 species)	206.3	7.2		49.9		
Total 82 species	461.7	20.9		100.0		

E, esciófita; *HD*, heliófita durable

were the most dominant, abundant, and best distributed in the area; together they contributed with 50% of the IVI, and the remaining 66 species contributed the other 50% (Table 14.16).

14.11.2 Forest Potential

The forests of the alluvial plain of the Nanay River is a special ecosystem used by man and fish during the flood of the river (Fig. 14.3). Due to its easy access during the growing season, the selective extraction of commercial woods is intense and oriented to domestic consumption, and the surplus is commercialized in Iquitos (Fig. 14.2). However, according to Kvist and Nebel (2000), the low Amazon of Peru has low population and few roads, so rivers constitute the largest available infrastructure. The alluvial plain of the Nanay River is flooded from January to May, with a water column that varies between 3 and 6 m. On the edge of the river and in the marginal zones of the lakes, trees and shrubs have been established in constant change without reaching the state of a climactic forest. Its soils do not present sedimentation due to the black waters, poor in silt (Fig. 14.4).

The density of trees/ha of this forest of 461.7/ha (Table 14.17) is in the range of 417 and 737 trees/ha for other areas of the Amazonian plain reported by Bongers et al. (1988), Nebel et al. (2000a,b,c). The result is greater than the 375 trees/ha reported by Burga (1994) for this forest, similarly to that reported by Pacheco et al. (1993) for the low terrace forest in the middle basin of the Nanay River (Santa María) whose range varied between 285 and 344 trees/ha; this suggests that the abundance of trees in the alluvial plain of the Nanay River does not present abrupt

Table 14.17 Trees (*N*), basal area (*G*), and volume (*V*) for the use of wood and diametric class in the alluvial forest of the Nanay River

			Clase diamétrica (cm)								
Uso madera	SP	Dato	10	20	30	40	50	60	70	≥80	Total
Aserrada	23	*N*	30.5	27.3	11.6	3.4	1.0	0.6	0.3	0.2	124.9
		G	1.4	1.2	1.1	0.5	0.2	0.2	0.1	0.1	4.8
		V	10.9	12.6	13.5	6.5	3.6	3.1	1.7	1.4	53.3
Leña y carbón	39	*N*	152.6	56.9	16.6	7.4	4.4	2.3	2.1	1.2	243.5
		G	2.6	2.6	1.5	1.1	1.0	0.7	0.9	1.2	11.7
		V	20.3	26.2	19.1	14.6	14.3	10.4	13.8	19.5	138.1
Otros	6	*N*	16.1	6.8	1.6	2.3	0.2	0.6			27.6
		G	0.3	0.3	0.1	0.4	0.1	0.2			1.3
		V	2.3	3.0	1.5	4.3	0.4	3.0			14.6
Postes	5	*N*	13.8	7.8	3.6	1.3	0.5	1.0		0.2	28.2
		G	0.2	0.4	0.3	0.2	0.1	0.3		0.1	1.6
		V	1.7	3.7	4.0	2.5	1.8	4.8		2.3	20.7
Redonda	10	*N*	24.6	10.3	0.4	1.2	0.8	0.3			37.5
		G	0.4	0.5		0.2	0.2	0.1			1.4
		V	3.2	4.9	0.4	2.5	2.6	1.2			14.7
Total		*N*	287.7	109.1	33.9	15.5	6.8	4.7	2.4	1.6	461.7
		G	4.8	5.0	3.1	2.3	1.6	1.5	1.0	1.5	20.9
		V	38.3	50.5	38.3	30.4	22.6	22.6	15.4	23.2	241.3

N, trees/ha; *G*, m^2/ha; *V*, m^3/ha; *SP*, number of species variations. However, according to Ruokolainen and Tuomisto (1998), it is sometimes possible to observe large areas that appear to be homogeneous, which are limited by areas where floristic and environmental characteristics change rapidly over short distances.

Similar occurs with the basal area of 20.9 m^2/ha and the estimated wood volume of 241.3 m^3/ha (Table 14.10), as Nebel et al. (2000a) reported volumes that varied between 59 and 240 m^3/ha in an alluvial forest of nutrient-rich white water; according to Parolin (2002), they have high primary production. These variations in volume can also be explained by the different levels of water stress of the trees during the flood. Due to the presence of restingas, not all the forest remains flooded during the same period; Paredes et al. (1998) found that the natural vegetation of the medium restingas is more diversified than that of the low restinga and is constituted by a taller tree forest with a layer of leaf litter on the surface. The average restingas are sporadically flooded, only in large floods. In that sense, Richards (1969) stated, as a general rule, that the locations with unfavorable increase in their conditions tend to make species less rich than those with optimal conditions.

On the other hand, Nebel et al. (2000b) reported a higher basal area in the tahuampa forests close to 28 m^2/ha in relation to the 20.9 m^2/ha of this study; it is less than that of the restinga forest (24 m^2/ha) and at 23.4 m^2/ha reported by Perea (1995). In the meander plain of the basin of the lower Amazon (DAP = 27.5), Rios and Burga (2005) reported a lower basal area (14.98 m^2/ha). The 20.9 m/ha of basal area found in this study is similar to the 21 m/ha reported by Melendez (2000) for plot XII arboretum near the study area (low terrace); however the volume of wood is minor (161 m^3/ha); this indicates that the height of the trees are different between zones or in any case it can be due to the bias produced in the ocular estimation.

The distribution by diametric class and the current wood volume reflects anthropogenic disturbances (Tables 14.11, 14.12, 14.13, and 14.14), as evidenced by the stumps observed in the forest (extraction of wood DAP = 40 cm); *Minquartia guianensis* has the harder wood and is scarce in the area; according to Louman et al. (2001), these anthropogenic disturbances constitute a regression in the succession.

The results of the alluvial plain of the Nanay River compared with those of the alluvial plain adjacent to the Ucayali River are lower. Nebel et al. (2000c) reported between 662 and 750 m^3/ha. This confirms the destruction of the Nanay forest and its impoverishment. The forestry potential from the logging point of view was represented by 23 species against the 60 species reported for Madre de Dios. With all the forestry potential is excellent, even the potential of wood for firewood and coal is considered as very good, which is why illegal loggers focus their operations in these areas.

Finally, Hernandez and Castellanos (2006) highlight the importance of tropical forests in terms of biodiversity and the generation of environmental goods and services, and considering their key role in the global carbon cycle, they argue that it is necessary to better understand the current and future response of tropical forests to global change.

For forest management, 12 scryophyte species are well adapted to the internal environment of the forest, constituting a guarantee for sustainable management. For firewood and charcoal: *Eschweilera coriacea*, *Campsiandra angustifolia*, *Licania harlingii*, *Hevea nitida*, *Inga* sp., *Couepia ulei*, and *Mabea elata* (Table 14.8); for sawn wood figure *Caraipa densifolia*, *Cariniana decandra*, *Sapium glandulosum*, and *Ocotea cernua* (Table 14.5); for other uses *Macrolobium angustifolium*. The species that probably require large clearings for their regeneration for firewood and charcoal are *Theobroma glaucum* and *Pouteria cuspidata* and for poles *Pouteria glomerata*; all are durable heliophytes.

14.11.3 Growth, Mortality, and Tree Recruitment

In the forest of the alluvial plain of Nanay, the cumulative diameter growth of the species by diametric classes showed a sigmoidal tendency (Fig. 14.30), determining that the financial maturity of the forest is achieved at 39.2 cm of DAP (Fig. 14.26), after which the growth rate rapidly decreases until the 75 cm of DAP, then, the growth of the trees is slower; in this case, it is advisable to cut the tree when it has reached at least 40 cm in diameter, as established in the INRENA legislation. The curve shows that between 10 and 20 cm of DAP, the growth is increasing (Fig. 14.26), apparently without any limitation, that is, it depends on the biotic potential of the individual, their photosynthetic activity, nutrient absorption, catabolic and anabolic processes, etc. (Zeide 1993); according to this author, in this section the weight of the positive factor is greater than the weight of the restrictions, so that growth shoots up; the increase of the restrictions is observed in the 30–40 cm stretch where the growth stops and then the growth decreases (Fig. 14.26). For Kiviste et al. (2002), this is due to the restrictions imposed by the environment as competition between the plants, limitation of resources, and stress and by their own condition of the plant, such as self-regulation mechanisms of growth and aging. It reflects the evolution of tree diameter throughout life, showing a sigmoid curve (Kiviste et al. 2002; Louman

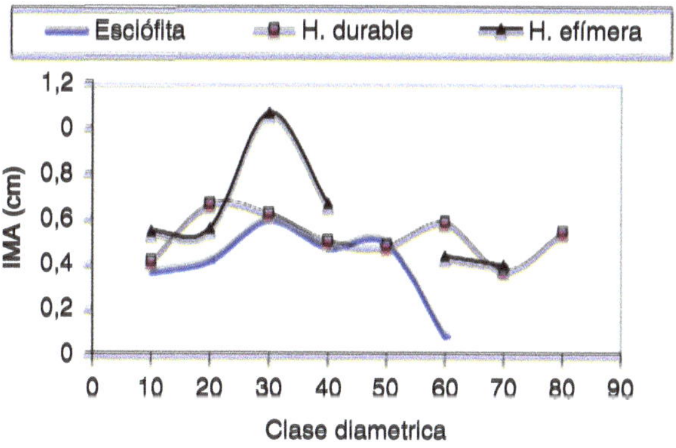

Fig. 14.30 Average annual accumulated increase by diametric class in the alluvial forest of the Nanay River

et al. 2001). So that the growth of the alluvial plain forest is stable, it happens independently of the species in question or of the changes and fluctuations of the environment in which it is developed, as they refer Kiviste et al. (2002).

In the heliophyte species, the diameter growth was between 0.54 and 0.56 cm/year. The lowest growth was registered in the species of undergrowth with 0.39 cm/year (Table 14.2). The ephemeral heliophyte species grow faster, and the IMA reached 1 cm in the class of 30–39.9 cm of DAP (Fig. 14.16), while the durable heliophytes and the sciophytes grew about 0.6 cm/year. Similar growth pattern was reported by Guardia (2004) demonstrating that the emergent species grew faster than the understory species. It is inferred that the species that grows in conditions of illumination different from that of its ecological guild affects its growth, which usually decreases; thus, Tello et al. (1992) reported that the diametric development of *Ormosia macrocalix* and *O. coccinea* decreased as they approach the primary forest, where there is less exposure to sunlight.

Under normal conditions, it has been observed that the average annual increment (AMI) of the trees of the alluvial plain of the Nanay River was 0.42 cm/year and the curve showed a gradual growth of the AMI as it increased the diameter of the tree, reaching the maximum peak in the diameter class of 30–39.9 cm with an IMA of 0.62 cm/year (Fig. 14.27); later, the IMA decreased gradually due to the maturity of the trees. This trend is natural; according to Louman et al. (2001), the class of 10–20 cm have minor increases, possibly because they do not have full access to the resources offered by the medium. This pattern in the growth of alluvial forest species is ideal and the curve is characterized by having a slow initial growth, followed by a phase of accelerated growth, which in most cases tends to be linear; in the final part of the curve, the growth returns to be slow when the tree is between mature and over mature stages.

In the most common species of this alluvial forest, the AMI was 0.42 cm/year. At the level of all species, this growth varied between 0.23 and 0.88 cm/year (Table 14.16). This growth rate fluctuated between 0.23 and 1.97 cm/year in the tropical moist forest in (El Magdalena) (Vallejo et al. 2005); according to this author, based on the information of Álvarez et al. 2002 cited by Vallejo et al. (2005), the average growth for all individuals was 0.39 cm/year, similar to the 0.42 cm/year of diametric growth of the Nanay forest. In addition, the diameter growth of the Nanay forest is in the range of growth in tropical forests that varies between 0.14 and 0.7 cm/year (Hernandez and Ortiz 2005). The rates of increase in diameter at breast height (DBH) vary significantly between tree species and among individuals of a tree species; they also vary in terms of their relation to age and microclimatic conditions (Ferri 1979).

The growth rate of this forest is reasonable. Nebel et al. (2000a) for forests flooded with white water reported an average annual increase in diameter between 0.4 and 0.45 cm/year in the plots not intervened and between 0.53 and 0.68 cm/year in the plots intervened. After reaching full development, the trees experience an increasingly slower growth until reaching the maximum diameter; this corresponds to the moment in which the IMA is null and they are framed within the law of growth. Biological growth is the result of a large number of complex processes (Zeide 1993).

In Bolivia's non-flood forests, growth rates according to Mostacedo et al. (2006) that are smaller vary between 0.18 and 0.38 cm/year and coincide with Hernandez and Castellanos (2006) that the diametric increase is not higher than 0.6 cm/year. The diameter growth rate of the managed forest trees was 0.83 cm/year (Linera 1996). Based on the results of the annual diametric increase that varied between 0.11 and 0.71 cm, Uslar (2003) supports the concept that tropical moist forests are more diverse and dynamic than tropical dry forests.

For the purposes of sustainable management to take advantage of their timber, it is necessary to intervene the forest to accelerate the growth of the remaining trees. The cutting of trees in itself constitutes a treatment, by forming new clearings, and increases the dynamics of the forest, favoring the diametric growth of forest species; Carvalho et al. (2004) show that 0.18 cm/year in an intervened forest passes to 0.42 cm in a forest when trees were extracted (DBH > 45 cm) and each species has a particular diametric development.

If the effect of the treatment were minimal, then, trees with 30 cm of DAP growing at a rate of 0.6 cm/year would reach the minimum cutting diameter (DAP = 40 cm) in 17 years, allowing the definition of utilization shifts 20 years old. This method, which is the simplest to estimate the allowable cut and the effect of the cutting cycle, is called mean time of passage (TMP), which is the number of years necessary for a tree to grow from the beginning of a diametric class at the start of the next major class (DFF 2006: 73). Still, with an IMA of 0.5 cm/year used by DFF (2006: 134), a short cycle of 20 years is obtained.

The Nanay forest is not in equilibrium, the rate of recruits whose value is 14.9 trees.ha^{-1}.year^{-1} exceeds the mortality rate of 14.16 trees.ha^{-1}.year^{-1} (Table 14.2), evidencing dynamic patterns corresponding to growing populations. However, this pattern in population dynamics that reflects the growth of young trees is favored by anthropogenic activities; in situ, 18.31 trees have been registered.ha^{-1}.year^{-1} caused by anthropogenic activities. What adds a mortality rate of 32.47 trees.ha^{-1}.year, so they present the forests near Iquitos, a pattern of declining populations; this constitutes an alert of the danger in which these forests are found if sustainable forest management policies are not defined and implemented.

If these forests are properly conserved, over time the pioneer trees die, and the vegetation enters the maturity stage. As a consequence, the floristic diversity diminishes, and gradually the more advanced successional species stimulate their growth by the formation of new clearings of small size, which leads to reach its position in the upper canopy.

14.11.4 Sustainable Forest Management and Environmental Services

The need to produce food, energy, and clothing for a growing population in the city of Iquitos increases the pressure on forest resources. The production of firewood and charcoal is a tala race, and agriculture is developed in highlands of forest aptitude, with this deforestation increases. Deforestation according to the cellular automata theory is a phenomenon driven mainly by the expansion of the network of routes and roads; in the satellite images, the environmental liability is observed, year after year the highland forests disappear; the remaining forests and fragments correspond to the alluvial forest of the Nanay River basin or the Itaya River. The sustainable management of this forest constitutes a new framework for the development of the Amazon, based on the participation of different social actors such as the riverbank population, illegal loggers, state institutions (IIAP, UNAP, Regional Government, municipal governments), and the organized civil society.

This new framework must include the environmental services to be achieved, in such a way that the environmental function becomes a service, necessary for the economic-social systems of value generation. This value consists of wood for sawing, roundwood, firewood and coal, and posts. Twelve species that are well adapted to the internal environment of the forest, which constitute a guarantee for sustainable management, were for firewood and coal *Eschweilera coriacea*, *Campsiandra angustifolia*, *Licania harlingii*, *Hevea nitida*, *Inga* sp., *Couepia ulei*, *Mabea elata*, *Theobroma glaucum*, and *Pouteria cuspidata*; for sawn wood *Caraipa densifolia*, *Cariniana decandra*, *Sapium glandulosum*, and *Ocotea cernua*; and for poles *Pouteria glomerata*. These species have a high density of trees/ha, except *Vochysia lomatophylla* (medium density) and *Pouteria glomerata* (low density), and can be managed in 20-year rotations.

The sustainable management of the Nanay forest would positively impact the quality of the water to be taken, and this forest is the main filter before the water reaches the main channel and the water intake by SEDALORETO; the sediments produced by the erosion of the upper parts remain in the forest. It will also allow conservation of the genetic resources of the useful species (wood, fruits, and resins), as well as the conservation of species of low density per hectare and scarce species such as *Ocotea argyrophylla*, *Pterocarpus amazoniens*, *Sloanea floribunda*, *Ocotea aciphylla*, and *Aniba guianensis*, among others. In passing, it also provides environmental services to fix carbon from atmospheric CO_2.

Also managing the forest allows maintaining the nutrient cycle. These remain in the place of extraction, providing essential resources to maintain biodiversity. Over time it allows to generate and improve the soil with organic matter and humus.

The vegetation acts as a climatic thermoregulator, because under the canopy, the light threshold changes, creating a suitable microclimate for the plants, maintaining the habitat for fish feeding, and, in passing, showing the scenic beauty and the possibility of recreation with fishing. All of the above, according to Hernandez and Castellanos (2006), enables environmental services; hence the importance of tropical forests in terms of biodiversity and the generation of environmental goods and services, and considering their key role in the global carbon cycle, is necessary to better understand the current and future response of tropical forests to global change.

Experiences on sustainable forest management in South America are many. The most successful occurs in Costa Rica and Bolivia. In Peru there are abandoned plans such as Alexander von Humboldt in Pucallpa and Pichis Palcazú in the Central Selva. The Santa Mercedes on the Putumayo River is in full execution. In addition, under the current legislation, we have forest concessions, which order and rationalize forest use, but silvicultural plans are not implemented.

14.12 Conclusions

In order to sustainably manage the alluvial forest of the Nanay River, it is necessary to know the characteristics of the structure, composition, potential, and growth of the forest. These characteristics are the following:

1. The structure of the Nanay forest is expressed in the following terms: In 15 species, the distribution of the number of trees by diametric classes was one (J-inverted), which characterizes the sciophyte species, and 37 were understory scyophyte species; together they represented 64.2% of the total species. The species with high demand for light represented 24.7% in durable heliophytes and 18.5% in ephemeral heliophytes; three species were scarce in the area.
2. The floristic composition is reflected in 23 families and 81 species, indicating high species richness. The species-area curve showed a species richness per unit area greater than 70%. This alluvial forest was characterized by the floristic association of *Eschweilera coriacea*, *Campsiandra angustifolia*, *Licania harlingii*, *Caraipa densifolia*, *Pouteria cuspidata*, and *Cariniana decandra*. The diversity of species according to the dominance of the species, calculated with the Simpson index, is 0.97, and the diversity calculated with measures of abundance with the Shannon-Wiever index was 3.88. The degree of similarity of the species between sampling strips was very high; the Morisita index showed 89% similarity between strips.
3. There is excellent wood potential in the alluvial forest with 241.3 m^3/ha. Even in wood for firewood and coal, the forest potential is considered as very good; the volume of wood reaches 138.1 m^3/ha in 39 species; for sawnwood, 23 species were registered with 53.3 m^3/ha; for other uses, 6 species; for posts, 5 species; and for roundwood, 10 species. The forest potential of the Nanay forest can be improved through sustainable management; for this, 12 species (scryophytes) are well adapted to the internal environment of the forest; for firewood and charcoal *Eschweilera coriacea*, *Campsiandra angustifolia*, *Licania harlingii*, *Hevea nitida*, *Inga* sp., *Couepia ulei*, and *Mabea elata*; for sawn wood *Caraipa densifolia*, *Cariniana decandra*, *Sapium glandulosum*, and *Ocotea cernua*; and for other uses *Macrolobium angustifolium*. Four species (durable heliophytes) require large clearings for regeneration, *Theobroma glaucum* and *Pouteria cuspidata* (firewood and coal) and *Pouteria glomerata* and *P. cuspidata* (poles).
4. The growth of the species is characterized by an average annual increase (AMI) of 0.43 cm/year and at the species level varied between 0.23 and 0.88 cm/year. The peak growth value of 0.62 cm/year was recorded between 30 and 39.9 cm of DAP. In the heliophyte species, the diameter growth varied between 0.54 and 0.56 cm/year; in the species of sotobosque, it was 0.39 cm/year.

5. The financial maturity of the forest is achieved at 39.2 cm of DAP; trees should be cut when they exceed 40 cm in diameter. The average time of passage for species to pass from the 30 cm class (30–39.9 cm) to the 40 cm class is 17 years; the rotation can be done every 20 years.
6. The recruitment rate was 14.91 trees.ha^{-1}.year^{-1}, and the mortality rate was 12.06 trees.ha^{-1}.year^{-1}. In the forest, dynamic patterns were observed corresponding to growing populations. In the vicinity of Iquitos by anthropogenic activities, it shows a pattern of decadent populations; that indicates the serious danger that these forests run if they do not define and implement policies of sustainable management, oriented to provide environmental services or participation in the voluntary market of avoided deforestation.

References

Baker TR, Phillips OL, Malhi Y, Almeida S, Arroyo L, Di Fiore A, Erwin T, Higuchi N, Killeen TJ, Monteagudo A, Neill DA, Nunez PV, Pitman NCA, Silva JNM, Martinez RV (2004a) Increasing biomass in amazonian forest plots. Philos Trans R Soc Lond B Biol Sci 359:353–365

Baker TR, Phillips OL, Malhi Y, Almeida S, Arroyo L, Di Fiore A, Erwin T, Killeen TJ, Laurance SG, Laurance WF, Lewis SL, Lloyd J, Monteagudo A, Neill DA, Patino S, Pitman NCA, Silva JNM, Martinez RV (2004b) Variation in wood density determines spatial patterns in Amazonian forest biomass. Glob Chang Biol 10:545–562

Bongers F, Pompa J, Del Castillo JM, Carabias J (1988) Structure and floristic composition of the lowland rain forest of Las Tuxtlas, Mexico. Vegetatio 74:55–80

Brenes G (1990) Parcelas de muestreo permanente, una herramienta de Investigación de nuestros bosques. Programa de restauración y selvicultura del Bosque Seco. Notas del Curso de Silvicultura de bosque natural. En http://www.acguanacaste.ac.cr/rothschildia/v1n1/textos/16.html.1p

Burga AR (1994) Determinación de la estructura diamétrica total y por especie en tres tipos de bosque en Iquitos - Perú. Documento Técnico. (Tesis Ingeniero Forestal). Iquitos. Universidad Nacional de la Amazonia Peruana, 16 p

Burga AR (2007) Inventarios forestales. Proyecto manejo de los recursos naturales en las cuencas de los ríos Pastaza y Morona. Fondo Nacional para áreas naturales protegidas por el estado (PROFONANPE)-UNAP, 96 p

Carrizosa J (2001) ¿Qué es ambientalismo? La visión ambiental compleja. PNUMA/IDEA/CEREC, Bogotá, p 17

Carvalho PJO, Silva JNM, Lopes CAA (2004) Growth rate of a *terra firme* rain forest in Brazilian Amazonian over an eight-year period in response to logging. Acta Amazon 34:209–217

Curtis JF, McIntosh RP (1950) The interrelations of certain analytic and synthetic phytosociological characters. Ecology 31:434–450

Darwin C (1859) The origin of species by means of natural selection, 1st edn. Murray, London Reeditado por E Mayr Harward University Press Cambridge Massachussets (1964). 502 p

De Lima PARA (2003) Antropogenización, dinámicas de ocupación del territorio y desarrollo en la amazonia brasileña: El caso del Estado de Amapá. Universidad Autónoma de Barcelona. Tesis Doctoral, Universidad de Bellaterra, España, pp 1–83

Delgado D, Finegan B, Zamora N, Meir P (1997) Efectos del aprovechamiento forestal y el tratamiento silvicultural en un bosque húmedo del noreste de Costa Rica: Cambios de la riqueza y composición de la vegetación. CATIE. Serie Técnica. Informe Técnico No 298. Colección manejo diversificado de bosques naturales 12:1–43

DFF-Direccion de Fomento Forestal (2006) Manejo Forestal: Elaboración de planes de manejo y planes operativos de aprovechamiento en bosques húmedos latifoliados. Departamento de fomento forestal. Nicaragua, pp 1–165

Encarnacion C, Salazar VA (2004) Marco conceptual inicial del enfoque sistémico para la conservación y uso sostenible de ecosistemas inundables de la amazonia peruana. Informe Técnico: Versión preliminar en edición. BIODAMAZ-IIAP, Iquitos, Perú, pp 1–90

FAO (1993) Conservation of genetic resources in tropical forest management. Principies and concepts. FAO, Roma. FAO Forestry Paper 107, pp 1–105

FAO (2001) Situación de los bosques del mundo. Depósitos de documentos de la FAO. Departamento de Montes, p 16. En www.fao.org/docrep/003/y0900s/y0900s06.htm

Ferreira O (1995) Manual de ordenación de bosques. Siguatepeque, Honduras, pp 1–128

Franco LJ, De La Cruz G, Cruz A, Rocha A, Navarrete N, Flores G, Kato E, Sanchez S, Sabarca LG, Bedia CM (2001) *Manual de ecología*, 2nd edn. Trillas, México 1266 p

Fredericksen T, Contreras F, Pariona W (2001) Guía de silvicultura para bosques tropicales de *Bolivia*. BOLFOR, Santa Cruz, Bolivia, pp 1–81

Gema-Servicios Geograficos y Medio Ambiente SAC (2007) EIA doce (12) pozos exploratorios – Lote 39. Línea base ambiental 3:1–99

GTZ/FUNDECO/IE (2001) Estrategia Regional de Biodiversidad para los Países del Trópico Andino: Conservación de ecosistemas transfronterizos y especies amenazadas. La Paz Bolivia, pp 1–203

Guardia VS (2004) Dinámica y efectos de un tratamiento silvicultural en el bosque secundario "Florencia". Tesis magíster Scientie, San Carlos, Turrialba, Costa Rica, pp 1–141

Hernandez L, Castellanos H (2006) Crecimiento diamétrico arbóreo en bosques de Sierra de Lema, Guayana Venezolana: Primeras Evaluaciones. Interciencia 31:787–793

Hernandez L, Ortiz J (2005) Avances del estudio sobre dinámica de bosques a lo largo de un gradiente climático entre sierra de Lema y la Gran Sabana. IV. Congreso Forestal Venezolano, Venezuela, 15 p

Higuchi N, Dos Santos J, Tribuzy ES, Lima A, Teixeira LM, Carneiro VMC, Felsemburgh CA, Pinto FR, Da Silva RP, Pinto DACM (2005) *Noçôes básicas sobre manejo florestal*. INPA, Manaus, pp 1–306

Hughell DA (1997) Optimización de inventarios forestales. Documento Técnico 59/1977. BOLFOR, Bolivia, pp 1–59

INRENA (2004) Planes de manejo en concesiones forestales con fines maderables. Lima-Perú, pp 1–107

IPCC (1996) Guideliness for national greenhouse gas inventaries: workbook and referent manual revised version 1996. UNEP, WMO, Module 1, 4, 5 p

Kiviste A, Álvarez GJG, Rojo AA, Ruiz GAD (2002) Funciones de crecimiento de aplicación en el ámbito forestal. Ministerio de Ciencia y Tecnología. Instituto Nacional de Investigación y Tecnología Agraria y alimentaria, Madrid, España, pp 1–190

Kvist LP, Nebel G (2000) Bosque de la llanura aluvial del Perú: ecosistemas, habitantes y uso de los recursos. Folia Amazon 10:5–56

Lamprecht H (1964) Ensayo sobre la estructura florística de la parte sur oriental del bosque universitario (El Caimital). Estado Barinas. Rev For Venez 6:77–106

Leano C (1998) Monitoreo de parcelas permanentes de medición en el bosque Chimanes Santa Cruz, Bolivia. Proyecto de Manejo Forestal Sostenible. BOLFOR. Documento Técnico 67/1998 p

Lewis SL, Phillips OL, Baker TR, Lloyd J, Malhi Y, Almeida S, Higuchi N, Laurance WF, Neill DA, Silva JNM, Terborgh J, Torres-Lezama A, Vasquez R, Martinez S, Brown J, Chave J, Kuebler C, Nunez-Vargas P, Vinceti B (2004) Concerted changes in tropic forest structure and dynamics: evidence from 50 South American long-term plots. Philos Trans R Soc B 359:421–436

Linera GW (1996) Crecimiento diamétrico de árboles caducifolios perennifolios del bosque mesófilo de montaña en alrededores de Xalapa. Madera Bosques 2:53–65

Loetsch F, Haller KE (1964) Forest inventor: volume I Statistof forest inventory and information from aerial photographs, 336 p

Louman B, Quiros D, Nilsson M (2001) Silvicultura de bosques latifoliados húmedos conénfasis en América Central. Serie Técnica. Turrialba, C.R.: CATIE. 46:1–265

Malleux OJ (1982) *Inventarios forestales en bosques* tropicales. UNA "La Molina", Lima-Perú, pp 1–414

Matteucci SD, Colma A (1982) *Metodología para el estudio de vegetación*. OEA, Washington, DC, pp 1–168

Melendez CJE (2000) Fitosociología de especies forestales en el arboretum del Centro de Investigación y Enseñanza Forestal (CIEFOR) Puerto Almendra. Iquitos-Perú. Tesis Ingeniero Forestal. Universidad Nacional de la Amazonia Peruana, Iquitos, pp 1–21

Moreno CE (2002) Métodos para medir la biodiversidad. Sociedad Entomológica Aragonesa (SEA), pp 1–83

Mostacedo B, Villegas Z, Pena M, Porter L, Licona JC, Alarcon A (2006) Fijación del carbono (biomasa aérea) en áreas de manejo forestal sujetas a diferentes intensidades de aprovechamiento: Implicancia a corto y a mediano plazo. Instituto Boliviano de Investigación Forestal (IBIF), Santa Cruz-Bolivia, pp 1–46

Namkoong G, Boyle T, Gregorius H, Joly H, Savolainen O, Ratnam W, Young A (1996) Testing criteria and indicators for assessing the sustainability of forest management: genetic criteria and indicators. CIFOR, Bogor 16680. CIFOR. Working Paper 10:1–15

Nebel G, Kvist LP, Vanclay J, Vidaurre H (2000a) Dinámica de los bosques de la llanura aluvial inundable de la amazonia peruana: efectos de las perturbaciones e implicancias para su manejo y conservación. Folia Amazon 11:65–90

Nebel G, Kvist LP, Vanclay JK, Cristensen H, Freitas L, Ruiz J (2000b) Dinámica de los bosques de la llanura aluvial inundable de la amazonia peruana: Estructura y composición florística del bosque de la llanura aluvial en la Amazonía peruana: I. el bosque alto: efectos de las perturbaciones e implicancias para su manejo y conservación. Folia Amazon 11:91–149

Nebel G, Gradsted J, Salazar VA (2000c) Depósito de detrito, biomasa y producción primaria neta en los bosques de la llanura aluvial inundable de la amazonia peruana. Folia Amazon 11:41–63

Nogueira ME, Nelson WB, Fearnside MF (2005) Wood density in dense forest in central amazonian. Brazil. For Ecol Manag 208:261–286

ONERN (1976) *Mapa Ecológico del Perú. Guía Descriptiva*. ONERN, Lima, Perú, pp 1–146

Pacheco GT, Tello RE, Burga AR (1993) Determinación de la estructura total de un bosque de Santa María río Nanay. Documento Técnico. UNAP. FIF. 17 p

Palacios WA (2004) Forest species communities in tropical rain forests of Ecuador. Lyonia 7:33–40

Panduro del AMY (1992) Diversidad arbórea de un bosque tipo Varillal, en Iquitos. Tesis Ingeniero Forestal, Universidad Nacional de la Amazonia Peruana, Iquitos-Perú, pp 1–105

Paredes AG, Kauffman S, Kalliola R (1998) Suelos aluviales recientes de la zona Iquitos-Nauta. 114: 231–251p. In: Kalliola RM, Flores Paitan S (eds) Geología y desarrollo amazónico: Estudio integrado de la zona de Iquitos, Perú. Annales Universitatis turkuensis Ser A II. Turun Yliopisto, Turku

Parolin P (2002) Bosques inundados en la amazonia central: su aprovechamiento actual y potencial. Universidad Nacional Agraria la Molina. Lima-Perú. Ecología Aplicada 1:111–114

Perea ZVM (1995) Caracterización por el método de las distancias del cuadrante errante de la vegetación arbórea de un bosque tipo Varillal de la zona de Puerto Almendras Iquitos-Perú. Tesis Ingeniero Forestal, Universidad Nacional de la Amazonia Peruana, Iquitos, 77 p

Peters CM (1994) Aprovechamiento sostenible de recursos no maderables en bosque húmedo tropical: Un manual ecológico. Instituto de Botánica Económica. Jardín Botánico de Nueva York, Bronx, NY 63 p

Ramirez GA (1999) Ecología aplicada: Diseño y Análisis Estadístico. Fundación Universitaria de Bogotá Jorge Tadeo Lozano. Colección Ecología. Santa Fé de Bogotá, pp 1–325

Rasanen M, Linna A, Irion G, Hernani LR, Vargas HR, Wesslingh F (1998) La geología y geoformas de la zona de Iquitos. 114:50–139

Richards PW (1969) Speciation in the tropical rain forest and the concept of the niche. Biol J Linn Soc 1:149–153

Rios ZR, Burga AR (2005) Tamaño óptimo de la unidad muestral para inventarios forestales en el sector Caballo cocha- Palo seco-Buen suceso, Provincia Mariscal Ramón Castilla, Loreto-Perú. Tesis magíster, Universidad Nacional de la Amazonia Peruana, Iquitos, pp 1–168

Ruokolainen K, Tuomisto H (1998) Vegetación natural de la selva de Iquitos. 114: 253–365 p. In: Kalliola RM, Flores Paitan S (eds) Geología y desarrollo amazónico: Estudio integrado de la zona de Iquitos, Perú. Annales Universitatis turkuensis Ser A II. Sociedad espanola de ciencias forestales (2005) *Diccionario forestal*. Ediciones Mundi Prensa, Madrid-Barcelona-México, pp 1–1314

Synnott TJ (1991) Manual de procedimientos de parcelas permanentes para bosque húmedo tropical. Instituto Tecnológico de Costa Rica. Departamento de Ingeniería Forestal. Cartago-Costa Rica. Serie de apoyo académico. 12:1–103

Tello ER, Burga R, Sevillano RH (1992) Desarrollo de *Ormosia macrocalix* ducke (huayruro negro) y de *Ormosia coccinea* Jack. (huayruro colorado), en el CIEFOR-Iquitos. Conocimiento UNAP. 2:57–66

Uslar YV (2003) Composición, estructura y dinámica de un bosque seco semideciduo en santa cruz, Bolivia. Documento Técnico. BOLFOR. 114:1–28

Vallejo JMI, Londono VAC, Lopez CR, Galeano G, Álvarez DE, Devia AW (2005) *Establecimiento de parcelas Permanentes en bosques de Colombia*. VOLUMEN I. Serie: Métodos para estudios ecológicos a largo plazo. Instituto de Investigación de Recursos Biológicos Alexander Von Humboldt, Bogotá D.C., Colombia, pp 1–306

Vicen CM, Vicen AC (1996) Diccionario de términos ecológicos. Editorial Paraninfo, S.A., Madrid, España, pp 1–169

Wittmann F, Anhuf D, Junk WJ (2002) Tree species distribution and community structure of Central Amazonian várzea forests by remote sensing techniques. J Trop Ecol 18:805–820

Worbes M, Klinge H, Revilla JD, Martius C (1992) On the dynamics, floristic subdivision and geographical distribution of várzea forests in Central Amazonia. J Veg Sci 3:553–564

Zeide B (1993) Analysis of growth equations. For Sci 39:594–616

Part VII
Human Impacts and Management

Chapter 15
Twenty-Five Years of Restoration of an Igapó Forest in Central Amazonia, Brazil

Fabio Rubio Scarano, Reinaldo Luiz Bozelli, André Tavares Corrêa Dias, Arcilan Assireu, Danielle Justino Capossoli, Francisco de Assis Esteves, Marcos Paulo Figueiredo-Barros, Maria Fernanda Quintela Souza Nunes, Fabio Roland, Jerônimo Boelsums Barreto Sansevero, Pedro Henrique Medeiros Rajão, André Reis, and Luiz Roberto Zamith

F. R. Scarano (✉)
Fundação Brasileira para o Desenvolvimento Sustentável, Departamento de Ecologia, Universidade Federal do Rio de Janeiro, Rio de Janeiro, RJ, Brazil

Departamento de Biologia Geral, Universidade Federal Fluminense, Niterói, RJ, Brazil
e-mail: fscarano@fbds.org.br

R. L. Bozelli
Universidade Federal do Rio de Janeiro, Rio de Janeiro, RJ, Brazil

A. T. C. Dias · M. F. Q. S. Nunes
Departamento de Ecologia, Universidade Federal do Rio de Janeiro, Rio de Janeiro, RJ, Brazil

A. Assireu · A. Reis
Instituto de Recursos Naturais, Universidade Federal de Itajubá, Itajubá, Minas Gerais, Brazil

D. J. Capossoli
Deloitte, Rio de Janeiro, RJ, Brazil

F. de Assis Esteves · M. P. Figueiredo-Barros
Núcleo de Pesquisas Ecológicas de Macaé, Universidade Federal do Rio de Janeiro, Rio de Janeiro, RJ, Brazil

F. Roland
Departamento de Biologia, Universidade Federal de Juiz de Fora, Juiz de Fora, MG, Brazil

J. B. B. Sansevero
Departamento de Ciências Ambientais, Universidade Federal Rural do Rio de Janeiro, Seropédica, RJ, Brazil

P. H. M. Rajão
Departamento de Ecologia e Evolução, Universidade Estadual do Rio de Janeiro, Rio de Janeiro, RJ, Brazil

L. R. Zamith
Departamento de Biologia Geral, Universidade Federal Fluminense, Niterói, RJ, Brazil

R. W. Myster (ed.), *Igapó (Black-water flooded forests) of the Amazon Basin*,
https://doi.org/10.1007/978-3-319-90122-0_15

15.1 Introduction

Bauxite mining in the *terra-firme* forest of Porto Trombetas (state of Pará, Central Amazonia, Brazil) produces tailings that were continuously discharged (from 1979 until 1989) into Lake Batata, which is located as a floodplain lake of the Trombetas River watershed. When discharge was halted, a bauxite tailing layer (2–5 m deep) buried ca. 600 ha of the lake (ca. 30% of the area of the lake during high-water period) and extensive marginal igapó vegetation. The company responsible for this impact, Mineração Rio do Norte, started in 1987 a complex limno-ecological monitoring program to mitigate the impact. As a prerequisite for the ecological recovery of the lake, a program to restore[1] the marginal igapó vegetation was launched (Bozelli et al. 2000). This experience is to our knowledge the only large-scale (both spatial and temporal sense) attempt to restore an igapó landscape. In this chapter, we review papers and theses that dealt with this specific challenge. Our main goal is to share lessons learned in both practice and theory.

Given the growing importance of landscape restoration in environmental diplomacy, we believe that reviewing this experience is timely. One of the Aichi Targets for 2020 of the United Nations (UN) Convention on Biological Diversity is to restore and rehabilitate 15% of degraded lands on the planet (Mittermeier et al. 2010). The Paris Agreement of the UN Framework Convention on Climate Change has country-level commitments that add up to a pledge to have 150 million hectares of degraded lands under restoration by 2020 and 350 million hectares by 2030 (Brancalion et al. 2016a). These targets converge to the UN Sustainable Development Goals, where restoration can play a role in helping achieve targets related to climate action, biodiversity conservation, and human well-being (Wood and DeClerck 2015).

Locally, Brazil has committed to restore 12 million hectares of degraded land by 2030 to the Paris Agreement. This commitment is backed up by a set of national policies. For instance, if farmers comply to the Native Vegetation Protection Law (that defines the proportion of land inside private properties that must be protected or restored), this climate goal will possibly be accomplished (Brancalion et al. 2016b; Scarano 2017; Soares-Filho et al. 2014). Furthermore, compliance means that all water bodies will have their terrestrial margins revegetated, which is particularly relevant for Amazonian flooded forests such as the igapó, despite criticisms about the lack of specificity of this legislation to Amazonian floodplains (Sousa Jr et al. 2011).

[1] We call the process undertaken here "ecological restoration," since the main goal of the interventions described next is to restore the ecological structure and function of the land-freshwater system in the long term. Although terms such as rehabilitation, afforestation, and recuperation can be applied to short- and midterm outcomes of some of the interventions described here, we will not use them, since the design of this process has a long-term focus.

15.2 The Study Site

The study site is located at the west margin of Lake Batata (1°25′–1°35' S and 56°15′–56°25' W), in the district of Porto Trombetas, municipality of Oriximiná, state of Pará, northern Brazil (Fig. 15.1). Local climate is *Am* in the *Köppen* system, mean annual rainfall in Porto Trombetas is ca. 2200 mm, and mean maximum and minimum temperatures are ca. 35 °C and 20 °C, respectively (Parrotta and Knowles 1999). Lake Batata is ca. 18 km long and ca. 3 km wide at its widest point. The lake lies on the right bank of the Trombetas River, and during the flooding season (from May to October), lake and river become a single water body since they are separated only by a narrow (50–120 m), low (3–4 m) stretch of land along most of its extension. Clear-water rivers, such as the Trombetas River, typically have a high transparency, low quantity of organic and inorganic suspended material, pH from low to neutral, and low nutrient availability. Forests seasonally flooded by clear-water rivers are locally called *igapó* (see Prance 1979 for terminology regarding Amazonian flooded forests). Local vegetation is subjected to 4–8 months of flooding per year, depending on topography.

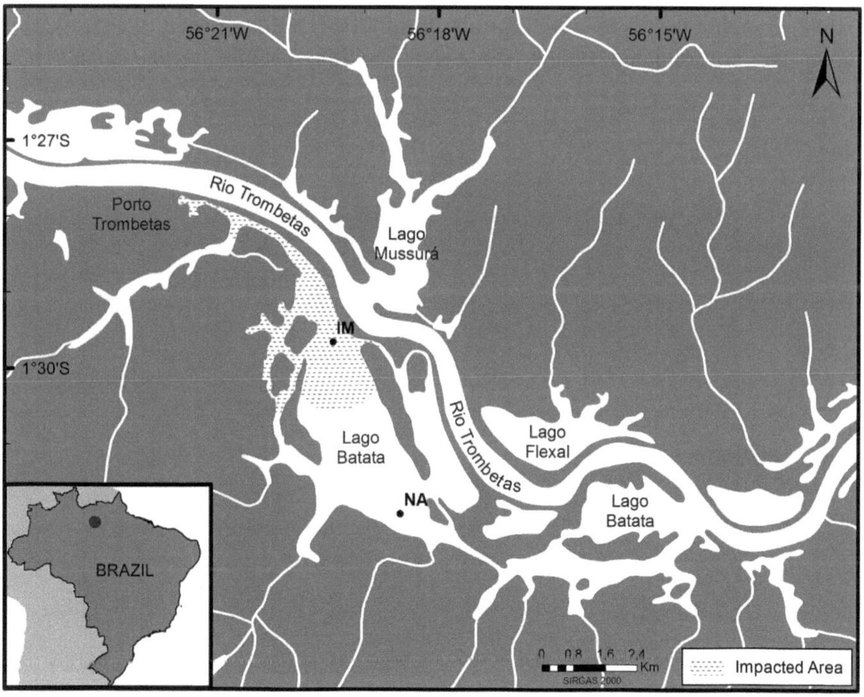

Fig. 15.1 Location of the study site. IM = area that received direct impact of the bauxite tailings; NA = natural portion of the lake, not directly affected by the impact

Washing bauxite produces tailings which consist of red, muddy clay composed primarily of iron and aluminum oxides (Lapa and Cardoso 1988). The bauxite tailing substrate consists of 75% clay, 21% silt, 3% fine sand, and 1% coarse sand soil. It differs from nonimpacted igapó soil in the proportions of clay (49%), silt (37%), and fine sand (13%). Frequent and prolonged exposure to full sunlight during the dry season has led to dehydration and consolidation of the bauxite tailings in most of the impacted area.

15.3 The Past and the Present

This new bauxite tailing substrate was located both on top of what used to be igapó forest (Fig. 15.2a) and on top of what used to be lake – even during the lowest water level – but was buried (Fig. 15.2b). The magnitude of the impact raised the question whether restoration would ever be viable. This concern was understandable because, to the long flood periods and low nutrient availability of the igapó, the impact imposed an extra burden to the ecosystem by creating this new sterile substrate – which at first was muddy and after sun exposure during dry season would become solid blocks (Dias et al. 2012). However, it soon became apparent that many water-dispersed or fish-dispersed seeds brought by annual flooding (see Barbieri et al. 2000; Mannheimer et al. 2003; Parolin et al. 2013) were capable of germinating upon the consolidated tailing substrate. Seedling establishment on this new

Fig. 15.2 Lake Batata region and neighboring igapó vegetation: (**a**) impacted igapó; (**b**) buried portion of the lake; (**c**) one of the restoration sites with initial growth of seedlings; (**d**) restoration site at more advanced stage of tree growth

substrate was dependent on local topography and consequent duration of flooding period but also on water currents and occasional litter traps that are formed and retain seeds and organic matter. As a result, there was a mosaic of areas undergoing natural regeneration and areas where intervention by man would be needed to introduce any vegetation cover.

15.3.1 Natural Regeneration

Current literature demonstrates great interest on the potential of tropical forest to spontaneously regenerate on abandoned or degraded land (Chazdon et al. 2016; Poorter et al. 2016). It can reduce the often high costs of producing and planting native trees to restore areas, and there is already evidence of the large scale that this can take place, for instance, in the Brazilian Atlantic rain forest (Rezende et al. 2015) and even in the Amazon (Brancalion et al. 2016c; Chazdon 2017). Given the scale of the impact, the level and frequency of inundation, and the relationship with the water currents in the igapó at the margins of Lake Batata, the fact that natural regeneration took place upon the sterile and compacted substrate of the bauxite tailings was indeed a surprise.

Ten years after the cessation of tailing entry into the lake, saplings, juveniles, and mature individuals of 51 plant species were growing in 1 ha of the impacted igapó between the river and the lake compared to 61 plant species in 1 ha of the adjacent natural forest (Barbieri 1995). This high initial reestablishment of species richness was then attributed mostly to transportation of hydrochoric seeds by the seasonal floodwater followed by germination upon the sterile tailing substrate under full sunlight (Scarano et al. 1998). Interestingly, however, these studies have also shown that despite the high number of tree species established, there was little floristic similarity between control and postimpact site: only two (*Buchenavia oxycarpa* (Mart.) Eichler and *Genipa americana* L.) out of ten dominant species were the same for impacted and nonimpacted sites.

In addition to water dispersal of seeds, fish dispersal was also relevant for regeneration success in many impacted areas. In the silted-up portion of the lake, fish species diversity has declined, and *Auchenipteris longimanus* became the most abundant species (Caramaschi et al. 2000). This auchenipterid species can attain 25 cm standard length and is widely distributed throughout the Amazon basin (Burgess 1989). Mannheimer et al. (2003) undertook germination tests with seeds found in the stomach and the intestine of this fish species and demonstrated its important role in seed dispersal. We then speculate that some of the differences found in dominance between native igapó and impacted sites could be at least partially due to the fruit diet of this fish and its dispersal role.

In another study, Dias et al. (2014) examined how natural regeneration would take place in a lower topography area, which used to be lake and that is now subjected to 7–8 months of flooding per year. The number of species found in this plant community increased from 15 in 2003 to 32 woody species (27 trees, 1 shrub, and

4 vines) in 2005. This was mostly due to a very marked increase in the number of regenerants in 2004 in relation to 2003 (fivefold) and in 2005 in relation to 2004 (fourfold). However, the number of recruited species remained rather constant (between 12 and 13). This is apparently more related to slow growth of these newcoming individuals than to mortality rates (only 3.5% for recruits as a whole over the 2 study years). The 32 species recorded in the 2005 survey belonged to ten different families, and the Fabaceae was particularly species rich (seven species, i.e., ca. 21% of the species total). There were no changes in the dominance structure of the community over the years, since the dominant species (those that concentrated 75% of the phytosociological importance value) were the same (*B. oxycarpa*, *G. americana*, and *Dalbergia inundata* Spruce ex Benth.) that also appeared as dominants in the previously mentioned studies of Barbieri (1995) and Scarano et al. (1998) for an impacted igapó site. However, while the site studied by these two papers did not need any human intervention, the site studied by Dias et al. (2014) does need tree planting to gain vegetation cover. After approximately 10 years since the substrate consolidated, only 2.12% of the area was covered by vegetation (as measured by canopy cover), which is a clear indication that human intervention was needed.

15.3.2 Interventions: Tree Planting

Planting of saplings was an annual practice in the impacted site between 1993 and 2005, in areas where natural regeneration did not take place (Fig. 15.2c, d). In total, 94.6 ha of igapó and buried lake have received planting of native tree species. Nearly 2/3 of the impacted area has now some degree of vegetation cover, and the other 1/3 remains as lake with impacted sediment, even when the water level is the lowest. Around 413,000 individuals of mostly typical igapó species were planted along lines spaced from one another by 1 m × 1 m, 1.3 m × 1.3 m, and 1.5 m × 1.5 m. Species composition and frequency varied between years due to sapling availability at the nursery, which is dependent on the amount and variety of seeds that local residents have to sell to the company and on the different responses of the species to the silvicultural practices in the nursery. Since nearly all the impacted site subject to a dry period during the flood pulse by now has been planted or has undergone natural regeneration, only complementary interventions to replace dead individuals in larger gaps are done since 2008, amounting to additional 180,000 planted seedlings.

This large-scale restoration effort was only possible due to a very effective operation undertaken by the company's nursery established in 1979 to produce seedlings to restore degraded areas of terra-firme forest where bauxite mining takes place – a necessary requirement for the compliance with environmental legislation (Parrotta and Knowles 2001). Since the early 1990s, the operations were expanded for the nursery to produce also seedlings of igapó species to attend to the demands of restoration of Lake Batata's impacted igapó. Seedlings of ca. 50 species of igapó have been produced in the nursery, between 1993 and 2017.

Capossoli (2013) has made a phytosociological and structural survey of planted trees that occurred in 5-year-old, 10-year-old, and 15-year-old planted sites and compared to controls (native, nonimpacted igapó vegetation neighboring the impacted sites). She used quadrats (out of regularly spaced transects) that totaled 0.1 ha on each of 12 sampling sites: 3 samples on each of the stands with different ages (5, 10, and 15 years old) and 3 on the control sites. In total, 43 species were found growing in the planted sites, while 54 were found in the control sites. This study also assessed natural regeneration underneath grown trees in the planted sites and control sites: 43 species were found in planted sites and 39 in control sites. In terms of species composition, there was a moderate to low similarity (sensu Müller-Dombois and Ellenberg 1974) between planted sites and control sites, which declined from stands with 5–10 years old (40% similarity) to stands with 15 years old (27%). However, species richness per unit area, diversity, and forest structure showed a stronger similarity between natural and planted forests (Fig. 15.3). The differences found between planted sites and control sites suggest that additional practices, such as improved species selection for replacement of dead individuals and litter and seed addition to promote regeneration and growth, could be relevant next steps.

15.3.3 Interventions: Selection of Framework Species

In the early years of the restoration operations, a large diversity of plant species was introduced on the impacted sites. This was possible due to the large number of species produced in the company's nursery. It is also the outcome of a trial-and-error attitude in face of an unknown but clearly great challenge, plus the requirements of Brazilian legislation at the time that demanded restoration efforts with a high diversity of species in response to environmental impact caused by mining and other types of enterprises. Moreover, our knowledge on the phenology and biology of igapó's tree species was too poor, at that time, to permit secure decisions about the right species. The results of Capossoli's (2013) thesis, however, showed that even native igapó was not as species rich per unit area as some of the early restoration stands (see Fig. 15.3g), and as a result, many unfit planted species died along the way. So, we started to aim for a selection of framework species with higher success rates (in terms of survival in the nursery and survival and growth in the field).

By planting a restricted number of native species, it is possible to reestablish a multilayered canopy structure and restore nutrient and carbon cycles. This new basic forest framework will then attract wildlife that can bring seeds (in addition to the annual flooding cycle) and slowly increase forest diversity and functionality. While this method has been successfully used to restore tropical forests in different parts of the world (Knowles and Parrotta 1995; Tucker and Murphy 1997; Wydhayagarn et al. 2009), a critical step is the selection of the appropriate framework species. The selection of framework species usually follows a set of criteria, namely, (i) high survival and growth of saplings planted in the degraded area, (ii)

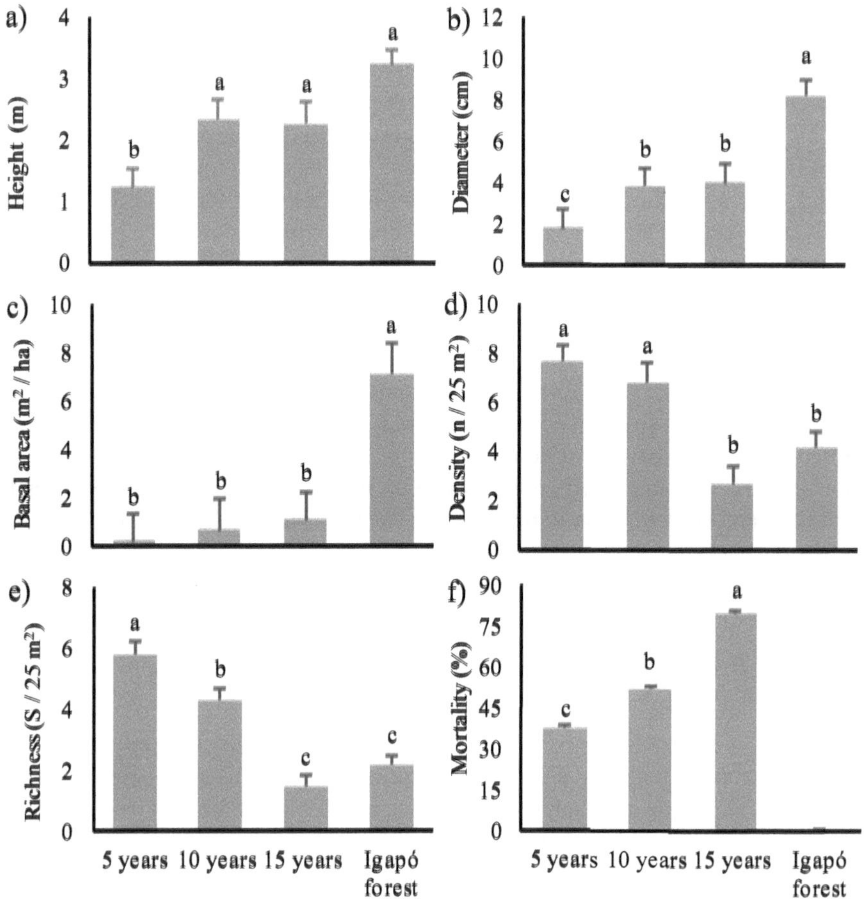

Fig. 15.3 Comparison of mean values of structural and diversity attributes measured for nonimpacted native igapó forests and planted igapó forests in impacted areas at the margins of Lake Batata. (**a**) Height (m); (**b**) diameter (cm); (**c**) basal area (m^2/ha); (**d**) plant density (ind/25m^2); (**e**) species richness (sp/25m^2); (**f**) mortality (%). Vertical bars indicate confidence intervals ($p < 0.05$). Means followed by the same letter do not differ statistically at 5% probability

easy propagation by seeds under nursery, (iii) crown architecture to shade out weeds or improve understory environmental conditions, and (iv) resistance to specific field harsh conditions (e.g., in this case, flooding and reduced penetrability of the substrate). The evaluation of these criteria is, however, extremely time- and labor-consuming as it requires experimental stands of planted vegetation to follow species performance in the field (Elliott et al. 2003).

Capossoli (2013), for instance, selected eight framework species for our study site based on functional-type analysis (Table 15.1). In Dias et al. (2014), however, we questioned Capossoli's selection when we found a lack of relationship of species survival and growth between planted and naturally recruiting plants of the same

Table 15.1 Matrix of the mean attributes of the optimal functional types identified at planted and native igapó forests by Lake Batata, according to a functional analysis undertaken at the software SYNCSA for Windows – Version 2.6.9 (©V.Pillar 1992–2010). This analysis searches for the group of plant traits and the number of plant functional types (PFTs) that maximize the congruence with environmental variables (Pillar and Sosinski 2003). PFTs were defined by UPGMA based on Gower similarity index after fuzzy weighting by the functional traits

			h	D	Lw	Lh	S	M	An	Zo	Hy	Ss	Ft	Nf
Optimal functional types	1	*Leptolobium nitens*	18.5	25	0	0	0	0	0.5	0	0.5	0.5	1	1
		Dalbergia inundata												
	2	*Couepia paraensis*	28.4	47.6	24.4	10.7	263.8	0.009	0	1	0	8	0.8	0.2
		Couepia paraensis subsp. *glaucescens*												
		Eschweilera ovata												
		Genipa americana												
		Macrolobium acacifolium (Benth)												
	3	*Swartzia polyphylla*	30	60	73.9	31.7	681.6	0.001	0	0	1	25	0	1

Legend: *h* maximum height (m), *D* maximum diameter (cm), *Lf* leaf width (cm), *Af* leaf height (cm), *S* specific leaf area ($cm^2\ g^{-1}$), *M* leaf specific mass ($g^{-1}\ cm^2$), *An* anemochory, *Zo* zoocory, *H* hydrochory, *Ss* seed size (cm^2), *Ft* flood tolerance, and *Nf* capacity for nitrogen fixation. An, Zo, Hy, Ft, and Nf are dichotomic data (0 = absence of the attribute; 1 = presence of the attribute)

selected species. Thus, plant surveys cannot replace experimental plantations for this purpose. Abundance during natural colonization of the bauxite tailings was also not a good indicator for performance of planted species. If, on the one hand, the abundant legumes *D. inundata* and *Leptolobium nitens* Vogel performed well as planted species, other abundant species either did not show good results after planting (e.g., *Couepia paraensis* (Mart. & Zucc.) Benth.) or were difficult to get seeds and cultivate in the nursery (e.g., *B. oxycarpa*, which can only be used infrequently). Moreover, *Myrciaria dubia* (Kunth) McVaugh showed some of the best results in terms of survival and growth in planted conditions – this species can be found in nearby nonimpacted areas but was never found naturally recruiting on the bauxite tailings. Currently, *D. inundata* is the main species used for replacement in the plantation sites, and *L. nitens* is the main species introduced via seed addition. However, we also use in smaller numbers and in higher areas not prone to longer-term flooding species such as *Burdachia prismatocarpa* Mart. ex A. Juss., *Eschweilera ovata* (Cambess.) Mart. ex Miers, *Myrciaria dubia*, *Panopsis rubescens* (Pohl) Rusby, and *Couepia paraensis* (Mart. & Zucc.) Benth.

15.3.4 Interventions: Seed and Litter Addition

One important characteristic of this restoration enterprise is that, along all years of planting, no chemical fertilizer was ever used. The main reason was our concern of not causing eutrophication of water bodies since the igapó is flood-prone. Given the characteristic slow growth, especially of some of the earlier planted stands, we decided to test whether litter addition could improve plant growth, and in particular canopy cover, and if seed addition results in improved regeneration in planted stands. Thus, we used a factorial experiment to test whether litter and seed addition could be used to increase the efficiency of ecological restoration on the 1993 planted stand (Dias et al. 2012). Our results clearly showed that the addition of litter collected from pristine igapó areas increased plant growth, as well as density and species richness of regenerants. The increase in individual plant growth was echoed at the community level by higher leaf area index values on litter addition plots compared to controls. Litter addition can enhance reaccumulation of nutrient pools during successional development, which has been proposed as an important feature to ensure self-sustainability of areas under restoration (Dias et al. 2012). These learnings have been transformed into regular practice of litter addition in the areas under restoration and now total some 7.0 hectares.

However, the success of the seed addition treatment depended on the species used. Of the seven sown species, only one of our framework species, *L. nitens* (a nitrogen-fixing legume), showed high establishment. The introduction of nitrogen-fixing species is also expected to build up the nitrogen pool in the system as has been reported for restoration programs in non-inundated forests. Since then, seed sowing for other species, such as *Ormosia excelsa* Benth. and *Parkia pendula* (Willd.) Benth. ex Walp., has also reached apparent success, although establishment seems slower than that found for *L. nitens*.

15.4 The Future

The study of Capossoli (2013) – which investigated floristic composition and plant functional types in planted forests in a chronosequence (5, 10, and 15 years old) compared to native mature igapó forest – prompted a modeling exercise on future successional trajectories of the planted igapó forest based on Markov chains. This approach has been used to model succession and estimate how long communities take to become restored as compared with a reference ecosystem (e.g., Balzter 2000; Peng et al. 2010; Chuang et al. 2011). The results showed a tendency of convergence of the planted forests with the native igapó; however, the modeling exercise indicates that the existing planted forest will not reach a state similar to that of native igapós even after 75 years, since their completion of 15 years. In other words, since a stand is first introduced, according to usual procedures, it takes longer than 90 years until the forest resembles the nonimpacted ones in species richness, composition, and structure.

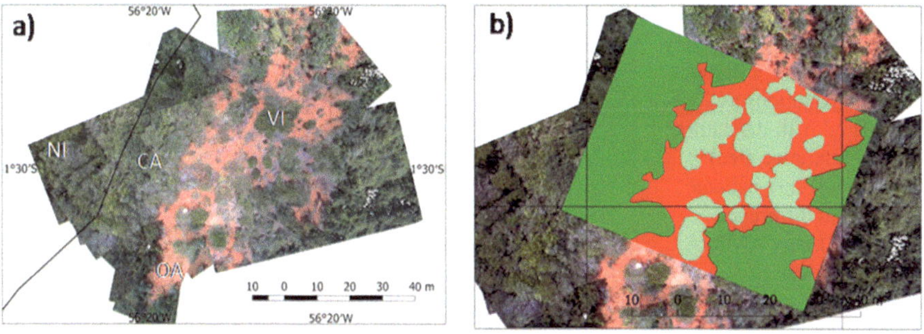

Fig. 15.4 Aerial images of one plantation at Lake Batata. Panel (**a**) shows the different landscape elements: natural igapó (NI), closed areas (CA), open areas (OA), and vegetation islands (VI). Panel (**b**) shows an example of the image treatment for quantifying the area of each landscape element within plantations

However, on a recent survey to examine vegetation recovery in this area, we found a large variation within plantation stands. Survival and growth of planted saplings show a strong spatial heterogeneity. We were able to identify three typical landscape elements within the planted stands: (i) closed areas, generally in higher topography in the interface with natural igapó areas; (ii) open areas, where saplings presented high mortality and low growth rate, leaving much bare ground; and (iii) vegetation islands, where nucleation process (as a possible result of topographic heterogeneity, influence of remaining large trees, management interventions with litter and seed additions, or a combination of all these factors) promoted high growth and survival of saplings forming dense vegetation clumps (Fig. 15.4a). For quantifying the recovery of these different landscape elements, we used vegetation cover as an indicator of vegetation structure and regenerant abundance and richness, as indicators of regeneration process. These indicators were recently suggested by the Secretary of State for the Environment of São Paulo as the first official governmental instrument for restoration monitoring in Brazil (Chaves et al. 2015). Our survey corroborated our visual expectation, showing marked differences in recovery among the landscape elements. Closed areas and vegetation islands showed similar vegetation cover and even higher values of regenerant abundance and richness as compared to natural igapó. On the other hand, open areas showed very low values of vegetation cover but still similar values of regenerant abundance and richness as compared to the reference areas (Fig. 15.5).

These results suggest that the speed of recovery will vary largely between and within planted stands, depending on topography, species introduced, nucleation process, and additional interventions (e.g., litter addition or seed sowing). We now trust that Capossoli's estimate of 75 years for recovery is probably maximum, since our analysis of spatial heterogeneity suggests that some areas can be considered as already recovered. Future evaluations of the entire impacted area under restoration should take into account this small spatial scale heterogeneity. We are now surveying aerial imagery, which will allow a quantification of the area of the different

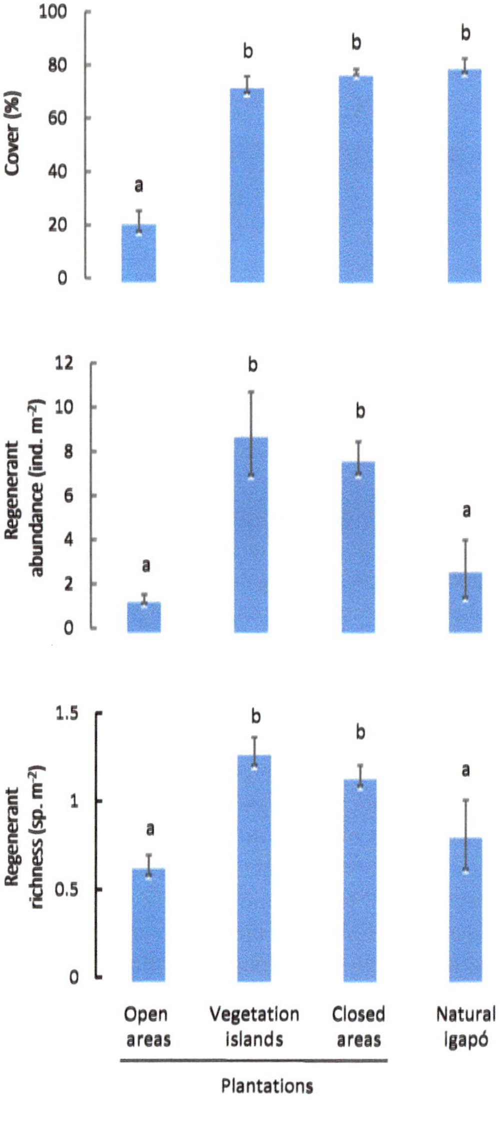

Fig. 15.5 Indicators used for evaluating the recovery of each landscape element within plantations as compared with the natural igapó (reference area). Different letters indicate significant difference ($P < 0.05$) after ANOVA and Tukey test

landscape elements in each planted stand (see Fig. 15.5b for an example). These numbers, together with in situ measurements of recovery indicators, will allow us to produce a more accurate estimate of the success of the restoration project as a whole and better quantify the results of different interventions. In parallel, rehabilitated sites (once areas of the lake) seem to provide similar ecological functions to aquatic communities such as fish (Erica Caramaschi, personal communication) and zooplankton (Josué 2017) when compared to natural areas of igapó in high-water period.

Clearly, not only the magnitude of the impact but also the ecological properties of the igapó system (slow growing, short growth season, nutrient poor) explain this relatively slow pace of recovery (although with much small-scale variation) despite all efforts undertaken. Continuous monitoring of this system will be required to calibrate the models we have developed and to inform which further actions should be taken to speed up the green coverage of the impacted sites.

15.5 Final Remarks

The main lessons learned from our 25-year experience of restoring a vast area impacted by a bauxite tailing spill over Lake Batata and its marginal igapó vegetation in Central Amazonia are the following:

1. Despite the sterility and difficult penetrability of the consolidated tailing substrate, igapó plants have been capable to establish and grow spontaneously in these areas, reaching reproductive stage and completing full life cycles.
2. Wherever natural regeneration was not possible, either due to topographic reasons or to water movements, human intervention by planting seedlings was also viable. In areas subjected to longer-term flooding (>8–9 months a year), planting is either physically not possible or, whenever possible, saplings die.
3. Natural regeneration and planting, altogether, were responsible for returning species richness and diversity to impacted areas, but floristic similarity with native, nonimpacted sites was moderate to low. This established vegetation produces shade, litter, and organic matter and thus creates new habitats and organic substrate for other species to establish naturally.
4. Recuperation is variable depending on site characteristics (annual length of flooding, species introduced, etc.) but often slow, and in worst-case scenarios, new forests may take more than 75 years to converge to similarity with native nonimpacted igapó. Thus, additional interventions are necessary to further forest growth, structure, and diversity.
5. Despite the challenges posed by the scale of this enterprise, relevant actions to speed up successional trajectory are feasible and necessary and include continuous monitoring and evaluation, litter and seed addition, and selection of framework species for replacements.
6. The processes of natural regeneration and human intervention are influenced decisively by the annual flood pulse that can provide matter and energy subsidy to the area but also impose limits as a function of the time of flooding, for example, for plants.

Acknowledgments We thank Mineração Rio do Norte for continuously supporting the studies reviewed in this chapter. We also thank the numerous students that collaborated in data collection; parabotanists that taught us a great deal about the local ecosystem and named the various plant species investigated in these studies (Delmo Ferreira, Heraldo Ferreira dos Santos, Pedro Ferreira);

and other researchers for on-site visits, collaboration of previous papers, and exchange of ideas (Erica Caramaschi, João Leal, Ricardo Barbieri, and their respective research teams). CNPq (Brazilian Scientific Council) and CAPES (Brazilian Council for Graduate Education) have provided research grants to all authors during these 25 years.

References

Balzter H (2000) Markov chain models for vegetation dynamics. Ecol Model 126:139–154

Barbieri R (1995) Colonização vegetal em habitats formados pela sedimentação do rejeito de bauxita em um lago amazônico (Lago Batata, PA). Ph.D. Thesis. Universidade Federal de São Carlos, São Paulo, Brasil, 104 pp.

Barbieri R, Esteves FA, Soares E (2000) Colonização por vegetação de igapó de novos habitats formados pela sedimentação do rejeito de bauxita. In: Bozelli RL, Esteves FA, Roland F (eds) Lago Batata – Impacto e Recuperação de um Ecossistema Amazônico. Universidade Federal do Rio de Janeiro and Sociedade Brasileira de Limnologia, Rio de Janeiro

Bozelli RL, Esteves FA, Roland F, (2000) Lago Batata – Impacto e Recuperação de um Ecossistema Amazônico. Universidade Federal do Rio de Janeiro and Sociedade Brasileira de Limnologia, Rio de Janeiro

Brancalion PHS, Pinto SR, Pugliese L, Padovezi A, Rodrigues RR, Calmon M, Carrascosa H, Castro P, Mesquita B (2016a) Governance innovations from a multi-stakeholder coalition to implement large-scale forest restoration in Brazil. World Dev Perspect 3:15–17

Brancalion PHS, Garcia LC, Loyola R, Rodrigues RR, Pillar VD, Lewinsohn TM (2016b) A critical analysis of the native vegetation protection law of Brazil (2012): updates and ongoing initiatives. Nat Conservação 14(Supplement):1–15

Brancalion PHS, Schweizer D, Gaudare U, Mangueira J, Lamonato F, Farah F, Nave A, Rodrigues RR (2016c) Balancing economic costs and ecological outcomes of passive and active restoration in agricultural landscapes: the case of Brazil. Biotropica 48:856–867

Burgess WE (1989) An atlas of freshwater and marine catfishes. A preliminary survey of the Siluriformes. T.F.H. Publications, Neptune City, p 783

Capossoli D (2013) Restauração Ecológica: Ligando Prática e Teoria. Ph.D. Thesis. Escola Nacional de Botânica Tropical, Instituto de Pesquisas Jardim Botânico do Rio de Janeiro, Rio de Janeiro, 211 pp

Caramaschi EP, Halboth DA, Mannheimer S (2000) Ictiofauna. In: Bozelli R, Esteves FA, Roland F (eds) Lago Batata: Impacto e Recuperação de um Ecossistema Amazônico. Universidade Federal do Rio de Janeiro and Sociedade Brasileira de Limnologia, Rio de Janeiro, pp 155–177

Chaves RB, Durigan G, Brancalion PHS, Aronson J (2015) On the need of legal frameworks for assessing restoration projects success: new perspectives from São Paulo state (Brazil). Restor Ecol 23:754–759

Chazdon RL (2017) Landscape restoration, natural regeneration, and the forests of the future. Ann Mo Bot Gard 102:251–257

Chazdon RL, Broadbent EN, Rozendaal DMA, Bongers F, Zambrano AMA, Aide TM, Balvanera P, Becknell JM, Boukili V, Brancalion PHS, Craven D, Almeida-Cortez JS, Cabral GAL, de Jong B, Denslow JS, Dent DH, DeWalt SJ, Dupuy JM, Durán SM, Espírito-Santo MM, Fandino MC, César RG, Hall JS, Hernández-Stefanoni JL, Jakovac CC, Junqueira AB, Kennard D, Letcher SG, Lohbeck M, Martínez-Ramos M, Massoca P, Meave JA, Mesquita R, Mora F, Muñoz R, Muscarella R, Nunes YRF, Ochoa-Gaona S, Orihuela-Belmonte E, Peña-Claros M, Pérez-García EA, Piotto D, Powers JS, Rodríguez-Velazquez J, Romero-Pérez IE, Ruíz J, Saldarriaga JG, Sanchez-Azofeifa A, Schwartz NB, Steininger MK, Swenson NG, Uriarte M, van Breugel M, van der Wal H, Veloso MDM, Vester H, Vieira ICG, Bentos TV, Williamson GB, Poorter L (2016) Carbon sequestration potential of second-growth forest regeneration in the Latin American tropics. Sci Adv 2:e1501639. https://doi.org/10.1126/sciadv.1501639

Chuang C-W, Lin C-Y, Chien C-H, Chou W-C (2011) Application of Markov-chain model for vegetation restoration assessment at landslide areas caused by a catastrophic earthquake in Central Taiwan. Ecol Model 222:835–845

de Rezende CL, Uezu A, Scarano FR, Araujo DSD (2015) Atlantic Forest spontaneous regeneration at landscape scale. Biodivers Conserv 24:2255–2272

Dias ATC, Bozelli RL, Darigo RM, Esteves FA, Santos HF, Figueiredo-Barros MP, Nunes MFQS, Roland F, Zamith LR, Scarano FR (2012) Rehabilitation of a bauxite tailing substrate in Central Amazonia: the effect of litter and seed addition on flood-prone forest restoration. Restor Ecol 20:483–489

Dias ATC, Bozelli RL, Zamith LR, Esteves FA, Ferreira P, Scarano FR (2014) Limited relevance of studying colonization in degraded areas for selecting framework species for ecosystem restoration. Nat Conservação 12:134–137

Elliott S, Navakitbumrung P, Kuarak C, Zangkum S, Anusarnsunthorn V, Blakesley D (2003) Selecting framework tree species for restoring seasonally dry tropical forests in northern Thailand based on field performance. For Ecol Manag 184:177–191

Josué IIP (2017) Zooplâncton como indicador da restauração de ambiente aquático impactado por atividades de Mineração: É possível recuperar funções. MSc Dissertation, Universidade Federal de Juiz de Fora, Juiz de Fora, Brasil, 92pp

Knowles OH, Parrotta JA (1995) Amazonian forest restoration: an innovative system for native species selection based on phenological data and field performance indices. Commonw For Rev 74:230–243

Lapa RP, Cardoso W (1988) Tailings disposal at the Trombetas bauxite mine. In: Larry J (ed) Proceedings of the TMS 117th annual meeting. Phoenix, Arizona, pp 65–76

Mannheimer S, Bevilacqua G, Caramaschi EP, Scarano FR (2003) Evidence for seed dispersal by the catfish *Auchenipterichthys longimanus* in an Amazonian lake. J Trop Ecol 19:215–218

Mittermeier R, Baião PC, Barrera L, Buppert T, McCullough J, Langrand O, Larsen FW, Scarano FR (2010) O protagonismo do Brasil no histórico acordo global de proteção à biodiversidade. Nat Conservação 8:197–200

Mueller-Dombois D, Ellenberg H (1974) Aims and methods of vegetation ecology. Wiley, New York

Parolin P, Wittmann F, Ferreira LV (2013) Fruit and seed dispersal in Amazonian floodplain trees – a review. Ecotropica 19:19–36

Parrotta JA, Knowles OH (1999) Restoration of tropical moist forests on bauxite-mined lands in the Brazilian Amazon. Restor Ecol 7:103–116

Parrotta JA, Knowles OH (2001) Restoring tropical forests on lands mined for bauxite: examples from the Brazilian Amazon. Ecol Eng 17:219–239

Peng S-L, Hou Y-P, Chen B-M (2010) Establishment of Markov successional model and its application for forest restoration reference in southern China. Ecol Model 221:1317–1324

Pillar VD, Sosinski EE (2003) An improved method for searching plant functional types by numerical analysis. J Veg Sci 14:323–332

Poorter L, Bongers F, Aide TM, Almeyda Zambrano AM, Balvanera P, Becknell JM, Boukili V, Brancalion PHS, Broadbent EN, Chazdon RL, Craven D, Almeida-Cortez JS, Cabral GAL, de Jong B, Denslow JS, Dent DH, DeWalt SJ, Dupuy JM, Durán SM, Espírito-Santo MM, Fandino MC, César RG, Hall JS, Hernández-Stefanoni JL, Jakovac CC, Junqueira AB, Kennard D, Letcher SG, Licona J-C, Lohbeck M, Marín-Spiotta E, Martínez-Ramos M, Massoca P, Meave JA, Mesquita R, Mora F, Muñoz R, Muscarella R, Nunes YRF, Ochoa-Gaona S, Orihuela-Belmonte E, Peña-Claros M, Pérez-García EA, Piotto D, Powers JS, Rodríguez-Velazquez J, Romero-Pérez IE, Ruíz J, Saldarriaga JG, Sanchez-Azofeifa A, Schwartz NB, Steininger MK, Swenson NG, Uriarte M, van Breugel M, van der Wal H, Veloso MDM, Vester H, Vicentini A, Vieira ICG, Bentos TV, Williamson GB, Rozendaal DMA (2016) Biomass resilience of Neotropical secondary forests. Nature 530:211–214

Prance GT (1979) Notes on the vegetation of Amazonia. 3. Terminology of Amazonian Forest types subject to inundation. Brittonia 31:26–38

Scarano FR (2017) Ecosystem-based adaptation to climate change: concept, scalability and a role for conservation science. Perspect Ecol Conserv 15:65–73

Scarano FR, Rios RI, Esteves FA (1998) Tree species richness, diversity and flooding regime: case studies of recuperation after anthropic impact in Brazilian flood-prone forests. Int J Ecol Environ Sci 24:223–235

Soares-Filho B, Rajão R, Macedo M, Carneiro A, Costa W, Coe M, Rodrigues H, Alencar A (2014) Cracking Brazil's forest code. Science 344:363–364

Sousa PT Jr, Piedade MTF, Candotti E (2011) Brazil's Forest code put wetlands at risk. Nature 458:478

Tucker NIJ, Murphy TM (1997) The effects of ecological rehabilitation on vegetation recruitment: some observations from the wet tropics of North Queensland. For Ecol Manag 99:133–152

Wood SLR, DeClerck F (2015) Ecosystems and human well-being in the sustainable development goals. Front Ecol Environ 13:123

Wydhayagarn C, Elliott S, Wangpakapattanawong P (2009) Bird communities and seedling recruitment in restoring seasonally dry forest using the framework species method in Northern Thailand. New For 38:81–97

Chapter 16
Conclusions, Synthesis, and Future Directions

Randall W. Myster

16.1 Conclusions

The Amazon is more than just a river, a basin, or a rain forest. For our species, it provides not just food and various ecosystem services and products (Fig. 16.1) but also nourishes our cultures, our emotions, and our psyches as part of our shared human consciousness (Fig. 16.2). In Chap. 1, I presented igapó forests as among the most stressed ecosystems in the Amazon that stress being caused primarily by being flooded by black water for months every year. Physical effects of that flooding include the depth and weight of the water column as well as the physical force of the moving water. Chemical effects include the water's acidity, lack of nutrients, and low oxygen content. After water recedes, igapó forests also receive stress, from drought facilitated by its sandy soil and from disturbances both natural (e.g., tree fall) and man-made (e.g., logging, fishing, agriculture). Given these conditions, it is not surprising that igapó forests have (compared to other Amazon forests) low diversity, a reduced physical structure, and slow growth. While the current biota in igapó forests lives and reproduces within these stressful conditions, evolution is always at work creating a future biota which must also exist within physical and chemical boundaries created by black-water flooding (Fig. 16.3).

Along with these results from my plots in igapó forests, in Chap. 1 I also related the difficulties of igapó recruitment. Seed load peaks in the early part of the year – near the end of the rainy season – and then decreases monotonically over the remainder of the year. Igapó forests have low seed species richness, and the seeds are not very evenly distributed among species. Given the high diversity and number of seeds coming from surrounding forests, however, the seed rain is probably not limiting igapó recruitment, regeneration, and plant community dynamics (Myster 2017a). What happens after dispersal is more limiting, and predation took most of

R. W. Myster (✉)
Oklahoma State University, Oklahoma City, OK, USA
e-mail: myster@okstate.edu

R. W. Myster (ed.), *Igapó (Black-water flooded forests) of the Amazon Basin*,
https://doi.org/10.1007/978-3-319-90122-0_16

Fig. 16.1 The Pasaje Paquito area of the Belén Market in Iquitos, Peru. Bark, roots, stems, leaves, flowers, fruits, and seeds taken from the Amazon are used to make items of both medicinal and magical value

those seeds, more seeds than was taken by pathogens or that germinated. Losses to predators decreased monotonically, and losses to pathogens increased monotonically, as months underwater increased.

Because of soil and black-water leaching, high mercury concentrations have been observed in black water and its biota even in regions without anthropogenic impacts. One consequence of this is biomagnification of mercury around one order of magnitude from one trophic level to other. In addition, land use can change the mercury cycle and increase concentrations in aquatic systems. To date, 298 arachnid species have been reported in igapó forests. Strategies to survive inundation include vertical migrations during flood events, remigration after the event, and timed reproduction. Horizontal migration is less frequent in igapó forest for most species. The five turtle species most common in the igapó – red-headed Amazon River turtle (*Podocnemis erythrocephala*), yellow-spotted Amazon river turtle (*Podocnemis unifilis*), giant South American river turtle (*Podocnemis expansa*), big-headed sideneck turtle (*Peltocephalus dumerilianus*), and matamata (*Chelus fimbriatus*) – have been historically exploited by human beings. The unique chemistry of the igapó promotes the uptake of mercury (Hg) by turtles in igapó more efficiently than any other environment of the Amazon. Igapó forests are widely used by primates, with only a very few species appearing to avoid them completely. For most species, use patterns are seasonal and based around exploitation of either seasonal abundances of fruits (during inundation peak) or of germinating seedlings (during the ebb). The few species, such as the golden-backed uakari (*Cacajao melanocephalus*) that occupy igapó year-round, have a highly flexible feeding ecology in order to survive the huge annual fluctuations in resource base availability. Mammal diversity does not differ between wet and dry years, species richness is higher in wet years, and evenness is low always. Diversity index differed by strata between sampling years but was always highest in the canopy. Alpha diversity was higher in the dry year compared to the wet year, and gamma and beta diversity were higher in the wet year compared to the dry year. During wet and dry year, frequency of sightings of species

Fig. 16.2 An artist's conception of a tropical forest ("Yellow birds" by Paul Klee)

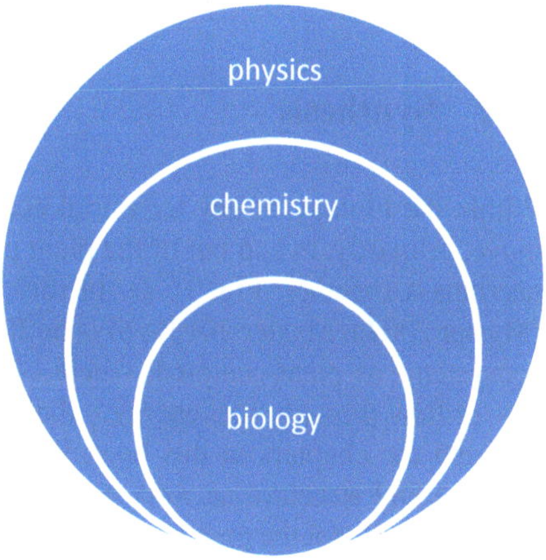

Fig. 16.3 The relationship between physics, chemistry, and biology in the known universe. Physics covers all known phenomena, within those phenomena are chemical reactions and within those reactions are those that are part of living organisms. For example, in this book, the igapó biota exists within boundaries set by physical and chemical conditions of black-water flooding

was influenced by time of day and varied by story level. There are an equal number of primate species in both igapó and *terra firme* habitats (11 sympatric primate species in each), and some primate species' use of igapó habitat is seasonal. These species may be moving into igapó during periods when fruit, especially that of palms like *Mauritia flexuosa*, is superabundant there although scarce in *terra firme*. Thus, during critical times of the year, igapó forests may be critical to maintaining sustainable primate populations and primate species richness in areas with flooded and unflooded forests.

A unique river of black water is the Nanay, located on the left bank of the Amazon River, between the Tigre River and the Napo River. Here trees reach maximum heights of 20 m and DBH up to 50 cm, but there are also endemic species as well as plants that have the great capacity to absorb chemical substances to counteract the contamination of the river and belong to the families Annonaceae, Apocynaceae, Burseraceae, Lauraceae, and Sapotaceae, among others. The forests of the flood plain of the Nanay River show a high productive potential and the conservation of biodiversity. Of the 81 species recorded in the forest of the alluvial plain of the Nanay River, the species that had the greatest ecological importance are characterized as an association of *Eschweilera coriacea, Theobroma glaucum, Campsiandra angustifolia, Licania harlingii, Caraipa densifolia, Pouteria cuspidata*, and *Cariniana decandra.* One author defined three possible tree species geographic patterns: AA, ample amazon; FP, floodplain area; and BW, black water. We checked 231 species with wide distribution on the Amazon (AA), 65 occur only on floodplain (FP), 61 occur only near black-water river (BW), 7 are restricted to one formation on Amazon, and 16 species are not registered on Species Link site. The majority of the species BW are considered rare (59%), and only 12 species (19.7%) occur in other Brazilian biome. We observed the same with floodplain species (FP); only 12.3% also occur in other Brazilian biomes. In general, the most constant/moderately frequent species in the igapó forest showed large geographic amplitude occurring in other Brazilian biomes. A great proportions of them also occurred in Cerrado (CE), followed by the Atlantic rain forest (MA).

16.2 Synthesis

Within the biology of any terrestrial ecosystem, both abiotic and biotic processes move, primarily, in and out of the plant phytomass (the biomass [Myster 2003] and necromass together, Fig. 16.4). In this volume and also in my first three books (Myster 2007a; Myster 2012b; Myster 2017b), I have presented a view of terrestrial ecosystems as plant-centered, where components of ecosystems cycle in and out of – or flow through, like energy – the total plant phytomass. The published literature and the chapters in this book have shown the phytomass is central to igapó function and structure as well. No other component or components of the igapó ecosystem, except the phytomass, can assume this central role as a conduit for physical, chemical, and biological parts of the ecosystem (Myster 2001a). Only the phytomass mediates and integrates between biogeochemical cycles (including cycles of productivity and decomposition; Myster 2003), conducting the vast majority of an ecosystem's energy and nutrient processing (Grime 1977). For example, carbon (C) in the atmosphere (as CO_2) is taken in by plants and converted to organic molecules (first glucose) which later either decomposes and returns C to the atmosphere (as CO_2) or is consumed by plants or animals which respire and (later) die and decompose all returning C to the atmosphere (as CO_2). Plants create C in the soil (as organic matter) which is consumed by the bacteria that take N out of the air and create nitrogen (N) compounds that plant take up in their roots (e.g., NO_3-, NH^4+). Also plants take up energy from the sun (as light) and change its form to chemical

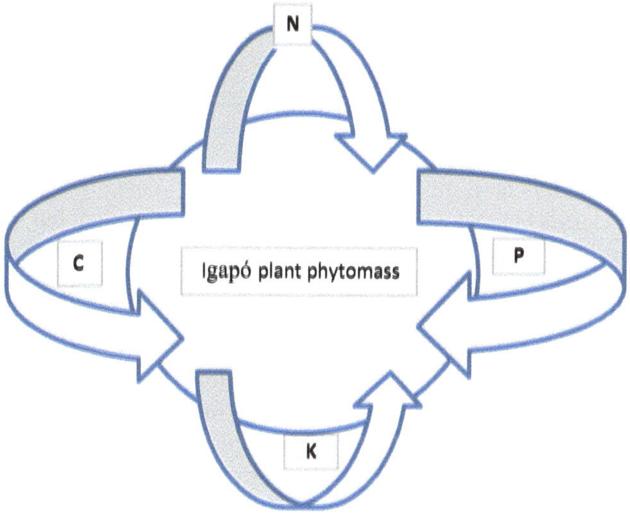

Fig. 16.4 A conceptual model of igapó forests with plant phytomass in the center showing how carbon (C), nitrogen (N), potassium (K), and phosphorus (P) cycle in and out

bonds in organic molecules. When those are consumed by living things, the large molecules are broken down to smaller molecules, and energy is released (ultimately as heat) which flows back into outer space.

The phytomass should, therefore, be put in the center in our conceptual models of terrestrial ecosystems, and I do that here for igapó (Fig. 16.4). My framework is in stark contrast to a more common way that ecosystems are conceptualized where the phytomass is presented as just another component – a green box – within ecosystems. For scientists who see ecosystems like that, their focus seems to be on chemistry or physics only without realizing the central part that vegetation plays in ecosystems. What I am suggesting with this perspective is that all ecologists need to think hard about the structure, function, and dynamics of the phytomass whenever they study any terrestrial ecosystem.

The phytomass exists as a collection of plant communities which are individualistic (Whittaker 1975; Pickett 1982; Myster and Pickett 1988), structurally build from the interactions between their species (Myster 2012a) and where permanent plots are necessary for their investigation (Myster and Malahy 2008). The dynamics of plant communities (and the phytomass) results from plant-plant replacements (Myster 2012c) among nine classes of replacement (Myster 2017a; Myster 2018). Seed and seedling mechanisms and tolerances after dispersal cause most of the replacements (Myster 2017a), and those replacements change plant spaces, both the space occupied by the phytomass – the phytospace – and the neighborhood space around plants. These plant spaces may overlap and change over a plant's life cycle. And so I extend the old concepts of replacement to include space created by *individual* plant tissue loss as well as by *individual* whole-plant mortality. While the neighborhood space is still being defined, it may change during a plant's life and overlap among different plants.

Table 16.1 Plant strategies based on disturbance x stress

	Low stress	High stress
Low disturbance	Competitive strategy	Stress-tolerant strategy
High disturbance	Ruderal strategy	No viable strategy

I propose that this new, more comprehensive model contains nine fundamental replacement classes (1) that don't result from the death of a plant but from a loss of biomass, or a freeing-up of space by other means, and thus a change in the phytospace and neighborhood space (replacements none = > one and none = > many), (2) where there is plant death but no resulting replacement (replacements one = > none and many = > none), for example, during the thinning phase of forest stand development, and (3) involving more than one plant at a time (replacements one = > many and many = > one) whose spaces may combine or overlap. While future models should include the possibility of these kinds of replacements, they should not necessarily assume that any species can be replaced by any other species.

The igapó forest phytomass is dynamic – changing through time and space – and those changes are determined by the plant-plant replacement process (Myster 2012c). In igapó, replacements obviously include trees but can also include herbs, shrubs, and other plants. Because stress is so important to igapó, we need to incorporate it within any models of plant community dynamics, for example, those that emphasize disturbance (Pickett and White 1985). All ecosystems are stressed to some degree, because none of them have the amounts of light, nutrients, etc. needed to maximize plant production (Grime 1977; Maestr et al. 2009). Among resources one is most limiting, and so there are different levels of limitation just as there are different levels of stress and flooding. What is most limiting in igapó forests? Inadequate supply of oxygen (with light and CO_2) to submerged tissues may be key (Jackson and Colmer 2005), and light/water interactions can be most limiting in the *terra firme* forests where many of the species originate (Myster 2012d). Also levels of the plant hormone ethylene and products of anaerobic metabolism by soil microorganisms (e.g., Mn^{2+}, Fe^{2+}, S^{2-}, H_2S, carboxylic acids) can build up in the soil (Jackson and Colmer 2005). And P is often limiting in tropical soils. Because of the physical aspects of flooding and stress, however, one could say that the *lack of water* is most limiting, and, if so, is the stress from water in igapó forests resource-related or non-resource-related (Maestr et al. 2009)?

The classic model of Grime (1977: Table 16.1) presents all plant species as having one of three strategies determined by *stress*, defined as anything within a plant community that limits plant primary production (e.g., less than optimal levels of light, water, or nutrients), and by *disturbance*, defined as anything that destroys plant biomass. This conceptualization leads to four types of habitats that present cues that plants are adapted to:

(1) Low stress/low disturbance habitats which have plants that compete well with other plants (i.e., late successional plant species, Bazzaz 1996)
(2) Low stress/high disturbance habitats which have fast-growing plants with small seeds (i.e., early successional plant species, Bazzaz 1996)

(3) High stress/low disturbance habitats which have plants that can tolerate stress first and foremost
(4) High stress/high disturbance habitats which do not allow reestablishment of the vegetation and thus have no plant species and no plant strategies.

Perhaps there are no plants with strategy (4) because under high stress, herbaceous plants and woody plants of low wood density (both common early successional plants that would normally establish after high disturbance) are just too frail to survive.

In igapó low stress means flooding that is shallow and of short duration, for example, away from black-water rivers and in areas of high relief. High stress means flooding that is deep and of long duration, for example, close to black-water rivers and in areas of low relief. The kind of flooding found in igapó cannot, however, be put into just two simple categories of low and high because flooding found in igapó creates gradients of stress. For example, as you walk away from black-water rivers, both flooding and stress decrease, and disturbances can occur at any point along these gradients (Myster 2001b). I have never observed areas along these gradients without vegetation which would be consistent with Grime's lack of a strategy (4). Instead I have sampled trees all along the igapó flooding gradient, both early and late successional (Myster 2007b, 2010, 2015). Indeed the individualistic distribution of species along the flooding/stress gradient suggests many possible strategies that igapó plants have (see discussion in Chap. 1).

Now while there is less richness and stem density as flooding/stress increases, how plants along the igapó flooding/stress gradient sort out in terms of successional strategy (perhaps indexed by seed size or wood density) is a future research topic of mine. Indeed, I have suggested (Myster 2007b, 2010, 2015) that flooding itself may be a disturbance because it removes plant biomass. If true, this would allow plant species to be placed on only one axis, which may illuminate or even reduce strategies. Clumping of tress along this axis would suggest strategies.

An important avenue for plant research in igapó forests is then how plant species are placed and adapted along this gradient. Since flooding stress is a strong driver of adaptive evolution (Jackson and Colmer 2005), what are the trade-offs, and which aspects of their life cycle are most critical in determining their distribution patterns along the gradient? Physiological studies may help interpret plant distribution patterns but cannot prove issues dealing with plant traits; only field sampling and controlled, focused field experiments can do that.

16.3 Future Directions

The chapters suggest the continued sampling of all ecosystem components in large forest plots with an emphasis on exploration of interactive links among ecosystem components. Experiments need to be designed to find these links and should have a special focus on the early parts of regeneration (i.e., recruitment of seeds and saplings).

In particular I suggest that researchers:

1. Set up permanent plots in various types of igapó in different locations and sample them for decades.
2. Sample soil and other environmental factors in the plots as well.
3. Conduct investigations into the ecophysiology of key tree species found in the plots.
4. Determine, with controlled field experiments, the plant-plant replacements which ultimately create forest structure, function, and dynamics.
5. Perform restoration experiments needed after logging and conversion to agriculture.

References

Bazzaz FA (1996) Plants in changing environments: linking physiological, population, and community ecology. Cambridge University Press, Cambridge

Grime JP (1977) Evidence for the existence of three primary strategies in plants and its relevance to ecological and evolutionary theory. Am Nat 111:1169–1194

Jackson MB, Colmer TD (2005) Response and adaptation by plants to flooding stress. Ann Bot 96:501–505

Maestr FT, Callaway RM, Valladares F, Lortie CJ (2009) Refining the stress-gradient hypothesis for competition and facilitation in plant communities. J Ecol 97:199–205

Myster RW (2001a) What is ecosystem structure? Caribb J Sci 37:132–134

Myster RW (2001b) Mechanisms of plant response to gradients and after disturbances. Bot Rev 67:441–452

Myster RW (2003) Using biomass to model disturbance. Community Ecol 4:101–105

Myster RW (2007a) Post-agricultural succession in the Neotropics. Springer-Verlag, Berlin

Myster RW (2007b) Interactive effects of flooding and forest gap formation on composition and abundance in the Peruvian Amazon. Folia Geobot 42:1–9

Myster RW (2010) Flooding duration and treefall interactive effects on plant community richness, structure and alpha diversity in the Peruvian Amazon. Ecotropica 16:43–49

Myster RW (2012a) A refined methodology for defining plant communities using data after sugarcane, banana and pasture cultivation in the Neotropics. Sci World J, vol. 2012, Article ID 365409, 9 pages. doi:https://doi.org/10.1100/2012/365409

Myster RW (2012b) Ecotones between forest and grassland. Springer-Verlag, Berlin

Myster RW (2012c) Plants replacing plants: the future of community modeling and research. Bot Rev 78:2–9

Myster RW (2012d) Spatial and temporal heterogeneity of light and soil water along a *terra firma* transect in the Ecuadorian Amazon. Can J For Res 42:1–4

Myster RW (2015) Flooding x tree fall gap interactive effects on black-water forest floristics and physical structure in the Peruvian Amazon. J Plant Interact 10:126–131

Myster RW (2017a) Does the seed rain limit recruitment, regeneration and plant community dynamics? Ideas Ecol Evolut 10:12–16

Myster RW (2017b) Forest structure, function and dynamics in western Amazonia. Wiley, Oxford

Myster RW (2018) The nine classes of plant-plant replacement. Ideas Ecol Evolut 11:29–34

Myster RW, Malahy MP (2008) Is there a middle-way between permanent plots and chronosequences? Can J For Res 38:3133–3138

Myster RW, Pickett STA (1988) Individualist patterns of annuals and biennials in early successional old fields. Vegetatio 78:53–60
Pickett STA (1982) Population patterns through twenty years of oldfield succession. Vegetatio 49:45–59
Pickett STA, White PS (1985) The ecology of natural disturbance and patch dynamics. Academic Press, Orlando, FL
Whittaker RH (1975) Communities and ecosystems. MacMillan, NY

Index

R. W. Myster (ed.), *Igapó (Black-water flooded forests) of the Amazon Basin*,
https://doi.org/10.1007/978-3-319-90122-0

MIX
Papier aus verantwortungsvollen Quellen
Paper from responsible sources
FSC® C105338

If you have any concerns about our products,
you can contact us on
ProductSafety@springernature.com

In case Publisher is established outside the EU,
the EU authorized representative is:
Springer Nature Customer Service Center GmbH
Europaplatz 3, 69115 Heidelberg, Germany

Printed by Libri Plureos GmbH
in Hamburg, Germany